INTRACELLULAR PROTEIN CATABOLISM

ADVANCES IN EXPERIMENTAL MEDICINE AND BIOLOGY

INTRACELLULAR PROTEIN CATABOLISM

Edited by

Koichi Suzuki
University of Tokyo
Tokyo, Japan

and

Judith S. Bond
Penn State University College of Medicine
Hershey, Pennsylvania

PLENUM PRESS • NEW YORK AND LONDON

Library of Congress Cataloging-in-Publication Data

Intracellular protein catabolism / edited by Koichi Suzuki and Judith
S. Bond.
        p.   cm. -- (Advances in experimental medicine and biology ; v.
389)
    "Proceedings of the tenth International Conference on
Intracellular Protein Catabolism, held October 30-November 3, 1994,
in Tokyo, Japan"--T.p. verso.
    Includes bibliographical references and index.
    ISBN 0-306-45201-4
    1. Proteins--Metabolism--Congresses. 2. Cell metabomism-
-Congresses.   I. Suzuki, Koichi, 1924-   . II. Bond, Judith S.
III. International Conference on Intracellular Protein Catabolism
(10th : 1994 : Tokyo, Japan) IV. Series.
QP551.I576   1996
574.1'33--dc20   95-26381
                                                              CIP

Proceedings of the Tenth International Conference on Intracellular Protein Catabolism,
held October 30–November 3, 1994, in Tokyo, Japan

ISBN 0-306-45201-4

© 1996 Plenum Press, New York
A Division of Plenum Publishing Corporation
233 Spring Street, New York, N. Y. 10013

10 9 8 7 6 5 4 3 2 1

PREFACE

The Tenth International Conference on Intracellular Protein Catabolism was held in Tokyo Japan, October 30-November 3, 1994, under the auspices of the International Committee on Proteolysis (ICOP). ICOP meetings have been held biennially in the USA, Europe, and Japan in turn. The previous three ICOP meetings (7th to 9th) were held in Shimoda, Japan, in 1988, in WildbadKreuth, Germany, in 1990, and in Williamsburg, Virginia, in 1992. Previous meetings were held in resort areas, this was the first meeting held in a large city. Attendance has grown every year so that nearly 400 participants from 19 different countries attended the Tokyo meeting. At the meeting, novel and updated results on the structure-function, physiology, biology, and pathology of proteases and inhibitors were discussed, together with cellular aspects of proteolysis and protein turnover. Thirty-nine invited papers and eight selected posters were presented orally and 171 poster presentations were discussed. This book documents almost all of the lectures and some selected posters. Since the world of proteolysis and protein turnover is expanding very rapidly, far beyond our expectations, it is impossible to cover all the new aspects of this field. However, this book will give an idea of the current status, trends, and directions of the field, and information necessary to understand what is and will be important in this field. Further, the editors hope that the novel ideas, approaches, methodologies, and important findings described in this book will stimulate further study on proteolysis and protein turnover.

We wish to express special thanks to Drs. Eiki Kominami, Tatsuya Samejima, and Sei-ichi Kawashima, and members of the organizing committee for their efforts in the preparation and organization of the meeting. Furthermore, we want to thank all the members of Dr. Suzuki's laboratory, especially Dr. Shoichi Ishiura, Dr. Hiroyuki Sorimachi, Ms. Kayoko Shimada and Ms. Keiko Tomita who devoted much time to the preparation of the meeting and publication of this book. Thanks are also due to Ms. Hisako Kamioka for her assistance in the editing process, especially retyping manuscripts following the Plenum Gamma Format, and to Ms. Deborah Atkinson-Davis at Penn State University for clerical assistance. Finally, we wish to acknowledge the helpful suggestions made by Ms. Eileen Bermingham of Plenum throughout the preparation of this book.

Koichi Suzuki
Judith Bond

ACKNOWLEDGMENTS

The meeting was supported by

The Commemorative Association for the Japan World Exposition (1970)

Thanks are due to the following organizations for their support

Foundation for Advancement of International Science
Pharmaceutical Society of Japan
The Japanese Agricultural Chemical Society of Japan
The Japanese Biochemical Society
The Japanese Society for Cell Biology
The Molecular Biology Society of Japan
The Protein Engineering Society of Japan

Special thanks are also due to the following sponsors

Asahi Breweries Ltd.
Asahi Denka Kogyo K.K.
Asahimatsu Foods Co.
Ciba-Geigy Foundation (Japan) for the Promotion of Science
Daiichi Pure Chemicals Co. Ltd.
Dinabot Co. Ltd.
Elizabeth Arnold Fuji Foundation
Food Research & Development Laboratories of Ajinomoto Co., Inc.
Fuji Flavor Co. Ltd.
Fuji Oil Co. Ltd.
Fujirebio, Inc.
Inoue Foundation for Science
International Reagents Corporation
Japan Tobacco, Inc., Food R & D Center
Japan Tobacco, Inc., Life Science Research Center
Kanebo Ltd. Cosmetics Laboratory
Kaneka Corporation
Kato Memorial Bioscience Foundation
Kenko Mayonnaise Co. Ltd.
Kikkoman Corporation
Kirin Brewery Co., Inc.

Kissei Pharmaceutical Co. Ltd.
Komatsugawa Chemical Engineering Co. Ltd.
Kowa Company Ltd., Kowa Research Institute
Life Science Foundation of Japan
Lotte Central Laboratory Co. Ltd.
Meiji Milk Products Co. Ltd.
Meiji Seika Kaisha Ltd.
Mercian Corporation
Minophagen Pharmaceutical Co. Ltd.
Morinaga & Co. Ltd.
Nakano Vinegar Co. Ltd.
Nestle Science Promotion Committee
Nichirei Corporation
Nihon Tokushu Garasu Kaihatsu Co. Ltd.
Nippi, Incorporated
Nippon Glaxo Ltd.
Nippon Meat Packers Inc. R & D Center
Nippon Roche K.K.
Nippon Roche Research Center
Nissin Flour Milling Co. Ltd.
Nitta Gelatin, Inc.
NOF Corporation
Novo Nordisk Pharma Ltd.
Osaka Pharmaceutical Manufacturers Association
Parke-Davis Pharmaceutical Research
Sankyo Foundation of Life Science
Sanwakagaku Kenkyusho Co. Ltd.
Seikagaku Corporation
Senju Pharmaceutical Co. Ltd.
Shimadzu Corporation
Shino-Test Corporation
Showa Scientific Co. Ltd.
Showa Sangyo Co. Ltd.
Soda Aromatic Co. Ltd.
Suntory Ltd.
Taiyo Kagaku Co. Ltd.
Takanashi Milk Products Co. Ltd.
T. Hasegawa Co. Ltd. Technical Research Center
Takasago International Corporation
Termo Life Science Foundation
The Asahi Glass Foundation
The Calpis Food Industries Co. Ltd. R & D Center
The Cell Science Research Foundation
The Chemo-Sero-Therapeutic Research Institute
The Kato Memorial Trust for Nambyo Research
The Naito Foundation
The Nissin Sugar Manufacturing Co. Ltd.
The Pharmaceutical Manufacturers' Association of Tokyo
The Research Foundation for Pharmaceutical Sciences
Toa Medical Electronics Co. Ltd.
Tone Chemical Co, Inc.

Tokyo Biochemical Research Foundation
Toray Industries, Inc.
Uehara Memorial Foundation
Upjohn Pharmaceuticals Ltd.
Wako Pure Chemicals Ltd.
Yamamori, Inc.
Yamaso Corporation
Yoshinaga Doi

CONTENTS

THE METZINCIN-SUPERFAMILY OF ZINC-PEPTIDASES

W. Bode, F. Grams, P. Reinemer, F. -X. Gomis-Rüth, U. Baumann,[1]
D. B. McKay,[2] and W. Stöcker[3]

Max-Planck-Institut für Biochemie
Am Klopferspitz 18, D-82152 Martinsried, Germany
[1] Institut für Organische Chemie und Biochemie
Universität Freiburg
Albertstraße 21
D-79104 Freiburg, Germany
[2] Beckman Laboratories for Structural Biology
Department of Cell Biology
Stanford University School of Medicine
Stanford, California 94305
[3] Zoologisches Institut der Universität Heidelberg
Im Neuenheimer Feld 230
D-69120 Heidelberg, Germany

INTRODUCTION

Over the past three years, the three-dimensional structures of a number of zinc proteinases that share the zinc-binding motif **HEXXHXXGXXH** have been elucidated. These proteinases comprise astacin, a digestive enzyme from crayfish [1,2,3], adamalysin II [4,5] and atrolysin C [6] from snake venom, the *Pseudomonas aeruginosa* alkaline proteinase [7] and serralysin from *Serratia marcescens* proteinase [8], the collagenases from human neutrophils [9,10,11]) and fibroblasts [12,13,14,15], human stromelysin 1 [16; K. Appelt, personal communication] and matrilysin [M. Browner, Keystone Symposia, March 5-12, 1994]. These enzymes represent four different families of zinc peptidases: the astacins [3,17], the bacterial serralysins [18], the adamalysins/reprolysins [19,20], and the matrixins (matrix metalloproteinases, MMPs) [21,22].

On the level of the amino acid sequences there is only very low similarity between these families [23]. However, a quantitative comparison of the three dimensional structures has uncovered the striking topological equivalence of their catalytic modules [24,25,26]. A sequence alignment based on topological constraints has revealed significant similarites indicating that these proteinases have evolved from a common ancestor. The designation "metzincins" has been coined for this superfamily of zinc peptidases [24,25,26].

Intracellular Protein Catabolism, Edited by Koichi Suzuki and Judith Bond
Plenum Press, New York, 1996

This chapter points out the common structural and functional features of the met-zincins as well as their more distant relationship with the thermolysin-like enzymes.

OVERALL THREE-DIMENSIONAL STRUCTURES

Astacin [1,2], adamalysin II [4,5], and the catalytic modules of neutrophil collagenase [9] and *Pseudomonas* alkaline proteinase [7] are globular entities, subdivided into two domains by the substrate binding cleft, with the zinc at its bottom. In the standard orientation the cleft lies horizontally in the paper plane (Fig.1).

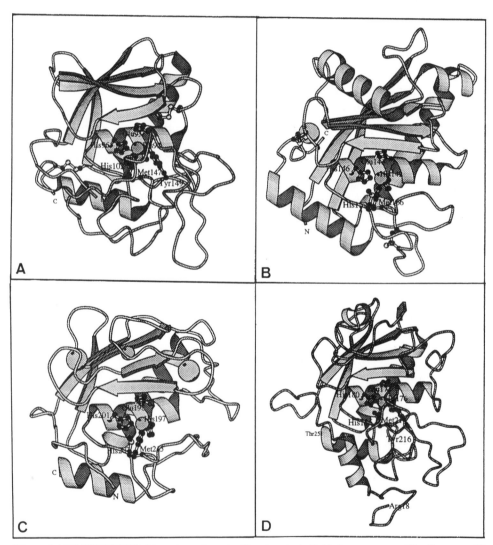

Figure 1. MOLSCRIPT [49] ribbon plots of astacin **(A)**, adamalysin II **(B)**, human neutrophil collagenase **(C)** and *Pseudomonas* alkaline protease **(D)**. Zinc and calcium ions are shown as small and large spheres, respectively. The zinc ligands, the methionine of the Met-turn, and the catalytic glutamic acid are labeled.

The upper domains (N-domains) comprise the N-terminal halves of these enzymes. Their core structure is made up from a twisted ß-sheet and two long α-helices (hA, hB). The sheet consists of four parallel (sI, sII, sIII, sV) and one antiparallel (sIV) strands (Fig.1). The sequential arrangement of these structural elements (sI - hA - sII - sIII - sIV - sV - hB), differs from their topological order (from the top: sII - sI - hA - sIII - sV - sIV - hB). The only antiparallel ß-strand sIV ("edge strand") forms the upper rim of the substrate binding region and is complementary to a bound substrate. In adamalysin [4,5] and in the alkaline proteinase [7], additional helices and strands are located on top of the ß-sheet (Fig.1).

Considerable variability between the four prototypical structures is also seen in the loop between strand sIII and the edge strand sIV (Fig.1). In collagenase, this segment is S-shaped enclosing a second, structural zinc ion and a structural calcium ion. Furthermore, this loop forms a bulge protruding into the active site cleft [9]. Similar, albeit smaller, bulge structures are present in adamalysin [4,5] and in the alkaline proteinase [7], but absent in astacin [1,2] (Fig.1).

The active site helix hB supplies the two histidine imidazoles of the **HEXXH** as zinc ligands. However, in contrast to thermolysin, the helix is terminated at a conserved glycine residue, three residues downstream of the second histidine zinc ligand. After the glycine, which adopts a conformation available to glycine residues only, the chain leaves the N-domain and forms the more irregularly folded C-terminal domain [1,2,4,5,7,9,24].

The C-domains vary considerably in size and conformation. Their only common regular secondary structure element is a lone α-helix close to the C-terminus (hC) (Fig.1).

Three residues downstream of the conserved glycine, the third histidine zinc ligand approaches the metal from below (Fig.1). After that point, the C-terminal chains run through spacers of different length before returning to the active site metal in a most remarkable conserved 1,4-tight turn arranged as a right handed screw. This turn has been designated as the "Met-turn", because of an invariant methionine residue, whose ϵ-methyl group is almost equally distant (4 Å) from the planar faces of the first and second histidine imidazoles [24,25,26].

In adamalysin and collagenase the polypeptide chain between the Met-turn and the C-terminal helix hC is arranged as a long loop running along the surface. In its initial part, this loop shapes the wall of the S_1'-subsite, which is especially large and deep in collagenase [27]. In astacin and the alkaline proteinase the chain subsequent to the Met-turn forms the right part of the lower domain before ending in the C-terminal helix hC (Fig.1). This long α-helix runs across the back of the molecule opposite to the active site thereby crosslinking the C- and N-domains. In astacin and adamalysin a disulfide bridge connects the end of helix hC to the N-terminal ß-sheet, and the polypeptide chain is terminated shortly after that point. In other metzincins, however, which are composed of multiple modules, it runs into the ß-sandwich module (serralysins), the hemopexin-like module (most matrixins), disintegrin-like structures (some adamalysins/reprolysins) or into other intervening domains (some astacins), respectively.

THE STRUCTURE OF THE ACTIVE SITE - IMPLICATIONS FOR CATALYSIS

The four prototypical metzincins exhibit strikingly similar zinc binding regions. The catalytic zinc is ligated by the three His-residues of the **HEXXHXXGXXH** zinc-binding motif with zinc-(N∈)-imidazole distances between 2.0 and 2.2 Å. In the free enzymes, a water molecule as a fourth zinc ligand is clamped between the catalytic Glu of the HEXXHXXGXXH-motif and the metal forming a trigonal pyramidal coordination sphere

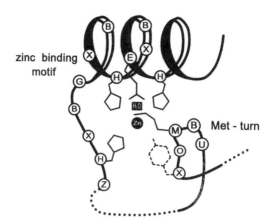

Figure 2. Zinc binding environment of the metzincins (see also text). Variable segments and the tyrosine residue of the astacins and serralysins are stippled.

around the zinc, with the N∈-nitrogen of the second histidine at the pyramid tip, and the zinc in the center of the triangular base formed by the other ligands (Fig.2). A special case are astacin and the alkaline proteinase, where, in the absence of a substrate or inhibitor, the hydroxyl oxygen of a conserved tyrosine residue following the Met-turn acts as a fifth zinc ligand resulting in a trigonal-bipyramidal coordination sphere [2,8,28].

Because of their smaller C-domains, the active site clefts of collagenase and adamalysin are more shallow than those of astacin and the alkaline proteinase. In all four, however, the northern wall of the cleft is built up in part by ß-strand sIV, the edge strand. In case of collagenase, adamalysin and also in the alkaline proteinase (albeit less pronounced), the wall is continued to the right by the protruding bulge segment. The bottom of the substrate binding region is lined by the Gly-turn, the catalytic zinc together with its imidazole ligands, and the S_1'-subsite, which is determined in collagenase and adamalysin to a large extent by the wall forming segment (see above). In astacin and in the alkaline proteinase the prominent southern cleft wall is formed by multiple turn structures preceding and following the Met-turn (Fig.1).

Structures of metzincins complexed with peptide inhibitors revealed that substrates are most likely bound in an extended conformation, which is different from thermolysin [9,11,12,13,25,27,29,30]. It appears that in the different structures the same kind of enzyme-substrate interactions are principally involved N-terminally of the cleavage spot, because the substrate aligns with the edge strand under formation of two (or three) hydrogen bonds in the known structures [9]. By contrast, the C-terminal half of the substrate is bound in different ways. In the matrixins [27] and probably in the adamalysins, it is fenced by the antiparallel bulge segment from above, and by the parallel lower wall-forming segment from below by two pairs of hydrogen bonds. Main chain interactions of that kind are not possible in astacin and less expressed in the alkaline proteinase [7].

The crystal structures of a transition state analog inhibitor complexed to astacin [30] suggests that the carbonyl group of the scissile peptide bond is polarized by the active site zinc. The zinc bound water molecule mediates between this carbonyl and the catalytic glutamate. Thereby, the water molecule acquires the nucleophilicity necessary for attack of the scissile peptide bond carbon resulting in a tetra-coordinate transition state. The catalytic Glu will then presumably serve as a proton shuttle to the cleavage products [27]. This scenario has been studied intensively, for example, in the catalytic mechanism of the zinc endopeptidase thermolysin; in that enzyme, a supporting function has been attributed to

Table 1. Amino acid sequence aligment of the catalytic modules of four representative metzincins. **AP** = *Pseudomonas* alkaline proteinase [45]; **ADA** = adamalysin II [5]; **COL** = human neutrophil collagenase [46]; **AST** = astacin [47]. s = ß-strand, h = α-helix. The alignment is based on topological constraints using the program OVRLAP [48]; topologically equivalent residues are in bold face

```
AP     1 GRSDAYTQVDNFLHAY-ARGGDELVNGHPSYTVDQAAEQILRE--QASWQ  47
ADA      -------------------------------------------------
COL   79 ----------------------------------------FMLTPG-NPKWE  89
AST    1 ----------------------------------------AAILGDEYLWS  12


AP    48 KAPGDSVLTLSYSFLTKPNDFFNTPWKYVSDIYSLGKFSAFSAQQQAQAK  97
ADA   -1 <EQNLPQRYIELVVVAD----RRVFMKYNSDLN-------IIRTRVHEIV  37
COL   90 R------TNLTYRIRNY--------TPQLS-----------EAEVERAIK 114
AST   13 G------GVIPYTFAGV-----------------------SGADQSAIL  31
         ssssssss                                hhhhhhhhh
         I                                       A


AP    98 LSLQSWSDVTNIHFVDAGSA------------------------------ 117
ADA   38 NIINGFYRSLNIRVSLTDLEIWSGQDFITIQSSSSNTLNSFGEWRERVLL  87
COL  115 DAFELWSVASPLIFTRIS-------------------------------- 132
AST   32 SGMQELEEKTCIRFVPR-------------------------------- 48
         hhhhhhhhhhhssssssss
         A          II


AP   118 --QQGDLTFGNFSS----------SVGGAAFAFLPDVPDALKGQSWYLIN 155
ADA   88 TRKRHDNAQLLTAI--------NFEGKIIGKAYTSSMCNPRSSVGIVKDH 129
COL  133 -QGEADINIAFYQRDHGDNSPFDGPNGILAHAFQPG--QGIGGDAHFDAE 179
AST   49 TTESDYVEIFTSGS--------------GCWSYVGR--ISGAQQVSLQAN  82
         ssssss                 sssssss        sssss
         III                    IV             V
```

zinc binding motif

```
AP   156 SSYSANVNPANGNYGRQTLTHEIGHTLGLSHPGDYNAGEGDPTYA----- 200
ADA  130 S-------PINL-LVAVTMAHELGHNLGMEHDGKD------------- 156
COL  180 ETWT--NT-SANYNLFLVAAHEFGHSLGLAHSSD-------------- 210
AST   83 G----------CVYHGTIIHELMHAIGFYHEHTRMDRDNYVTINYQNVD 121
                     hhhhhhhhhhhhhhh
                     B
```

Met-turn

```
AP   201 -----------DATYAEDTRAYSVMSYWEEQNTG--------------- 223
ADA  157 -----------CLRGAS----LCIMRPGL------------------- 170
COL  211 -------------------PGALMYPNY------------------- 219
AST  122 PSMTSNFDIDTYSRYVGEDYQYYSIMHYGKYSFSIQWGVLETIVPLQNGI 171


AP   224 ----QDFKG-AYSSAPLLDDIAAIQKLYGANLTT--------------- 252
ADA  171 -----TPGR---SYEFSDDSMGYYQKFLNQYKPQCILNKP---------- 202
COL  220 -----AFRE-TSNYSLPQDDIDGIQAIYGLSSNPIQP------------ 250
AST  172 DLTDPYDKAH-----MLQTDANQINNLYT---NECSLRH---------- 202
              hhhhhhhhhhhh
              C
```

His231 in stablizing the transition state during catalysis [29]. A residue comparable to His231 of thermolysin has not been observed to be involved in the catalytic mechanism of the metzincins. However, in astacin [30] and in the alkaline proteinase [7] the tyrosine zinc ligand plays a special role, because this tyrosine side chain is displaced by a transition state analog inhibitor and the phenolic oxygen becomes engaged in a hydrogen bond to the inhibitor PO_2H-group mimicking the tetracoordinate carbon during catalysis. Concomitantly with the tyrosine shift the lower domain moves upward and the cleft gets narrower.

Several metzincins, like the collagenases, gelatinases [22], snake venom proteinases [9] and astacins [31] are able to cleave proteins of the extracellular matrix like type I or type IV collagen, laminin, fibronectin and others. Correspondingly, there is a preference for Pro in P_3, and large, mainly hydrophobic residues in P_1' [27,32]. Crayfish astacin is an exception in preferring small aliphatic P_1' residues [3,31]. These similar specificities are in good accordance with structural constraints. For example, the S_1' pockets in general are rather large [e.g. 27], but blocked in astacin [3]. A grove in the northern wall of each of the four prototypes corresponds to a specific S_3-subsite, and P_2 residues are harbored in shallow depressions at the bottom of the cleft, whereas the P_1 side chains are projecting from the cleft and are in contact with the southern wall.

THE CONFORMATION OF THE N-TERMINI AND THE ACTIVATION OF PROMETZINCINS

In adamalysin the N-terminal section is rather short and freely accessible to the solvent, whereas astacin, collagenase and the alkaline proteinase have similar N-terminal segments running parallel to helix hC before they enter the molecular body at a conserved tryptophan residue (Fig.1, Table 1). In the alkaline proteinase there are additional structural elements preceding this region, which are in contact with the C-terminal ß-sandwich module [7].

Two versions of collagenase are known, a Met80-form in which the first six residues are unordered, and a one residue longer Phe79-form with a perfectly ordered N-terminal region [9,33]. The N-terminal ammonium group of the Phe79-form and the side chain of the conserved Asp232 of helix hC form a salt bridge [10], and the next residue, another invariant aspartate in the same helix, Asp233, is indirectly linked via a hydrogen bonding network to the Met-turn and to the first histidine zinc ligand. Hence, the N-terminus of collagenase and of other matrixins is in touch with the active center, which might explain the 3.5-times higher "superactivity" of the Phe-form compared to the truncated Met-form [9,10]. Recently, a membrane bound matrix metalloproteinase has been found with an Asp232/Tyr substitution indicating an alternative arrangement of the N-terminus [34].

In human prostromelysin 1, the only promatrixin of known structure, Pro86 is the first residue which has a topologically equivalent counterpart in the mature enzyme. A cysteine residue as part of a conserved motif **PRCGXPD**, the "cysteine switch", is coordinating the catalytic zinc [K. Appelt, personal communication]. The cysteine-containing propeptide can be removed proteolytically or chemically to yield the active enzymes [35,36].

Similarly as in collagenase, helix hC of the alkaline proteinase contains a topologically equivalent conserved tandem of aspartates (Table 1). The first aspartate, Asp237, is salt bridged to the side chain of Arg18 linking the N-terminal stretch to the catalytic module [7] (Fig.1). The second aspartate, Asp238, is likewise engaged in an equivalent hydrogen bonding system with the active site[7]. A special role for the activation of the proserralysins is attributed to the calcium-binding of the calcium ß-sandwich module [7].

The single conserved aspartate residue (Asp186) in the helix hC of the astacins, is structurally and functionally equivalent to the second Asp in collagenase and alkaline proteinase (Table 1). However, activation of proastacins is most likely achieved in a completely different fashion. In the mature form of astacin the first three residues are submerged in a water-filled cavity of the C-domain [1,2,3]. There, the N-terminal ammonium group is salt bridged to Glu103, the residue directly following the third zinc histidine ligand, His102. It has been shown that latent astacins, which carry elongated N-termini cannot form this active conformation, but are activated upon proteolytic release of an N-terminal propeptide [e.g. 37].

An aspartate tandem present in the helix hC of adamalysin is shifted by one residue (Table 1) and is not conserved in the adamalysin-family [5]. However, the activation of proadamalysins may occur in a similar fashion as in the collagenases, since they also contain a cysteine-switch-like propeptide (**PKMCGV**) in their N-terminal regions [38,39].

TOPOLOGY DERIVED SEQUENCE ALIGNMENT

The three-dimensional structures of astacin, adamalysin, the alkaline proteinase and neutrophil collagenase were superimposed and quantitatively analyzed. This allowed for the assignment of topologically equivalent regions and for the adjustment of a sequence alignment [24,25,26] (Table 1). There are considerably low overall sequence identities between the four prototypical metzincins (Fig.3) with similarity scores between 13% and 27% identity (Table 1; Fig.3). The closest relationship is observed for the pairs alkaline proteinase/collagenase (26.5% identity) and alkaline proteinase/astacin (26.2% identity), whereas the score is lower for the pair astacin/collagenase (15.8% identity) (Fig.3). These values reflect a basic evolutionary position for the bacterial serralysins and a more distant position for the adamalysins (Fig.3) [24,25,26].

The metzincins are distinguished by several criteria, including a conserved overall topology, and, furthermore, two most remarkable sequence motifs, which may serve as signatures [24,25,26]. In the consensus sequence **HEBXHXBGBXHZ**, which includes the

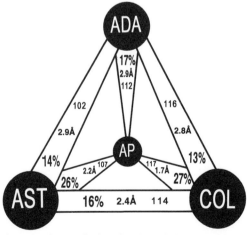

Figure 3. Relationships between the metzincins (for abbreviations see Table 1). Indicated in Å: root mean square deviations of topologically equivalent Cα-backbone atoms; unlabeled: number of equivalent sequence positions (bold faced in Table 1); indicated in %: identity scores within topologically equivalent regions.

three histidine zinc ligands and the catalytic Glu, the **B** residue is bulky and hydrophobic anchored in the hydrophobic core of the protein, and **Z** labels a conserved residue, which is typical for each of the families (Table 1; Fig.2). In the astacins, **Z** is the Glu-residue, that is salt-bridged to the N-terminus. The equivalent serine or threonine of the matrixins presumably helps to stabilize the mature N-terminus. In the serralysins and adamalysins the Z-position is occupied by a conserved proline [7,8], or an aspartate [5], respectively, which presumably play structural parts. The aspartate is even present in some adamalysins/reprolysins like PH30ß [8], in which other catalytically essential residues have been replaced by non-functional ones. In the Met-turn consensus sequence (**UBMO**X), **B** is a hydrophobic side chain embedded in the hydrophobic core, while **U** and **O** are conserved family-specific residues, and **X** is the tyrosine zinc-ligand of the astacins and serralysins (Fig.2, Table 1).

DISTANT RELATIONSHIPS TO THE THERMOLYSINS

The similarity scores between the metzincins and the thermolysin-like zinc peptidases are extremely low. However, the topologies of the metzincin N-domains and the N-terminal third of thermolysin are clearly equivalent. The similar structural elements comprise the zinc-binding active site helix, the ß-strands sI, sIII, sV and sIV and to the helix hA [2,5,25,24,26]. Hence, the metzincins and the thermolysin-like enzymes may have arisen by divergent evolution from a common ancestor [24,25]. We have previously introduced the term "zincins" for these proteins [24], which contain a **HEXXH**-zinc-binding motif, and, most importantly, are related due to the similar fold of their polypeptide chains.

CONCLUSIONS

The increase of sequence information on zinc-peptidases in the past few years has revealed that this class of proteolytic enzymes is made up by a variety of distinct protein families [1,3,5,23,40,41,42,43,44]. The knowledge of the corresponding three-dimensional structures now may allow recognition of new facets of the catalytic mechanism of metalloproteases as a basis for drug design.

A common topological pattern has crystallized for various families, which could hardly be expected from pure sequence alignments. A striking observation in this context has been furnished by E. Schlagenhauf, R. Etges and P. Metcalf at the European Molecular Biology Laboratory (Heidelberg) who solved the crystal structure of leishmanolysin, the surface protease of *Leishmania*. This enzyme has turned out to be a member of the metzincins. However, in leishmanolysin the second and the third histidine zinc ligands are separated by a long intervening spacer precluding the detection of an HEXXHXXGXXH motif [Peter Metcalf and Robert Etges, personal communication].

ACKNOWLEDGEMENTS

We thank Drs Robert Zwilling, Francesc X. Aviles, and Robert Huber for encouragement and support, and Dr. Krzysztof Appelt for making unpublished results available to us. This work was supported by grants from the Deutsche Forschungsgemeinschaft: Sto 185/3-1 to W.S. and SFB207 to W.B.

REFERENCES

1. Bode, W., Gomis-Rüth, F.-X., Huber, R., Zwilling, R., and Stöcker, W., 1992, Structure of astacin and implications for the activation of astacins and zinc ligation of collagenases, *Nature (London)* 358: 164-167.
2. Gomis-Rüth, F.-X., Stöcker, W., Huber, R., Zwilling, R., and Bode, W., 1993, The refined 1.8 Å X-ray crystal structure of astacin, a zinc-endopetidase from the crayfish *Astacus astacus* L. Structure determination, refinement, molecular structure, and comparison to thermolysin, *J. Mol. Biol.* 229: 945-968.
3. Stöcker, W., Gomis-Rüth, F.-X., Bode, W., and Zwilling, R., 1993, Implications of the three-dimensional structure of astacin for the structure and function of the astacin family of zinc-endopeptidases, *Eur. J. Biochem.* 214: 215-231.
4. Gomis-Rüth, F.-X., Kress, L.F., and Bode, W., 1993, First structure of a snake venom metalloproteinase: a prototype for matrix metalloproteinases/collagenases, *EMBO J.* 12: 4151-4157.
5. Gomis-Rüth, F.-X., Kress, L.F., Kellermann, J., Mayr, I., Lee, X., Huber, R., and Bode, W., 1994, Refined 2.0 Å X-ray crystal structure of the snake venom zinc endopeptidase adamalysin II. Primary and tertiary structure determination, refinement, molecular structure and comparison with astacin, collagenase and thermolysin, *J. Mol. Biol.* 239: 513-544.
6. Zhang, D., Botos, I., Gomis-Rüth, F.-X., Doll, R., Blood, C., Njoroge, F.G., Fox, J.W., Bode, W., and Meyer, E.F., 1994, Structural interaction of natural and synthetic inhibitors with the venom metalloproteinase atrolysin C (form d), *Proc. Natl. Acad. Sci. USA* 91: 8447-8451.
7. Baumann, U., Wu, S., Flaherty, K.M., and McKay, D.B., 1993, Three-dimensional X-ray crystallographic structure of the alkaline protease of *Pseudomonas aeruginosa. EMBO J.* 12: 3357-3364.
8. Baumann, U., 1994, Crystal structure of the 50 kDa metallo protease from *Serratia marcescens, J. Mol. Biol.* 242: 244-251.
9. Bode, W., Reinemer, P., Huber, R., Kleine, T., Schnierer, S., and Tschesche, H., 1994, The crystal structure of human neutrophil collagenase inhibited by a substrate analog reveals the essentials for catalysis and specificity, *EMBO J.* 13: 1263-1269.
10. Reinemer, P., Grams, F., Huber, R., Kleine, T., Schnierer, S., Pieper, M., Tschesche, H., and Bode, W., 1994, Structural implications for the role of the N terminus in the 'superactivation' of collagenases, *FEBS Lett.* 338: 227-233.
11. Stams, T., Spurlino, J.C., Smith, D.L., Wahl, R.C., Ho, T.F., Qorronfleh, M.W., Banks, T.M., and Rubin, B., 1994, Structure of human neutrophil collagenase reveals large S₁' specificity pocket, *Nature (London) Struct. Biol.* 1: 119-123.
12. Borkakoti, N., Winkler, F.K., Williams, D.H., D'Arcy, A., Broadhurst, M.J., Brown, P.A., Johnson, W.H., and Murray, E.J., 1994, Structure of the catalytic domain of human fibroblast collagenase complexed with an inhibitor, *Nature (London) Struct. Biol.* 1: 106-110.
13. Lovejoy, B., Cleasby, A., Hassell, A.M., Longley, K., Luther, M.A., Weigl, D., McGeehan, G., McElroy, A.B., Drewry, D., Lambert, M.H., and Jordan, S.R., 1994, Structure of the catalytic domain of fibroblast collagenase complexed with an inhibitor, *Science* 263: 375-377.
14. Lovejoy, B., Hassell, A.M., Luther, M.A., Weigl, D., and Jordan, S.R., 1994, Crystal structure of recombinant 19-kDa human fibroblast collagenase complexed to itself, *Biochemistry* 33: 8207-8217.
15. Spurlino, J.C., Smallwood, A.M., Carlton, D.D., Banks, T.M., Vavra, K.J., Johnson, J.S., Cook, E.R., Falvo, J., Wahl, R.C., Pulvino, T.A., Wendolski, J.J., and Smith, D.L., 1994, 1.56 Å structure of mature truncated human fibroblast collagenase, *Proteins* 19: 98-109.
16. Gooley, P.R., O'Connell, J.F., Marcy, A.I., Cuca, G.C., Salowe, S.P., Bush, B.L., Hermes, J.D., Esser, C.K., Hagmann, W.K., Springer, J.P., and Johnson, B.A., 1994, The NMR structure of the inhibited catalytic domain of human stromelysin-1, *Nature (London) Struct. Biol.* 1: 111-118.
17. Dumermuth, E., Sterchi, E.E., Jiang, W., Wolz, R.L., Bond, J.S., Flannery, A.V., and Beynon, R.J., 1991, The astacin family of metalloendopeptidases, *J. Biol. Chem.* 266: 21381-21385.
18. Häse, C.C., and Finkelstein, R.A., 1994, Bacterial extracellular zinc-containing metalloproteases, *Microbiological Reviews* 57: 823-837.
19. Wolfsberg, T.G., Bazan, J.F., Blobel, C.P., Myles, D.G., Primakoff, P., and White, J.M., 1993, The precursor region of a protein active in sperm-egg fusion contains a metalloprotease and a disintegrin domain: Structural, functional, evolutionary implications, *Proc. Natl. Acad. Sci. USA* 90: 10783-10787.
20. Bjarnason, J.B., and Fox, J.W., 1994, Hemorrhagic metalloproteinases from snake venoms, *Pharmacol. Ther.* 62: 325-372.
21. Woessner, J.F. jr., 1991, Matrix metalloproteinases and their inhibitors in connective tissue remodeling, *FASEB J.* 5: 2145-2154.

22. Birkedal-Hansen, H., Moore, W.G.I., Bodden, M.K., Windsor, L.J., Birkedal-Hansen, B., DeCarlo, A., and Engler, J.A., 1993, Matrix metalloproteinases: a review, *Crit. Rev. Oral. Biol. Med.* 4: 197-250.

23. Rawlings, N.D., and Barrett, A.J., 1993, Evolutionary families of peptidases, *Biochem. J.* 290: 205-218.

24. Bode, W., Gomis-Rüth, F.-X., and Stöcker, W., 1993, Astacins, serralysins, snake venom and matrix metalloproteinases exhibit identical zinc-binding environments (HEXXHXXGXXH and Met-turn) and topologies and should be grouped into a common family, the "metzincins", *FEBS Lett.* 331: 134-140.

25. Stöcker, W., Grams, F., Baumann, U., Reinemer, P., Gomis-Rüth, F.-X., McKay, D.B., and Bode, W., 1995, The metzincins - topological and sequential relations between the astacins, adamalysins, serralysins, and matrixins (collagenases) define a superfamily of zinc-peptidases, *Protein Science*,4: 825 - 840.

26. Stöcker, W., and Bode, W. (1995) Structural features of a superfamily of zinc-endopeptidases, the metzincins, *Current Opinion in Structural Biology*, accepted for publication.

27. Grams, F., Reinemer, P., Powers, J.C., Kleine, T., Pieper, M., Tschesche, H., Huber, R., and Bode, W., 1995, X-ray structures of human neutrophil collagenase complexed with peptide hydroxamate and peptide thiol inhibitors - implications for substrate-binding and rational drug design, *Eur. J. Biochem.*, 228:830-841.

28. Gomis-Rüth, F.-X., Grams, F., Yiallouros, I., Nar, H., Küsthardt, U., Zwilling, R., Bode, W., and Stöcker, W., 1994, Crystal structures, spectroscopic features and catalytic properties of cobalt(II)-, copper(II)-, nickel(II)- and mercury(II)-derivatives of the zinc-endopeptidase astacin. A correlation of structure and proteolytic activity, *J. Biol. Chem.* 269: 17111-17117.

29. Matthews, B.W., 1988, Structural basis of the action of thermolysin and related zinc peptidases, *Accts. Chem. Res.* 21: 333-340.

30. Grams F, Stöcker W, Dive V, and Bode W, in preparation.

31. Stöcker, W., and Zwilling, R., 1995, Astacin, *Meth. Enzymol.* 248: 305-325.

32. Netzel-Arnett, S., Fields, G., Birkedal-Hansen, H., and Van Wart, H., 1993, Sequence specificities of human fibroblast and neutrophil collagenases, *J. Biol. Chem.* 266: 6747-6755.

33. Knäuper, V., Osthues, A., DeClerk, Y.A., Langley, K.E., Bläser, J., and Tschesche, H., 1993, Fragmentation of human polymorphonuclear leucocyte collagenase, *Biochem. J.* 291: 847-854.

34. Sato, H., Takino, T., Okada, Y., Cao, J., Shinagawa, A., Yamamoto, E., and Seiki, M., 1994, A matrix metalloproteinase expressed on the surface of invasive tumour cells, *Nature (London)* 370: 61-65.

35. Springman, E.B., Angleton, E.L., Birkedal-Hansen, H., and Van Wart, H.E., 1990, Multiple modes of activation of latent human fibroblast collagenase: Evidence for the role of a Cys_73 active site zinc complex in latency and a "cysteine switch" mechanism for activation, *Proc. Natl. Acad. Sci. USA* 87: 364-368.

36. Nagase, H., Enghild, J.J., Suzuki, K., and Salvesen, G., 1990, Stepwise activation mechanisms of the precursor of matrix metalloproteinase 3 (stromelysin) by proteinases and (4-aminophenyl) mercuric acetate, *Biochemistry* 29: 5783-5789.

37. Corbeil, D., Milhiet, P.-M., Simon, V., Ingram, J., Kenny, A.J., Boileau, G., and Crine, P., 1993, Rat endopeptidase-24.18 Å subunit is secreted into the culture medium as a zymogen when expressed in COS-1 cells, *FEBS Lett.* 335: 361-366.

38. Hite, L.A., Shannon, J.D., Bjarnasson, J.B., and Fox, J.W., 1992, Sequence of a cDNA clone encoding the zinc metalloproteinase hemorrhagic toxin e from *Crotalus atrox*: evidence for a signal, zymogen, and disintegrin-like structures, *Biochemistry* 31: 6203-6211.

39. Grams, F., Huber, R., Kress, L.F., Moroder, L., and Bode, W., 1994, Activation of snake venom metalloproteinases by a cysteine-switch-like mechanism, *FEBS Lett.* 335: 76-80.

40. Jongeneel, C.V., Bouvier, J., and Bairoch, A., 1989, A unique signature identifies a family of zinc-dependent metallopeptidases, *FEBS Lett.* 242: 211-214.

41. Murphy, G.J.P., Murphy, G., and Reynolds, J.J., 1991, The origin of matrix metalloproteinases and their familial relationships, *FEBS Lett.* 289: 4-7.

42. Vallee, B.L., and Auld, D.S., 1990, Zinc coordination, function and structure of zinc enzymes and other proteins, *Biochemistry* 29: 5647-5659.

43. Jiang, W., and Bond, J.S., 1992, Families of metallopeptidases and their relationships, *FEBS Lett.* 312: 110-114.

44. Hooper, N.M., 1994, Families of zinc metalloproteases, *FEBS Lett.* 354: 1-6.

45. Okuda, K., Morihara, K., Atsumi, Y., Takeuchi, H., Kawamoto, S., Kawasaki, H., Suzuki, K., and Fukushima, J., 1990, Complete nucleotide sequence of the structural gene for alkaline proteinase from *Pseudomonas aeruginosa*, *Infect. Immun.* 58: 4083-4088.

46. Hasty, K.A., Pourmotabbed, T.F., Goldberg, G.I., Thompson, J.P., Spinella, D.G., Stephens, R.M., and Mainardi, C.L., 1990, Human neutrophil collagenase. A distinct gene product with homology to other matrix metalloproteinases, *J. Biol. Chem.* 265: 11421-11424.

47. Titani, K., Torff, H.-J., Hormel, S., Kumar, S., Walsh, K.A., Rödl, J., Neurath, H., and Zwilling, R., 1987, Amino acid sequence of a unique protease from the crayfish *Astacus fluviatilis, Biochemistry* 26: 222-226.
48. Rossman, M.G., and Argos, P., 1975, A comparison of the heme binding pocket in globulins and cytochrome b5, *J. Biol. Chem.* 250: 7525-7532.
49. Kraulis, P.J., 1991, MOLSCRIPT: A program to produce both detailed and schematic plots of protein structures, *J. Appl. Cryst.* 24: 946-950.

2

STRUCTURE AND BIOSYNTHESIS OF MEPRINS

P. Marchand and J. S. Bond

Department of Biochemistry and Molecular Biology
The Pennsylvania State University
500 University Drive
Hershey, Pennsylvania 17033

INTRODUCTION

Meprins are cell surface and secreted metalloendopeptidases that are expressed in kidney and intestine of mice, rats, and humans (Jiang *et al.*, 1992; Gorbea *et al.*, 1993; Corbeil *et al.*, 1992; Johnson and Hersh, 1992; Sterchi *et al.*, 1988; Yamaguchi *et al.*, 1994). The enzymes are capable of hydrolyzing a variety of proteins and biologically active peptides, such as bradykinin, glucagon, transforming growth factor-α, and parathyroid hormone (Butler *et al.*, 1987; Choudry and Kenny, 1991; Wolz *et al.*, 1991). Meprins are implicated in the degradation of extracellular and membrane-bound proteins and peptides, and it has been suggested that they are involved in the processing and inactivation of peptide hormones *in vivo* (Kenny *et al.*, 1989; Yamaguchi *et al.*, 1994).

Meprins were first discovered in 1981 as enzymes of high azocaseinase activity in BALB/c mouse kidney (Beynon *et al.*, 1981). Soon thereafter it was found that two meprin phenotypes are present in the mouse population (Bond *et al.*, 1983). While meprin accounts for the great majority (about 90%) of azocaseinase activity of the kidneys of many mouse strains (e.g., ICR, C57BL/6), certain inbred strains of mice (e.g., C3H/He, CBA) are deficient in this activity. The "low meprin activity" kidney phenotype is inherited as an autosomal recessive trait (Bond *et al.*, 1984). Biochemical characterization and molecular cloning of the enzymes revealed that the molecular basis for the high and low activity phenotypes in different strains of mice is the differential expression of two types of subunits, referred to as α and β subunits. The investigation of meprins from inbred mouse strains has been very useful for the elucidation of meprin structure and expression.

Meprins are highly regulated at the transcriptional and post-translational level. The enzymes comprise a group of oligomeric glycoprotein complexes composed of α and/or β subunits (Gorbea *et al.*, 1991). Both types of subunits are catalytic. They differ in substrate specificity and are differentially expressed in a species-, strain- and tissue-specific manner. The α and β subunits are also regulated by differential post-translational modifications *in vivo*, which affect the structure of the mature subunits and thereby influence the activities

Intracellular Protein Catabolism, Edited by Koichi Suzuki and Judith Bond
Plenum Press, New York, 1996

and subcellular localization. In order to better understand the regulation and activities of this group of enzymes it thus became important to study their structure and biosynthesis. The recent cloning and heterologous expression of the meprin subunit cDNAs provided new insights into meprin structure and biosynthesis, as well as their evolutionary relationship to other metalloendopeptidases.

MEPRINS AND THEIR RELATIONSHIP TO OTHER METALLOPROTEASES

Cloning of the meprin α and β subunit cDNAs revealed that the subunits are the products of evolutionarily related genes (Jiang *et al.*, 1992; Gorbea *et al.*, 1993; Corbeil *et al.*, 1992; Johnson and Hersh, 1992; Dumermuth *et al.*, 1993). In addition, several other proteins with partial sequence homology were identified in the data bases, and it became apparent that these proteins are part of a newly identified family of metalloproteases, the "astacin" family (Dumermuth *et al.*, 1991). To date more than 20 members of this family have been identified by protein or DNA sequence analyses. The domain structures of representative members of the astacin family are shown in Fig. 1. These proteins include the crayfish protease astacin (Titani *et al.*, 1981), meprin α and β subunits, blastula protein 10 (BP10) of sea urchins (Lepage *et al.*, 1992), *Drosophila* and mouse *tolloid* (Shimell *et al.*, 1991; Takahara *et al.*, 1994), and human bone morphogenetic protein, BMP-1 (Wozney *et al.*, 1988). Other astacin family proteases not shown include proteins from nematode, hydra, frog, fish, and quail (Maéno *et al.*, 1993; Yasamasu *et al.*, 1992; Elaroussi and DeLuca, 1994). These proteins are expressed in developing and/or mature organisms, generally in a very tissue-specific manner. They are implicated in diverse functions including digestion (astacin), processing of membrane-bound and extracellular proteins (meprins), embryonic

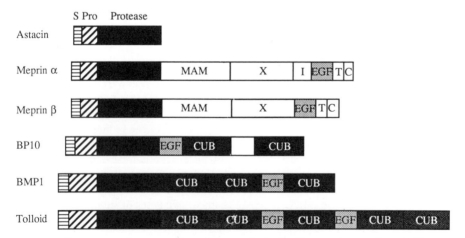

Figure 1. Domain structure of some members of the astacin family. The proteins are aligned with respect to the protease domains. All proteins shown contain signal sequences (horizontal striped boxes), prosequences (diagonal striped boxes) and protease domains (black boxes). Additional domains shared by several proteins in this family are epidermal growth factor (EGF)-like domains and C1r and C1s repeats (CUB domain). The deduced amino acid sequence for meprin subunits predict additional domains referred to as MAM, X, transmembrane-spanning (T), and cytoplasmic (C) domains. Meprin α subunits contain an inserted (I) domain not found in β subunits (Marchand *et al.*, 1995).

pattern formation (tolloid), mesoderm differentiation (BP10), and cartilage and bone formation (BMP-1).

The defining common feature of all proteinases in the astacin family is that they contain a catalytic protease domain that is approximately 200 amino acids in length and has sequence homology to the protease astacin from the crayfish *Astacus fluviatilis* . Sequence homologies between the protease domains vary between 30% and 95% identity; the protease domains of meprin subunits and astacin are approximately 32-35% identical. Though astacin family members are generally believed to have proteolytic activtity *in vivo*, this activity has been demonstrated *in vitro* only for a few proteins, and the physiological substrates are largely unknown.

A comparison of the deduced primary sequences of the proteins in the astacin family shows that these enzymes have many similarities. In addition to the protease domain, meprin subunits contain several other domains that are common to all or some astacin family members. For example, all astacin-like proteases are synthesized with a signal sequence and prosequence preceding the catalytic domain. Removal of the prosequence is a prerequisite for proteolytc activity of meprin subunits, and this is probably true for all other members in this family (Kounnas *et al.*, 1991). A model for the activation mechanism of astacin family proteases has been proposed based on the x-ray structure of astacin (Bode *et al.*, 1992). The x-ray structure showed that the NH_2-terminal amino acid of the protease formed an internal water-linked salt-bridge with a conserved Glu residue (Glu-103 in astacin) near the active site. Computer-assisted modelling indicates that proforms cannot exhibit such a conformation. Activation liberates the mature NH_2-terminus, presumably allowing the new NH_2-terminal residue to fold into the active site, thus triggering the conformational transition from the proform to the active state.

The meprin subunits are much larger proteins than the crayfish protein astacin, and like most proteins in the astacin family contain additional domains COOH-terminal to the protease domain that may regulate their function. Meprin subunits, BP10, BMP-1, and *tolloid* contain one or more epidermal growth factor (EGF)-like domains. BP-10, BMP-1, and *tolloid*, but not meprins, contain complement-like C1r/C1s repeats, also referred to as CUB domains. Meprins are the only astacin family members with MAM domains; MAM domains are recently identified extracellular domains shared by several cell surface proteins including meprin, A5 protein, protein tyrosine phosphatase μ, and enteropeptidase (Beckmann and Bork, 1993; Kitamoto *et al.*, 1994). EGF-like, CUB, and MAM domains are believed to modulate the function of the molecules by promoting interactions with substrates or other proteins. Meprin subunits are the only members of the family that contain COOH-terminal hydrophobic sequences capable of anchoring the proteins to membranes. All other members of the family described thus far are secreted from cells.

The proteases in the astacin family can be distinguished from other metalloendopeptidases, such as those of the thermolysin, serralysin, matrixin, and snake venom families, by the presence of a highly conserved 18 amino acid astacin family signature sequence which contains three histidines involved in zinc binding (Jiang and Bond, 1992). A second conserved region in astacin family proteases contains a tyrosine that coordinates the zinc and a methionine involved in a 'Met-turn' of the peptide chain (Bode *et al.*, 1993). The Met-turn has also been identified in crystal structures of adamalysins, serralysins, and matrixins, and these have been classified collectively as "metzincins" (Stöcker *et al.*, 1995). Meprins can further be distinguished from some other metalloproteases by the geometry and identity of the ligands involved in zinc binding (Jiang and Bond, 1992). In meprins, and the other astacin family proteases, the active-site zinc is coordinated by three histidines, a tyrosine, and a water molecule. The penta-coordination of zinc is also found in the serralysins, and is quite different from that of the thermolysin family, in which two histidines and a glutamic acid residue are ligands to the active-site zinc ion (Gomis-Rüth *et al.*, 1994).

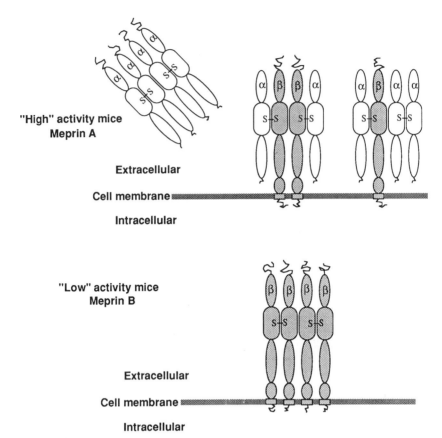

Figure 2. Model for the meprin isoforms expressed in mouse kidney proximal tubule cell membranes. The meprin subunits are shown schematically, α subunits in white and β subunits in grey. Intersubunit disulfide bonds are indicated by horizontal bars. Each isoform consists of two disulfide linked dimers which associate non-covalently to form tetramers. "High" activity mice contain disulfide-linked αβ dimers and αα dimers; membrane-bound tetramers of meprin A are formed by the association of two αβ dimers, or one αβ dimer and one αα dimer. In either instance, the β subunit anchors all subunits to the membrane. In addition, homotetramers of α subunits are secreted from the cells into the tubule lumen (urine) in these mice. "Low" activity mice contain $β_4$ homotetramers (meprin B) formed by the association of two disulfide-linked ββ dimers. The meprin β subunits and the secreted meprin α subunits retain the prosequences; the membrane-associated α subunits do not contain the prosequences.

STRUCTURE OF THE MEPRIN SUBUNITS

The cDNA-inferred amino acid sequences predict primary translation products of the mouse α and β subunits of 760 and 704 amino acids, respectively (Jiang *et al.*, 1992; Gorbea *et al.*, 1993). The primary sequences and the arrangement of functional domains are similar in α and β (Fig.1). The two subunits share approximately 42% overall sequence identity, although the sequence similarity differs markedly in different domains of the protein. For example, the highest sequence similarity is found between the protease domains (56% identity); the MAM domains of α and β are 44% identical, and the EGF-like domains are 34% identical. A striking difference between α and ß is that α subunits contain an additional inserted domain (I domain) that is completely absent from β subunits. The I domain is located between a spacer domain (designated as X domain) and the EGF-like domain (Marchand *et al.*, 1995).

While the primary sequences of meprin α and β predict many similarities, the mature proteins differ significantly in activity, structure, and membrane association (Fig. 2). The differences are determined by differential post-translational modifications of α and β subunits *in vivo* . Characterization of purified meprin subunits from mouse kidney indicated that the post-translational modifications include glycosylation, and NH_2-terminal and COOH-terminal proteolytic processing. The mature subunits in mouse kidney are glycoproteins of 90 (α) and 110 kDa (β). They have a carbohydrate content of about 30%, and both subunits contain primarily N-linked carbohydrates. Alpha and β subunits are differentially glycosylated, and there is evidence for gender-specific differences in glycosylation (Stroupe *et al.*, 1991).

NH_2-terminal proteolytic processing determines the expression of active and latent forms. NH_2-terminal sequencing of the detergent-solubilized meprin subunits from mouse kidney membranes revealed that the signal sequence and the prosequence are removed from the NH_2-terminus in α, whereas the mature ß subunit has the signal sequence removed but retains the prosequence. Secreted α subunits isolated from mouse urine also appear to retain the prosequence (Bond and Beynon, in press). The membrane-associated α subunits, with the prosequence removed, are in a fully active conformation. Retention of the prosequence in β correlates with latency of activity; β subunit activation requires proteolytic removal of the prosequence by trypsin-like proteases (Kounnas *et al.*, 1991). It is possible that while the β subunits are inactive at the cell surface under normal conditions, certain conditions trigger activation. The expression of latent β subunits in mouse kidney is in contrast to the intestinal mouse β subunits, and kidney and intestinal rat β subunits, which lack the prosequence (Gorbea *et al.*, 1994; Johnson and Hersh, 1992).

Meprin α and β subunits also differ with respect to COOH-terminal processing, which affects membrane association and secretion of the subunits (Marchand *et al.*, 1995). Beta subunits are not proteolytically processed at the COOH-terminus. They are type I integral membrane proteins, anchoring to the cell membrane through a COOH-terminal hydrophobic transmembrane-spanning domain. Meprin β subunits cannot be solubilized by high salt concentrations, phospholipase C treatment, reducing agents, or urea, but can be solubilized by detergents, such as octylglucoside. In contrast, a fragment of 10-12 kDa is proteolytically removed from the COOH-terminus of the meprin α subunit during maturation, consequently mature α subunits contain neither the hydrophobic membrane-spanning nor the EGF-like domain encoded by the message (Marchand *et al.*, 1994). Meprin α subunits can be solubilized by reducing agents and urea, indicating that their attachment to the membrane is dependent on S-S bridges and noncovalent interactions.

OLIGOMERIC ORGANIZATION AND MEMBRANE ASSOCIATION OF MEPRIN SUBUNITS

Analyses of purified meprin A (EC 3.4.24.18) from mouse kidney by nonreducing SDS-PAGE and analytical ultracentrifugation indicated that meprins consist of disulfide-linked dimers, and that under non-denaturing conditions dimers form tetramers (Marchand *et al.*, 1994). Immunoblot analyses of kidney brush border membrane extracts with subunit-specific antibodies detected several isoforms, and the expression of isoforms differed in the mouse strains examined. ICR mouse (a high meprin activity strain) kidneys contain αα and αβ dimers, but no ββ dimers can be detected; C3H/He mice (a low meprin activity strain) contain no α subunits, and only β oligomers. The proposed quaternary structure for meprin

A in ICR mouse kidney is indicated in Fig.2. The disulfide-linked dimers are proposed to form various tetrameric complexes at the cell surface. αα/αβ complexes and αβ/αβ complexes remain cell-associated through the transmembrane domain of β. There is clearly an excess of α subunits over β subunits in the renal brush border membranes, indicating that the αα/αβ complex may be the predominant form in the membranes. In addition to the membrane-anchored complexes, tetramers containing only α subunits are secreted into the urine. The oligomeric structure of meprin B (EC 3.4.24.63) in C3H mouse kidney is proposed to be that of a tetramer containing β subunits (Fig.2). The oligomeric structure of meprin A and B is unique among cell surface proteinases.

BIOSYNTHESIS AND PROTEOLYTIC PROCESSING

The biosynthesis of mouse meprin subunits has been studied in kidney organ culture, and more recently in transfected COS-1 and 293 cells (Hall *et al.*, 1993; Marchand *et al.*, 1995). A model for the biosynthesis of meprin A in mouse kidney is shown in Fig. 3. The

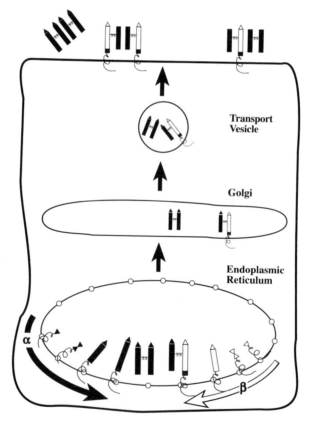

Figure 3. Diagram of the biosynthesis of meprin A in mouse epithelial cells. Alpha subunits are indicated by black symbols, β subunits are shown in white. The triangles indicate signal sequences (which are removed in the ER) and prosequences. Both subunits are initially synthesized with EGF-like domains (indicated by the grey boxes) and transmembrane-spanning and cytoplasmic domains. Alpha subunits are proteolytically processed in the ER at a cleavage site NH_2-terminal to the EGF-like domains. Beta subunits and secreted α subunits retain the prosequences, but α subunits at the cell surface lose the prosequence.

biosynthesis of meprin subunits begins in the rough endoplasmic reticulum (ER), where the subunits are translated, have their signal sequences removed, dimerize to disulfide-linked homo- and heterodimers, and obtain core N-linked glycosylation. Also in the ER the COOH-terminal fragment of the α subunit is proteolytically removed. The subunits then obtain complex glycosylation in the Golgi apparatus, and are transported to the cell surface through the constitutive secretory pathway. The mature membrane and extracellular forms are higher oligomers, most likely tetramers, composed of αα and/or αβ dimers. They either remain associated with the cell membrane (complexes containing β subunits) or are secreted (homo-oligomeric complexes of α subunits). In kidney, cell-associated α subunits lose the prosequence, presumably after they have reached the cell surface, whereas secreted α subunits retain the prosequence.

The forces that drive dimerization have not been determined. An unexpected finding is the apparent lack of ββ dimer formation in ICR mouse kidney. Alpha subunits are synthesized at twice the rate of β (Hall *et al.*, 1993); thus an excess of α subunits over β subunits might explain why αα and αβ dimers are favored, but undetermined additional factors may contribute to the preferential formation of αβ heterodimers over ββ dimers.

The enzymes involved in the proteolytic processing of meprins have not been identified. The COOH-terminal processing of α occurs in kidneys *in vivo* as well as in transfected COS-1 cells, and transfected cells have been used to study this cleavage event. Pulse-chase studies of expressed meprin subunits with or without a COOH-terminal epitope tag indicate that the COOH-terminal cleavage of α occurs after covalent dimerization and core glycosylation, but before the subunits leave the ER and acquire complex glycosylation (Marchand *et al.*, 1995). Expression of chimeric meprin subunits further established that the inserted domain of α is essential for processing. Insertion of the α subunit I domain into the β subunit sequence was sufficient to cause COOH-terminal proteolysis of β and secretion of the otherwise membrane-anchored subunits. Deletion of the I domain from the α subunit resulted in the retention of unprocessed α subunit precursors in the ER, indicating that removal of the COOH-terminal extension in α subunits may be a prerequisite for their transport out of the ER (Marchand *et al.*, 1995). Initial evidence from the expression of additional meprin α/β chimeras indicates that the transmembrane domain of α may be responsible for the retention of unprocessed α subunits (unpublished observation). Furin-like enzymes are not involved in the processing of the α COOH-terminus.

The biosynthesis of α, but not β, in mouse kidney involves the removal of a prosequence, and the subcellular site and the identity of the proprotein processing enzyme are not known. It is clear that the activation does not occur by autoactivation, a mechanism that is seen for example for furin-like processing enzymes (Steiner *et al.*, 1992). As in most other proenzymes, the cleavage of meprin α subunits occurs at a basic residue. The proprotein cleavage in α occurs following the residues R-T-R in α; the corresponding sequence in the β subunit is N-G-K. It is thus possible that the prosequence cleavage depends on a basic residue in position P3, or that the processing enzyme requires arginine in the P1 position. The latter hypothesis is further supported by the observation that the prosequence is proteolytically removed *in vivo* in mature rat kidney β subunits; rat and mouse β subunits differ in that there is an arginine residue preceding the cleavage site in rat β. It would be helpful to characterize the requirements for proprotein cleavage using site-directed mutagenesis, but no cell expression system has been identified to date that will process meprin subunit prosequences. Transfected COS-1 cells, 293 cells, or MDCK cells express both meprin α and β subunits as inactive zymogens.

Interestingly, secreted forms of α in mouse urine also retain the prosequence. This indicates that NH$_2$-terminal processing may occur at the cell surface by a membrane-associated protease, and that those subunits that have only transient interactions with the cell surface leave the cells as unprocessed zymogens. Notably, the amino acid sequence preced-

Intracellular Protein Catabolism
Koichi Suzuki and Judith S. Bond, eds.

ing the cleavage site in α closely resembles the cleavage site in the proform of another bursh border enzyme, lactase-phlorizin hydrolase (Oberholzer *et al.*, 1993). In the latter protein, cleavage following the sequence K-T-R occurs in a post-Golgi compartment, either in the trans-Golgi network, in transport vesicles, or at the cell surface (Lottaz *et al.*, 1993). It is thus possible that there is a common processing enzyme for meprins, lactase-phlorizin hydrolase, and other brush border enzymes. Current effort concentrates on the identification of this novel processing enzyme using a kidney organ culture system, and may provide insights not only into meprin biosynthesis but into the processing and regulation of other membrane proteins as well.

CONCLUSIONS

The identification of two types of meprin subunits, which are evolutionarily related but differ in tissue-specific expression, activities, structures, and membrane association, provides an opportunity to increase our understanding of the regulation of cell surface proteinases. In addition to the regulatory mechanisms at the transcriptional and post-translational level, there is the possibility that the activities of the subunits may be modulated by subunit interactions in the oligomers. The identification of interaction domains and the possible influence of oligomerization on proteolytic activity are being studied.

The work presented here also relates to the regulation of the intracellular transport and processing of membrane and secreted proteins. Future work will likely concentrate on the identification of membrane-bound proprotein processing enzymes, about which generally very little is known. Finally, the retention of uncleaved α subunits in the ER and the specific COOH-terminal cleavage of meprin α subunits by an unidentified resident ER proteinase make meprins an excellent model system to identify components of the proteolytic machinery in the endoplasmic reticulum.

ACKNOWLEDGMENT

This work was supported in part by USPHS Grant DK 19691.

REFERENCES

Beckmann, G., and Bork, P., 1993, An adhesive domain detected in functionally diverse receptors, *TIBS* 18:40-41.

Beynon, R. J., Shannon, J. D., and Bond, J. S., 1981, Purification and characterization of a metalloendoproteinase from mouse kidney, *Biochem. J.* 199:591-598.

Bode, W., Gomis-Rüth, F. X., Huber, R., Zwilling, R., and Stöcker, W., Structure of astacin and implications for activation of astacins and zinc-ligation of collagenases, *Nature* 358:164-166.

Bode, W., Gomis-Rüth, F. X., and Stöcker, W., 1993, Astacins, serralysins, snake venom and matrix metalloproteinases exhibit identical zinc-binding environments (HEXXHXXGXXH and Met-turn) and topologies and should be grouped into a common family, the 'metzincins', *FEBS Lett* 331:134-140.

Bond, J. S., Shannon, J. D., and Beynon, R. J., 1983, Certain mouse strains are deficient in a kidney brush-border metalloendopeptidase activity, *Biochem. J.* 209:251-255.

Bond, J. S., Beynon, R. J., Reckelhoff, J. F., and David, C. S., 1984, *Mep-1* gene controlling a kidney metallopeptidase is linked to the major histocompatibility complex in mice. *Proc. Natl. Acad. Sci. USA* 81:5542-5545.

Bond, J. S. and Beynon, R. J., 1995 The astacin family of metalloendopeptidases, *Protein Science*, 4: 1247-1261.

Butler, P. E., McKay, M. J., and Bond, J. S., 1987, Characterization of meprin, a membrane-bound metalloen-dopeptidase from mouse kidney. *Biochem. J.* 241:229-235.

Choudry, Y., and Kenny, A. J.,1991, Hydrolysis of transforming growth factor-alpha by cell surface peptidases *in vitro, Biochem. J.* 280:57-60.

Corbeil, D., Milhiet, P.-E., Simon, V., Ingram, J., Kenny, A. J., Boileau, G., and Crine, P., 1993, Rat endopeptidase-24.18 alpha subunit is secreted into the culture medium as a zymogen when expressed by COS-1 cells, *FEBS Lett* 335:361-366.

Dumermuth, E., Sterchi, E. E., Jiang, W., Wolz, R. L., Bond, J. S., Flannery, A.V., and Beynon, R. J., 1991, The astacin family of metalloendopeptidases, *J. Biol. Chem.* 266:21381-21385.

Dumermuth, E., Eldering, J. A., Grünberg, J., Jiang, W., and Sterchi, E. E., 1993, Cloning of the PABA peptide hydrolase α subunit (PPHα) from human small intestine and its expression in COS-1 cells, *FEBS Lett* 335:367-375.

Elaroussi, M. A., and DeLuca, H. F.,1994, A new member to the astacin family of metalloendopeptidases: A novel 1,25-dihydroxyvitamin D-3-stimulated mRNA from chorioallantoic membrane of quail, *Biochim. Biophys. Acta* 1217:1-8.

Gomis-Rüth, F. X., Grams, F., Yiallouros, I., Nar, H., Küsthardt, U., Zwilling, R., Bode, W., and Stöcker, W., 1994, Crystal structures, spectroscopic features, and catalytic properties of cpbalt (II), copper (II), nickel (II), and mercury (II) derivatives of the zinc endopeptidase astacin, *J. Biol. Chem.* 269:17111-17117.

Gorbea, C. M., Flannery, A. V., and Bond, J. S.,1991, Homo- and heterotetrameric forms of the membrane-bound metalloendopeptidases meprin A and B, *Arch. Biochem. Biophys.* 290:549-553.

Gorbea, C. M., Marchand, P., Jiang, W., Copeland, N. G., Gilbert, D. J., Jenkins, N. A., and Bond, J. S., 1993, Cloning, expression, and chromosomal localization of the mouse meprin β subunit, *J. Biol. Chem.* 268:21035-21043.

Gorbea, C. M., Beynon, R. J., and Bond, J. S.,1994, Meprins: activation, subunit interactions, and anchoring, in *Mammalian Brush Border Membrane Proteins II* (Lentze, M. J., Naim, H. Y., and Grand, R. J., eds.) pp 89-97, Thieme Medical Publishers, New York.

Hall, J. L., Sterchi, E. E., and Bond, J. S., 1993, Biosynthesis and degradation of meprins, kidney brush border proteinases, *Arch. Biochem. Biophys.* 307:73-77.

Jiang, W., Gorbea, C. M., Flannery, A. V., Beynon, R. J., Grant, G. A., and Bond, J. S., 1992, The alpha subunit of meprin A. Molecular cloning and sequencing, differential expression in inbred mouse strains, and evidence for divergent evolution of the alpha and beta subunits, *J. Biol. Chem.* 267:9185-9193.

Jiang, W., and Bond, J. S., 1992, Families of metallopeptidases and their relationships, *FEBS* 312:110-114.

Johnson, G. D., and Hersh, L.B., 1992, Cloning a rat meprin cDNA reveals the enzyme is a heterodimer, *J. Biol. Chem.*.267:13505-13512.

Kenny, A. J., O'Hare, M. J., and Gusterson, B. A., 1989, Cell-surface peptidases as modulators of growth and differentiation, Lancet 2:785-787.

Kitamoto, Y., Yuan, X., Wu, Q., McCourt, D. W., and Sadler, J. E., 1994, Enterokinase, the initiator of intestinal digestion, is a mosaic protease composed of a distinctive assortment of domains, *Proc. Natl. Acad. Sci. USA* 91:7588-7592.

Kounnas, M. Z., Wolz, R. L., Gorbea, C. M., and Bond, J.S., 1991, Meprin-A and -B: Cell surface endopeptidases of the mouse kidney, *J. Biol. Chem.* 266:17350-17357.

Lepage, T., Ghiglione, C., and Gache, C., 1992, Spatial and temporal expression pattern during sea urchin embryogenesis of a gene coding for a protease homologous to the human protein BMP-1 and to the product of the *Drosophila* dorsal-ventral patterning gene *tolloid, Development* 114, 147-164.

Lottaz, D., Oberholzer, T., Bahler, P., Semenza, G., and Sterchi, E. E., 1993, Maturation of human lactase-phlorizin hydrolase. Proteolytic cleavage of precursor occurs after passage throught Golgi complex, *FEBS* 313:270-276.

Maeno, M., Xue, Y., Wood, T., Ong, R. C., and Kung, H., 1993, Cloning and expression of cDNA encoding *Xenopus laevis* bone morphogenetic protein-1 during early embryonic development, *Gene* 134:257-261.

Marchand, P., Tang, J., and Bond, J. S., 1994, Membrane association and oligomeric organization of the α and β subunits of mouse meprin A, *J. Biol. Chem.* 269:15388-15393.

Marchand, P., Tang, J., Johnson, G. D., and Bond, J. S., 1995, COOH-terminal proteolytic processing of secreted and membrane forms of the α subunit of the metalloprotease meprin A: Requirement of the I domain for processing in the endoplasmic reticulum, *J. Biol. Chem.* 270:5449-5456.

Oberholzer, T., Mantei, N., and Semenza, G., 1993, The prosequence of lactase-phlorizin hydrolase is required for the enzyme to reach the plasma membrane. An intramolecular chaperone?, *FEBS Lett* 333:127-131.

Shimell, M. J., Ferguson, E. L., Childs, S. R., and O'Connor, M. B., 1991, The *Drosophila* dorsal-ventral patterning gene tolloid is related to human bone morphogenetic protein 1, *Cell* 67:469-481.

Steiner, D. F., Smeekens, S. P., Ohagi, S., and Chan, S. J., 1992, The new enzymology of precursor processing endoproteases, *J. Biol. Chem.* 267:23435-23438.

Sterchi, E. E., Naim, H. Y., Lentze, M. J., Hauri, H.-P., and Fransen, J. A., 1988, N-benzoyl-L-tyrosyl-p-aminobenzoic acid hydrolase: a metalloendopeptidase of the human intestinal microvillus membrane which degrades biologically active peptides, *Arch. Biochem. Phys.* 265:105-118.

Stöcker, W., Grams, F., Baumann, U., Reinemer, P., Gomis-Rüth, F. X., McKay, D. B., and Bode, W., 1995 The metzincins - topological and sequential relations between the astacins, adamalysins, serralysins, and matrixins (collagenases) define a superfamily of zinc-peptidases, *Protein Science,* 4: 823-840.

Stroupe, S. T., Butler, P. E., and Bond, J. S., 1989, Proteolytic activation of a plasma membrane proteinase, in *Mechanism and Regulation of Intracellular Proteolysis* (Katumuma, N., and Kominami, E., eds.) pp 178-187, Japan Scientific Soc. Press.

Takahara, K., Lyons, G. E., and Greenspan, D. S., 1994, Bone morphogenetic protein-1 and a mammalian *tolloid* homologue (mTld) are encoded by alternatively spliced transcripts which are differentially expressed in some tissues, *J.Biol. Chem.* 269:32572-32578.

Titani, K., Torff, H.-J., Hormel, S., Kumar, S., Walsh, K. A., Rödl, J., Neurath, H., and Zwilling, R., 1987, Amino acid sequence of a unique protease from the crayfish *Astacus fluviatilis, Biochemistry* 26:222-226.

Wolz, R. L., Harris, R. B., and Bond, J. S., 1991, Mapping the active site of meprin-A with peptide substrates and inhibitors, *Biochemistry* 30:8488-8493.

Wozney, J. M., Rose, V., Celeste, A. J., Mitsock, L. M., Whitters, M. J., Kriz, R. W., Hewick, R. M., and Wang, E. A., 1988, Novel regulators of bone formation: Molecular clones and activities, *Science* 242:1528-1533.

Yamaguchi, T., Fukase, M., Sugimoto, T., Kido, H., and Chihara, K., 1994, Purification of meprin from human kidney and its role in parathyroid hormone degradation, *Biol. Chem. Hoppe-Seyler* 375:821-824.

Yasamasu, S., Yamada, K., Akasaka, K., Mitsunaga, K., Iuchi, I., Shimada, H., and Yamagami, K., 1992, Isolation of cDNAs for LCE and HCE, two constituent proteases of the harching enzyme of *Oryzias latipes,* and concurrent expression of their mRNAs during development, *Dev. Biol.* 153:250-258.

INVOLVEMENT OF TISSUE INHIBITORS OF METALLOPROTEINASES (TIMPS) DURING MATRIX METALLOPROTEINASE ACTIVATION

H. Nagase,[1] K. Suzuki,[1] Y. Itoh,[1] C. -C. Kan,[2] M. R. Gehring,[2] W. Huang,[3] and K. Brew[3]

[1] Department of Biochemistry and Molecular Biology
University of Kansas Medical Center
Kansas City, Kansas 66160
[2] Agouron Pharmaceuticals, Inc
San Diego, California 92121-1122
[3] Department of Biochemistry and Molecular Biology
University of Miami School of Medicine
Miami, Florida 33101

INTRODUCTION

Matrix metalloproteinases (MMPs), also termed "matrixins", constitute a family of zinc metalloendopeptidases that participate in breakdown of extracellular matrix macromolecules (Woessner, 1991). These enzymes are considered to play an important role in many biological processes such as in reproduction, embryogenesis, tissue resorption, and in the control of cell behavior. Overproduction of matrixins is associated with various connective tissue diseases such as arthritis, periodontitis, glomerulonephritis, tissue ulceration as well as being connected with tumor cell invasion and metastasis (Woessner, 1991; Birkedal-Hansen et al., 1993).

Currently, eleven distinct members of the matrixin family have been identified. They include three collagenases [interstitial collagenase (MMP-1), neutrophil collagenase (MMP-8) and collagenase 3 (MMP-13) (Freije et al., 1994)]; two gelatinases [gelatinase A (MMP-2) and gelatinase B (MMP-9)]; two stromelysins [stromelysin 1 (MMP-3) and stromelysin 2 (MMP-10)]. Other members that have diverged activity from the above subgroups in structure and function include matrilysin (MMP-7), stromelysin 3 (MMP-11) (Murphy et al., 1993; Pei et al., 1994), metalloelastase (MMP-12) (Shapiro et al., 1993) and membrane-type MMP (MT-MMP/MMP-14) (Sato et al., 1994). MMPs are synthesized as preproenzymes and secreted as proenzymes from many cell types. They are homogenous with each other as indicated by significant sequence similarity and all, except matrilysin (MMP-7),

Intracellular Protein Catabolism, Edited by Koichi Suzuki and Judith Bond
Plenum Press, New York, 1996

contain three characteristic domains: a propeptide domain (77-87 residues), a catalytic domain (162-173 residues) and a C-terminal hemopexin/vitronectin-like domain (202-213 residues). MMP-7 is the smallest member and lacks a C-terminal domain. MMP-2 (gelatinase A) and MMP-9 (gelatinase B) have three repeats of 58 residues that are homologous to the fibronectin type II domain. These repeats are responsible for the ability of the two gelatinases to bind to gelatin (Collier *et al.*, 1992; Bányai *et al.*, 1994) and collagen (Murphy *et al.*, 1994). MMP-9 has an additional proline-rich collagen-like sequence (Wilhelm *et al.*, 1989), but its function is not known. MT-MMP (MMP-14) contains a 24-hydrophobic residue transmembrane domain after the C-terminal hemopexin/vitronectin-like domain (Sato *et al.*, 1994).

The importance of matrixins in extracellular matrix breakdown is emphasized by the following features: 1) all of the MMPs are secreted or plasma membrane-bound and they act on substrates outside of the cells at a physiological pH; 2) the synthesis of many MMPs is transcriptionally enhanced by a number of growth factors (e.g., epidermal growth factor, basic fibroblast growth factor, platelet-derived growth factor, nerve growth factor), inflammatory cytokines (e.g., interleukin 1 and tumor necrosis factor α), oncogenic cellular transformation, phorbol esters and various other agents (see Woessner, 1991; Birkedal-Hansen *et al.*, 1993 for reviews). The elevated synthesis of MMPs, on the other hand, may be negatively regulated by transforming growth factor β, retinoic acid, interferon γ, glucocorticoid, progesterone, and estrogen. (Woessner, 1991; Birkedal-Hansen *et al.*, 1993).

While the temporal gene expression of matrixins in various cell types is critical for matrix turnover, the biological activities of matrixins are further controlled extracellularly by the regulated activation of proMMPs and the inhibition of activated MMPs by endogenous inhibitors such as tissue inhibitors of metalloproteinases (TIMPs) and α_2-macroglobulin. Three TIMPs (TIMP-1, TIMP-2 and TIMP-3) have been identified and they are about 40% identical to each other in sequence including 12 conserved cysteines (Wilde *et al.*, 1994). TIMPs inhibit MMPs by forming 1:1 molar complexes.

ProMMPs are activated *in vitro* by proteinases, thiol-modifying agents (e.g., mercurial compounds, iodoacetic acid, N-ethyl maleimide, and oxidized glutathione), chaotropic agents, SDS, HOCl, and heat treatment (Woessner, 1991; Koklitis *et al*, 1991). The latency of proMMPs is maintained by interaction of an unpaired cysteine in the conserved sequence PRCG[V/N]PD in the propeptide with the zinc atom at the active site of the enzyme (Salowe *et al.*, 1992; Holz *et al*, 1992), which prevents the formation of a water-zinc complex required for the catalytic activity of the enzyme. To explain the diverse means of activation of proMMPs VanWart and colleagues proposed a "cysteine switch" model (Springman *et al.*, 1990; VanWart and Birkedal-Hansen, 1990). A critical step in activation of proMMPs involves disruption of the Cys-Zn interaction. One activation method is by treatment of proMMPs with non-proteolytic reagents which either induce the conformational changes in the zymogen or react with the SH group of the unique cysteinyl residue in the propeptide. Thiol-reagents interact with the dissociated Cys, which prevents the reassociation of Cys and Zn^{2+}. When proMMPs are treated with a proteinase, the proteinase attacks a region near the middle of propeptide resulting in a formation of an intermediate which retains a portion of propeptide including the PRCG[V/N]PD sequence. The removal of the N-terminal half of the propeptide destabilizes the Cys-Zn interaction and activates the proMMPs. This renders the final activation site to be susceptible to a second proteolysis. This reaction is conducted by an intermolecular reaction of the intermediates or by activated MMPs (Nagase *et al.*, 1990; Crabbe *et al.*, 1992; 1994). MMP-3 plays a critical role in the generation of fully active collagenases (MMP-1 and MMP-8) (Suzuki *et al.*, 1990; Knäuper *et al.*, 1993). This unique "stepwise" proteolytic activation mechanism applies to most proMMPs, which led us to hypothesize that TIMPs may regulate the MMP activities by interfering with activation processes.

Additional complexities are observed in the activation of progelatinase as their precursors (progelatinase A and progelatinase B) form 1:1 molar stoichiometry complexes with TIMP-2 and TIMP-1, respectively. Progelatinase/TIMP complexes are formed through the interaction of the C-terminal hemopexin/vitronectin-like domain of the zymogen and the C-terminal domain of the inhibitor (Murphy *et al.*, 1991; 1992; Howard and Banda, 1991; Fridman *et al.*, 1992; Goldberg *et al.*, 1992; Willenbrock *et al.*, 1993). The activation of progelatinase/TIMP complexes does not readily occur as the N-terminal inhibitory domain of TIMP in the complex is exposed and capable of inhibiting other MMPs.

In this communication we discuss a possible role of TIMPs during activation processes of proMMP-3 and progelatinase/TIMP complexes.

INTERACTION OF AN MMP-3 INTERMEDIATE WITH TIMP-1

A number of proteinases can activate proMMP-3 including trypsin, chymotrypsin, neutrophil elastase, cathepsin G, plasmin, plasma kallikrein, thermolysin, tryptase and chymase (Oakda *et al.*, 1988; Okada and Nakanishi, 1990; Suzuki *et al.*, 1995). The studies with neutrophil elastase, plasma kallikrein and chymotrypsin have demonstrated that all these proteinases initially attack the sequence $F^{34}VRRKD^{39}$ in the propeptide (Nagase *et al.*, 1990). The resulting intermediates undergo further processing and hydrolysis of the H^{82}-F^{83} bond occurs by an intermolecular reaction of the intermediates. To test whether the initial intermediate can bind to TIMP, we have constructed a variant of cDNA for proMMP-3 which lacks the signal peptide, the first 34 propeptide, and the C-terminal domain ([V^{35}]proMMP-3(ΔC); the N-terminal of this protein is V^{35}. The recombinant protein was expressed in *E. coli*. [V^{35}]proMMP-3(ΔC) was isolated from the soluble fraction of the *E. coli* extract in the presence of 5 mM EGTA. The addition of 5 mM $CaCl_2$ and 50 μM $ZnCl_2$ to [V^{35}]proMMP-3(ΔC) rapidly converted to MMP-3(ΔC) of M_r 24,000 in a dose dependent manner, suggesting that the removal of the 48-residue propeptide is due to the bimolecular reaction of [V^{35}]proMMP-3(ΔC). We have then tested whether TIMP-1 can form a complex with [V^{35}]proMMP-3(ΔC) prior to its processing to MMP-3(ΔC). Since TIMP-1 (28 kDa) and MMP-3(ΔC) have similar mobilities on SDS/polyacrylamide gel, we used a recombinant N-TIMP-1 terminal domain (N-TIMP-1) of 14.5 kDa expressed in *E. coli*. The K_i values of N-TIMP-1 against MMP-3 is 1.2 nM, about 1/10 of that of the full-length TIMP-1.

Incubation of [V^{35}]proMMP-3(ΔC) with N-TIMP-1 at a 1:1 molar ratio inhibited the conversion to MMP-3(ΔC). The formation of a complex between [V^{35}]proMMP-3(ΔC) and N-TIMP-1 was confirmed by isolating the complex by anti-MMP-3 affinity chromatography. However, upon prolonged incubation of the complex (>2 h incubation at 37°C at a concentration of 10μM) [V^{35}]proMMP-3(ΔC) was gradually converted to the 25-kDa MMP-3(ΔC). These results suggested that while the initial MMP-3 intermediate generated by proteolytic activation can form a complex with TIMP, their interaction is relatively weak (Fig. 1). Nonetheless, during this process, little proteolytic activity was detected in MMP-3. These observations indicate that TIMP-1 can not only inhibit the activated enzyme but also interferes with the activation process of proMMP-3.

The activation processes of proMMP-2 (progelatinase A) and proMMP-9 (progelatinase B) and their complexes with respective TIMPs by 4-aminophenylmercuric acetate and proteinases are summarized in Figs. 2 and 3. Both proMMP-2 and proMMP-9 are activated by treatment with 4-aminophenylmercuric acetate (APMA), but proMMP-2 is resistant to activation by many proteinases (Okada *et al.*, 1990). ProMMP-9 is activated by MMP-3, trypsin and partially by plasmin, chymotrypsin and cathepsin G (Morodomi *et al.*, 1990; Goldberg *et al.*, 1992; Ogata *et al.*, 1992; Okada *et al.*, 1992).

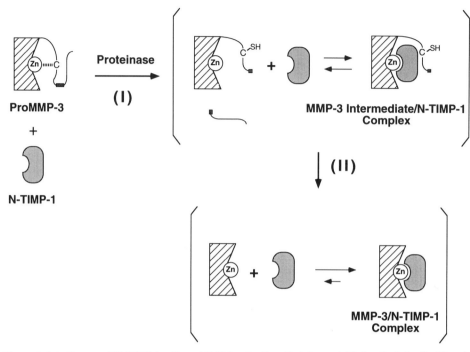

Figure 1. Interference of N-TIMP-1 with proMMP-3 activation. Proteinases initially attack the "bait" region in the propeptide indicated by a hatched box and generate an intermediate. The binding of the intermediate with TIMP-1 was demonstrated by interaction of the recombinant N-TIMP-1. Interaction of the MMP-3 intermediate and N-TIMP-1 is not tight and ACTIVATION OF PROGELATINASE/TIMP COM-PLEXES[V^{35}]proMMP-3(ΔC) with the recombinant [V^{35}]proMMP-3(ΔC) is gradually converted to the mature MMP-3 (ΔC) which also forms complexes with N-TIMP-1 consequently.

When proMMP-2 formed a complex with TIMP-2, treatment of the complex with APMA at 37°C caused conformational changes in proMMP-2 and dissociated the Cys-Zn interaction of the zymogen, but gelatinolytic activity did not develop. The dissociation of the Cys-Zn interaction, which was demonstrated by the availability of the SH group of Cys 75 for alkylation upon APMA treatment, allowed the N-terminal domain of TIMP-2 to interact with the catalytic site of the "activated" proMMP-2 without processing the propeptide. The proMMP-2/TIMP-2 complex can bind to another molecule of the active MMP and inhibit its activity. Once a ternary proMMP-2/TIMP-2/MMP complex was formed, treatment of the complex with APMA generated the 65-kDa active MMP-2, and the majority of active MMP-2 remained as a ternary complex (Fig. 2).

A physiological activator of proMMP-2 is found on the plasma membrane of fibroblasts and neoplastic cells treated with concanavalin A or a phorbol ester (Overall *et al.*, 1990; Brown *et al.*, 1990). The recently discovered MT-MMP has been characterized as a proMMP-2 activator (Sato *et al.*, 1994). Activation of proMMP-2 by the membrane-bound activator requires the C-terminal domain of the zymogen (Murphy *et al.*, 1992). We, therefore, postulated that the formation of proMMP-2/TIMP-2 complex may prevent the activation of proMMP-2. To test this we have prepared the plasma membrane fraction from human uterine cervical fibroblasts treated with con-

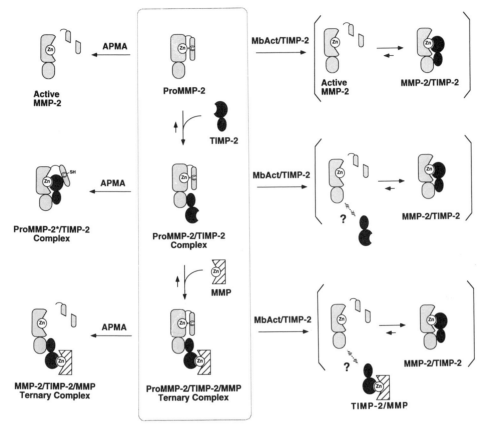

Figure 2. Activation steps of proMMP-2 and the proMMP-2/TIMP-2 complexes. See text for details. MbAct, plasma membrane fraction prepared from the concanavalin A-treated human uterine cervical fibroblasts. AbAct contains endogenous TIMP-2.

canavalin A. Incubation of the proMMP-2/TIMP-2 complex with the plasma membrane converted proMMP-2 to the 65-kDa form, although the conversion is not as effective as for free proMMP-2 (Fig. 2). This suggests that the site in the C-terminal domain that interacts with the membrane activator is probably different from that interacting with TIMP-2. When the proMMP-2/TIMP-2 complex was reacted with MMP-3 and then with the membrane, proMMP-2 was more readily converted to the 65-kDa form. All of the 65-kDa forms generated by plasma membranes from the concanavalin A-treated human uterine cervical fibroblasts did not exhibit much proteolytic activity, suggesting that the activated MMP-2 is inhibited by an endogenous inhibitor (Fig. 2). Reversed zymographic analysis of the membrane preparation showed that TIMP-2 was associated to the membrane whereas the membranes from the untreated cells lacked TIMP-2. It is, therefore, concluded that TIMP-2 present on the membrane reacted with the activated MMP-2 as in the case of free proMMP-2. It is, however, not known whether the inhibition of the membrane-activated proMMP-2/TIMP-2 complex is due to the TIMP-2 bound to proMMP-2 or that present on the membrane. Nonetheless, the interaction of the membrane-bound TIMP-2 and MMP-2 was relatively slow: a complete inhibition of MMP-2 by the membrane-bound TIMP-2 requires in-

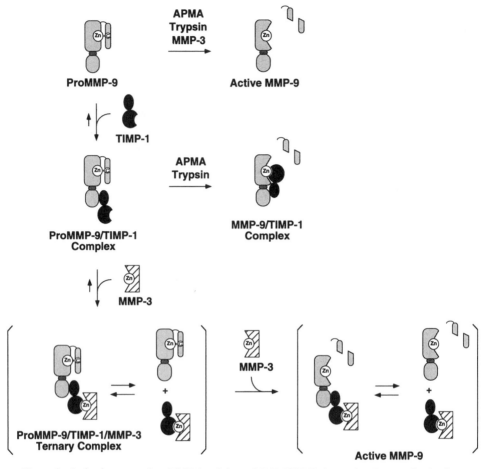

Figure 3. Activation steps of proMMP-9 and the proMMP-9/TIMP-1 complex. See text for details.

cubation at 37°C for 6 h. Therefore, when proMMP-2 is activated on the cell surface, MMP-2 may be active for a certain period of time.

ProMMP-9 (progelatinase B) forms a complex with TIMP-1 also through the C-terminal domains. Activation of the proMMP-9/TIMP-1 complex with trypsin or APMA generated an active MMP-9 by converting the zymogen to a low M_r species, but it was also inhibited by TIMP-1. A catalytic quantity of MMP-3 activates free proMMP-9 (Okada *et al.*, 1992; Goldberg *et al.*, 1992; Okada *et al.*, 1992) but not the proMMP-9/TIMP-1 complex (Goldberg *et al.*, 1992). In the latter case, MMP-3 binds to TIMP-1 and forms a ternary complex, proMMP-9/TIMP-1/MMP-3. The formation of the ternary complex weakened the interaction between proMMP-9 and TIMP-1 and it partially dissociated into free proMMP-9 and a TIMP-1/MMP-3 complex (Fig. 3). When MMP-3 exceeded a molar stoichiometry to the complex, excess MMP-3 was capable of activating proMMP-9. Under these conditions, since TIMP-1 in the complex was masked by MMP-3, the activated MMP-9 exhibited full proteolytic activity (Fig. 3). Thus, when proMMP-9 is bound to TIMP-1, masking of TIMP-1 by an active MMP is critical for the expression of MMP-9 activity.

CONCLUSIONS

The ability of the recombinant MMP-3 intermediate $[V^{35}]$proMMP-3(ΔC) to bind to N-TIMP-1 before its conversion to the mature MMP-3(ΔC) suggests that the endogenous inhibitors of MMPs can not only inhibit active MMPs but also interfere with steps in their activation mechanisms. The unique stepwise activation mechanisms of proMMPs therefore provide greater opportunities for endogenous inhibitors to regulate MMP activities. Such mechanisms might have evolved to control the balance between degradation and synthesis of extracellular matrix that takes place during tissue remodeling and reorganization.

The abilities of progelatinases A and B to bind to TIMP-2 and TIMP-1, respectively, further regulate the MMP activities and progelatinase activation. In the case of the proMMP-2/TIMP-2 complexes, proMMP-2 is activated by APMA, but it does not display gelatinolytic activity as it is readily inhibited by TIMP-2 bound to the zymogmen. In contrast, the membrane-bound activator can convert the proMMP-2 component of the complex to the 65-kDa active form as well as activating proMMP-2. However, activated MMP-2 is inhibited by TIMP-2 when the plasma membrane was prepared from concanavalin A-treated human uterine cervical fibroblasts. Although the rate of interaction of the cell-surface activated MMP-2 and TIMP-2 could not be determined using crude plasma membrane preparations, it is not very rapid. The activated MMP-2, therefore, probably escapes from the inhibition by TIMP-2 and exhibits its proteolytic activity for a certain period of time.

The proMMP-2/TIMP-2 complex inhibits other active MMPs by forming a ternary complex (Fig. 2). Such a complex can be activated by APMA and the membrane activator. In the latter case, activated MMP-2 is again inhibited by TIMP-2, which presumably is derived from the membrane. Interestingly, TIMP-2 on the membranes does not inhibit MMP-3. Therefore, the association of TIMP-2 with the plasma membrane is an additional control mechanism for MMP-2 activity. It is not known, however, if the expression of the membrane-bound activator and the binding of TIMP-2 to the membrane are under coordinated regulation.

The activation of the proMMP-9/TIMP-1 complex required saturation of TIMP-1 with active MMPs. This mode of activation results in a shift of the predominant MMP activity from one type to another: free, active MMP is first inhibited by the TIMP-1 in the complex, which in turn allows proMMP-9 to be activated by a small amount of MMP-3.

Taken together, these observations emphasize that the expression of required MMP activities during tissue catabolism is not only under the control of the temporal expression of specific MMP and TIMP genes, but also it is closely related to the involvement of TIMPs during and after proMMP activation processes. Shifts of any of these balances therefore may result in tissue destruction or abnormal matrix deposition.

ACKNOWLEDGEMENTS

We thank Denise Byrd for typing the manuscript. This work was supported by NIH Grants AR39189 and AR40994.

REFERENCES

Bányai, M. J., Tordai, H., and Patthy, L., 1994, The gelatin-binding site of human 72 kDa type IV collagenase (gelatinase A), *Biochem. J.* 298:403-407.

Birkedal-Hansen, H., Moore, W. G. I., Bodden, M.D., Windsor, L. J., Birkedal-Hansen, B., DeCarlo, A., and Engler, J.A., 1993, Matrix metalloproteinases: a review, *Crit. Rev. Oral Biol. Med.* 4:197-250.

Brown, P.D., Levy, A.T., Margulies, I. M. K., Liotta, L. A., and Stetler-Stevenson, W. G., 1990, Independent expression and cellular processing of M_r 72,000 type IV collagenase and interstitial collagenase in human tumorigenic cell lines, *Cancer Res.* 50:6184-6191.

Collier, I. E., Krasnov, P. A., Strongin, A. Y., Birkedal-Hansen, H., and Goldberg, G. I., 1992, Alanine scanning mutagenesis and functional analysis of the fibronectin-like collagen-binding domain from human 92-kDa type IV collagenase, *J. Biol. Chem.* 267:6776-6781.

Crabbe, T., Willenbrock, F., Eaton, D., Hynds, P., Carne, A. F., Murphy, G., and Docherty, A. J. P., 1992, Biochemical characterization of matrilysin. Activation conforms to the stepwise mechanisms proposed for other matrix metalloproteinases, *Biochemistry.* 31:8500-8507.

Crabbe, T., O'Connell, J. P., Smith, B. J., and Docherty, A. J. P., 1994, Reciprocated matrix metalloproteinase activation: A process performed by interstitial collagenase and progelatinase A, *Biochemistry* 33:14419-14425.

Freije, J. M. P., Diez-Itza, I., Balbin, M., Sánchez, L. M., Blasco, R., Tolivia, J., and López-Otín, C., 1994, Molecular cloning and expression of collagenase-3, a novel human matrix metalloproteinase produced by breast carcinomas, *J. Biol. Chem.* 269:16766-16773.

Fridman, R., Fuerst, T. R., Bird, R. E., Hoyhtya, M., Oelkuct, M., Kraus, S., Komarek, D., Liotta, L. A., Berman, M. L., and Stetler-Stevenson, W. G., 1992, Domain structure of human 72-kDa gelatinase/type IV collagenase. Characterization of proteolytic activity and identification of the tissue inhibitor of metalloproteinase-2 (TIMP-2) binding regions, *J. Biol. Chem.* 267:15398-15404.

Goldberg, G. I., Strongin, A., Collier, I. E., Genrich, L. T., and Marmer, B. L., 1992, Interaction of 92-kDa type IV collagenase with the tissue inhibitor of metalloproteinases prevents dimerization, complex formation with interstitial collagenase, and activation of the proenzyme with stromelysin, *J. Biol. Chem.* 267:4583-4591.

Holz, R. C., Salowe, S. P., Smith, C. K., Cuca, G. C., and Que L., Jr., 1992, EXAFS evidence for a "cysteine switch" in the activation of prostromelysin. *J. Am. Chem. Soc.* 114:9611-9614.

Howard, E. W., and Banda, M. J., 1991, Binding of tissue inhibitor of metalloproteinases 2 to two distinct sites on human 72-kDa gelatinase, *J. Biol. Chem.* 266:17972-17977.

Knäuper, V., Wilhelm, S. M., Seperack, P. K., DeClerck, Y. A., Langley, K. E., Osthues, A., and Tschesche, H., 1993, Direct activation of human neutrophil procollagenase by recombinant stromelysin, *Biochem. J.* 295:581-586.

Koklitis, P., Murphy, G., Sutton, C., and Angel, S., 1991, Purification of recombinant human prostromelysin. Studies on heat activation to give high-M_r and low-M_r active forms, and a comparison of recombinant with natural stromelysin activities, *Biochem. J.* 276:217-221.

Morodomi, T., Ogata, Y., Sasaguri, Y., Morimatsu, M., and Nagase, H., 1992, Purification and characterization of matrix metalloproteinase 9 from U937 monocytic leukemia and HT1080 fibrosarcoma cells, *Biochem. J.* 285:603-611.

Murphy G., Houbrechts, A., Cockett, M.I., Williamson, R. A., O'Shea, M., and Docherty, A. J. P., 1991, The N-terminal domain of tissue inhibitor of metalloproteinases retains metalloproteinase inhibitory activity, *Biochemistry* 30:8097-8102.

Murphy, G., Willenbrock, F., Ward, R. V., Cockett, M. E., Eaton, D., and Docherty, A. J. P., 1992, The C-terminal domain of 72 kDa gelatinase A is not required for catalysis, but is essential for membrane activation and modulates interactions with tissue inhibitors of metalloproteinases, *Biochem. J.* 283:637-641.

Murphy, G., Segain, J.-P., O'Shea, M., Cockett, M., Ioannoum C., Lefebvre, O., Chambon, P., and Basset, P., 1993, The 28-kDa N-terminal domain of mouse stromelysin-3 has the general properties of a weak metalloproteinase, *J. Biol. Chem.* 268:15435-15441.

Murphy, G., Nguyen, Q., Cockett, M. I., Atkinson, S. J., Allen, J. A., Knight, G. C., Willenbrock, F., and Docherty, A. J. P., 1994, Assessment of the role of the fibronectin-like domain of gelatinase A by analysis of a deletion mutant, *J. Biol. Chem.* 269:6632-6636.

Nagase, H., Enghild, J. J., Suzuki, K., and Salvesen G., 1990, Stepwise activation mechanisms of the precursor of matrix metalloproteinase 3 (stromelysin) by proteinases and (4-aminophenyl) mercuric acetate, *Biochemistry* 39:5783-5789.

Ogata, Y., Enghild, J. J., and Nagase, H., 1992, Matrix metalloproteinase 3 (stromelysin) activates the precursor for the human matrix metalloproteinase 9, *J. Biol. Chem.* 267:3581-3584.

Okada, Y., and Nakanishi, I., 1989, Activation of matrix metalloproteinase 3 (stromelysin) and matrix metalloproteinase 2 ("gelatinase") by human neutrophil elastase and cathepsin G, *FEBS Lett* 249:353-356.

Okada, Y., Harris, E. D., Jr., Nagase,H., 1988, The precursor of a metalloendopeptidase from human rheumatoid synovial fibroblasts. Purification and mechanisms of activation by endopeptidases and 4-aminophenylmercuric acetate, *Biochem. J.* 254:731-741.

Okada, Y., Morodomi, T., Enghild, J.J., Suzuki, K., Yasui, A., Nakanishi, I., Salvesen, G., and Nagase, H., 1990, Matrix metalloproteinase 2 from human rheumatoid synovial fibroblasts. Purification and activation of the precursor and enzymic properties, *Eur. J. Biochem.* 194:721-730.

Okada, Y., Gonoji, Y., Naka, K., Tomita, K., Nakanishi, I., Iwata, K., Yamashita, K., and Hayakawa, T., 1992, Matrix metalloproteinase 9 (92-kDa gelatinase/type IV collagenase) from HT 1080 human fibrosarcoma cells. Purification and activation of the precursor and enzymic properties, *J. Biol. Chem.* 267:21712-21719.

Overall, C. M., and Sodek, J., 1990, Concanavalin A produces a matrix-degradative phenotype in human fibroblasts. Induction and endogenous activation of collagenase, 72-kDa gelatinase and Pump-1 is accompanied by the suppression of the tissue inhibitor of matrix metalloproteinases, *J. Biol. Chem.* 265:21141-21151.

Pei, D., Majmudar, G., and Weiss, S. J., 1994, Hydrolytic inactivation of a breast carcinoma cell-derived serpin by human stromelysin-3, *J. Biol. Chem.* 269:25849-25855.

Salowe, S. P., Marcy, A. I., Cuca, G. C., Smith, C. K., Kopka, I. E., Hagmann, W. K., and Hermes, J. D., 1992, Characterization of zinc-binding sites in human stromelysin-1: stoichiometry of the catalytic domain and identification of a cysteine ligand in the proenzyme, *Biochemistry* 31:4535-4540.

Sato, H., Takino, T., Okada, Y., Cao, J., Shinagawa, A., Yamamoto, E., and Seiki, M., 1994, A matrix metalloproteinase expressed on the surface of invasive tumor cells, *Nature* 370:61-65.

Shapiro, S. D., Kobayashi, D. K., and Ley, T. J., 1993, Cloning and characterization of a unique elastolytic metalloproteinase produced by human alveolar macrophages. *J. Biol. Chem.* 268:23824-23829.

Springman, E. B., Angleton, E. I., Birkedal-Hansen, H., and VanWart, H. E., 1990, Multiple modes of activation of latent human fibroblast collagenase: Evidence for the role of a Cys[73] active-site zinc complex in latency and a "cysteine switch" mechanism for activation, *Proc. Natl. Acad. Sci. USA* 87:364-368.

Suzuki, K., Enghild, J. J., Morodomi, T. Salvesen, G., and Nagase, H., 1990, Mechanisms of activation of tissue procollagenase by matrix metalloproteianse 3 (stromelysin), *Biochemistry* 29:10261-10270.

Suzuki, K., Lees, M., Newlands, G. F. J., Nagase, H., and Woolley, D. E., 1995, Activation of precursors for matrix metalloproteinases 1 (interstitial collagenase) and 3 (stromelysin) by rat mast-cell proteinases I and II, *Biochem. J.* 305:301-306.

VanWart, H. E., and Birkedal-Hansen, H., 1990, The cysteine switch: A principle of regulation of metalloproteinase activity with potential applicability to the entire matrix metalloproteinase gene family, *Proc. Natl. Acad. Sci. USA* 87:5578-5582.

Wilde, Craig G., Hawkins, P. R., Coleman, R. T., Levine, W. B., Delegeane, A. M., Okamoto, P. M., Ito, L. Y., Scott, R. W., and Seilhamer, J. J., 1994, Cloning and characterization of human tissue inhibitor of metalloproteinases-3, *DNA Cell Biol.* 13:711-718.

Wilhelm, S. M., Collier, I. E., Marmer, B. L., Eisen, A. Z., Grant, G. A., and Goldberg, G. I., 1989, SV40-transformed human lung fibroblasts secrete a 92-kDa type IV collagenase which is identical to that secreted by normal human macrophages, *J. Biol. Chem.* 264:17213-17221.

Willenbrock, F., Crabbe, T., Slocombe, P. M., Sutton, C. W., Docherty, A. J. P., Cockett, M. I., O'Shea, M., Brocklehurst, K., Philips, I. R., and Murphy, G., 1993, The activity of the tissue inhibitors of metalloproteinases is regulated by C-terminal domain interactions: A kinetic analysis of the inhibition of gelatinase A, *Biochemistry* 32, 4330-4337.

Woessner, J. F., Jr., 1991, Matrix metalloproteinaes and their inhibitors in connective tissue remodeling, *FASEB J.* 5:2145-2154.

4

STRUCTURE AND FUNCTION OF A NOVEL ARGININE-SPECIFIC CYSTEINE PROTEINASE (ARGINGIPAIN) AS A MAJOR PERIODONTAL PATHOGENIC FACTOR FROM *PORPHYROMONAS GINGIVALIS*

Kenji Yamamoto[1], Tomoko Kadowaki[1], Kuniaki Okamoto[1],
Masahiro Yoneda[1], Koji Nakayama,[1] Yoshio Misumi,[2] and Yukio Ikehara[2]

[1] Departments of Pharmacology and Microbiology
Kyushu University
Faculty of Dentistry
Fukuoka 812-82, Japan
[2] Department of Biochemistry
Fukuoka University School of Medicine
Fukuoka 814-20, Japan

INTRODUCTION

Progressive periodontal disease is characterized by acute progressive lesions of gingival connective tissues, excessive leukocyte infiltration, and occurrence of a characteristic microflora (1). *Porphyromonas gingivalis* (formerly *Bacteroides gingivalis*), a Gram-negative anaerobic rod-shaped bacterium, is frequently isolated from cases of advanced periodontitis in humans and implicated in the etiology of the disease (review in references 2 and 3). This organism is known to produce a number of potential virulence factors, including fimbriae, hemagglutinins, lipopolysaccharides, and various hydrolytic enzymes (review in reference 4 and 5). Among the candidate virulence factors, the proteolytic enzymes are of special importance since they have the capacity closely related to the virulence of the organism. For example, the ability to degrade physiologically important proteins, including collagens and fibronectin, is likely to be associated with the direct periodontal tissue breakdown. Also, the ability to degrade immunoglobulins and complement factors appears to be involved in the disruption of normal host defense mechanisms. Further, these proteases may contribute to the bacterial nutrition during infection. A number of *P. gingivalis* proteinases have so far been described in both the cell-associated and secretory forms (review in reference 4 and 5). Recent molecular genetic approaches have also revealed that a variety of distinct proteinases are expressed by the organism (6-11). However, there is little agreement

Intracellular Protein Catabolism, Edited by Koichi Suzuki and Judith Bond
Plenum Press, New York, 1996

as yet on the number and molecular relationship of individual proteinases produced by the organism . Like other putative virulence factors reported, little is known about the precise mechanisms by which the proteinases function as virulence factors.

Proteases from this organism include collagenolytic enzymes (12-17), trypsin-like enzymes (review in reference 5), IgA proteases (18-20), and glycylprolyl dipeptidylami-nopeptidase (review in reference 5). Several groups have recently reported the isolation of trypsin-like proteinases from *P. gingivalis*, all of which are thiol-dependent (review in reference 5). More recent data have also suggested that several of the cysteine proteinases, which are specific for arginine and/or lysine peptide bonds, are present in processed forms of high molecular mass precursors. For example, the sequence of an arginine-specific cysteine proteinase (argingipain) has revealed that the 50 kDa extracellular mature enzyme from the strain 381 is derived from a gene coding for a 109 kDa translation product (21). Likewise, the Prt protease has been thought to be initially synthesized as a 98 kDa precursor and then processed to the 53 kDa mature enzyme (22). However, little information is available about the question of whether or not the processing is autocatalytic or results from the activity of other proteases.

Previous studies from our laboratory have shown that *P. gingivalis* 381 secretes the unique cysteine proteinase, termed "argingipain" due to its peptide cleavage specificity at arginine residue at the P1 position, the source from *P. gingivalis* species, and the classification as a cysteine proteinase related to papain (23). In this paper we describe the enzymatic and structural characteristics of this enzyme and its possible roles as a periodontal pathogenic factor in the pathogenic process of progressive periodontal disease.

ENZYMATIC PROPERTIES

A major cysteine proteinase was purified from the culture supernatant of *P. gingivalis* 381 by conventional chromatographic techniques, involving ammonium sulfate fractionation, chromatography on DEAE-Sephacel, CM-Toyopearl 650S, TSKgel G2000SW, and isoelectric focusing (23). The molecular mass of the purified enzyme was estimated to be 44 and 50 kDa by sodium dodecyl sulfate/polyacrylamide gel electrophoresis under reduced and non-reduced conditions, respectively, and to be 50 kDa by gel filtration, indicating that the enzyme is composed of a single polypeptide chain with a molecular mass of 44-50 kDa. Analysis of the enzymatic properties has revealed several distinctive features for this enzyme. The proteolytic activity is absolutely thiol-dependent, but the enzyme also has in part characteristics of both metallo and serine endopeptidases, as shown by the inhibition of activity by metal chelators, chymostatin, and the chloromethyl ketones of tosyl-L-lysine and tosyl-L-phenylalanine. However, protease inhibitors, such as cystatins (e.g., human cystatin S), serpins (e.g., α1-antichymotrypsin), and tissue inhibitor of metalloproteinases (e.g., human TIMP-1 and TIMP-2), had no effects on the activity, suggesting its evasion from normal host defense systems in vivo. Studies with synthetic peptide substrates and the NH$_2$-terminal sequence analysis of its autodegradation products have revealed the most restricted cleavage specificity of the enzyme at the peptide bonds containing Arg in the P1 position and nonpolar amino acid residues at P2, P3, P5, P6, and P'1 positions (Fig. 1). In particular, the presence of hydrophobic amino acid residues at the P2 and/or P3 positions was essential for efficient cleavage.

The best substrates for the enzyme are found to be Boc-Phe-Ser-Arg-MCA and Z-Phe-Arg-MCA which are preferred substrates for trypsin-like serine proteinases and cathepsins B- and L-like cysteine proteinases, respectively. The optimal pH for hydrolysis of the two synthetic substrates by the enzyme was around 7.5. The optimal pH for hydrolysis

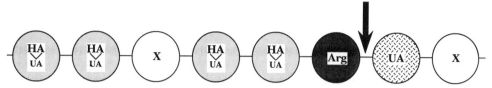

Figure 1. Substrate cleavage specificity of argingipain. The arrow indicates the cleavage site by the enzyme. The positions of P2, P3, P5, and P6 showed the predominance of hydrophobic amino acids rather than uncharged amino acids for efficient cleavage. HA: hydrophobic amino acid, UA: uncharged amino acid.

of protein substrates, including hemoglobin and casein, is also found to be the same as that obtained with the synthetic substrates.

It should be noted that argingipain has the ability to degrade acid-soluble human type I and IV collagens by incubation at 37°C for 2 h (23). The extensive degradation of the type I collagen is observed even by incubation at 20°C. Since the type collagen I is a major component of the connective tissue in gingiva, periodontal ligament, and alveolar bone, and since the type IV collagen is a major component of the basement membrane, it is more likely that the enzyme plays an important role in the direct destruction of periodontal tissue components, which is known as one of the most important pathological conditions in periodontal disease. It is also noted that human IgG and IgA are efficiently degraded by the enzyme, suggesting its ability to disrupt a part of the humoral immunity system. In addition, argingipain has shown another important ability related to the virulence of *P. gingivalis*. The chemiluminescence (CL) response of polymorphonuclear leukocytes (PMNs), which is known to be closely related with the bactericidal activity, is strongly suppressed by the enzyme in the presence of cysteine (23). The observed suppression of the CL response by the enzyme is abolished by the presence of leupeptin or by the addition of the anti-argingipain IgG. The results indicate that the enzyme is important in relation to disruption of the normal host defense mechanisms.

STRUCTURAL CHARACTERIZATION

To know the details of structural and functional features of the enzyme and to clarify the mechanism for its extracellular translocation across the two membranes of the organism, it is essential to determine the entire structure of the enzyme. For this, we have isolated and sequenced the gene for the enzyme (12). A DNA fragment for the enzyme was selectively amplified by polymerase chain reaction using mixed oligonucleotide primers designed from the NH$_2$-terminal amino acid sequence of the purified enzyme. The nucleotide sequence of the isolated DNA and the deduced amino acid sequence are shown in Fig. 2. The nucleotide sequence (3094 nucleotides) includes the complete coding region and parts the of 5'- and 3'-noncoding regions. The open reading frame consisting of 2973 nucleotides encodes 991 amino acid residues with a calculated mass of 108,780 Da, which is much larger than that of the purified extracellular enzyme (44 - 50 kDa).

The deduced amino acid sequence exhibited no significant similarity to the sequences of representative members of the cysteine protease family. The sequence data indicate that the primary translation product of argingipain comprises at least four functional domains: signal peptide, NH$_2$-terminal prosequence, proteinase domain, and COOH-terminal prosequence. The NH$_2$-terminal sequence, the first 24 residues, has characteristics of a signal sequence. The domain from residues 3 to 24 comprises uncharged amino acid residues with

MKNLNKFVSI *ALCSSLLGGM* *AFAQQTELGR* NPNVRLLEST QQSVTKVQFR MDNLKFTEVQ   60

TPKGMAQVPT YTEGVNLSEK GMPTLPILSR SLAVSDTREM KVEVVSSKFI EKKNVLIAPS   120

KGMIMRNEDP KKIPYVYGKS YSQNKFFPGE IATLDDPFIL RDVRGQVVNF APLQYNPVTK   180

TLRIYTEITV AVSETSEQGK NILNKKGTFA GFEDTYKRMF MNYEPGRYTP VEEKQNGRMI   240

VIVAKKYEGD IKDFVDWKNQ RGLRTEVKVA EDIASPVTAN AIQQFVKQEY EKEGNDLTYV   300

LLVGDHKDIP AKITPGIKSD QVYGQIVGND HYNEVFIGRF SCESKEDLKT QIDRTIHYER   360

NITTEDKWLG QALCIASAEG GPSADNGESD IQHENVIANL LTQYGYTKII KCYDPGVTPK   420

NIIDAFNGGI SLVNYTGHGS ETAWGTSHFG TTHVKQLTNS NQLPFIFDVA CVNGDFLFSM   480

PCFAEALMRA QKDGKPTGTV AIIASTINQS WASPMRGQDE MNEILCEKHP NNIKRTFGGV   540

TMNGMFAMVE KYKKDGEKML DTWTVFGDPS LLVRTLVPTK MQVTAPAQIN LTDASVNVSC   600

DYNGAIATIS ANGKMFGSAV VENGTATINL TGLTNESTLT LTVVGYNKET VIKTINTNGE   660

PNPYQPVSNL TATTQGQKVT LKWDAPSTKT NATTNTARSV DGIRELVLLS VSDAPELLRS   720

GQAEIVLEAH DVWNDGSGYQ ILLDADHDQY GQVIPSDTHT LWPNCSVPAN LFAPFEYTVP   780

ENADPSCSPT NMIMDGTASV NIPAGTYDFA IAAPQANAKI WIAGQGPTKE DDYVFEAGKK   840

YHFLMKKMGS GDGTELTISE GGGSDYTYTV YRDGTKIKEG LTETTYRDAG MSAQSHEYCV   900

EVKYAAGVSP KVCVDYIPDG VADVTAQKPY TLTVVGKTIT VTCQGEAMIY DMNGRRLAAG   960

RNTVVYTAQG GYYAVMVVVD GKSYVEKLAV K   991

Figure 2. The amino acid sequence of the argingipain precursor deduced from the nucleotide sequence of its cDNA. Amino acid sequences determined by Edman degradation of the purified enzyme and derived peptides are indicated by a underline and dotted lines, respectively. Residues in the putative signal sequence are shown by italics. The vertical arrow, the closed and open arrowheads indicate the residues of NH_2-terminal ends of NH_2-terminal propeptide, proteinase domain, and COOH-terminal hemagglutinin domain, respectively.

a high content of hydrophobic amino acids. A lysine residue at position 2 has been frequently found in signal sequences. Further, this region has the most probable site of signal sequence cleavage (between Gln^{24} and Gln^{25}). It is most likely that the initially synthesized preproenzyme is translocated across the inner membrane with the aid of the signal peptide.

The following sequence comprising 203 residues is considered to be the NH_2-terminal prosequence. This domain appears to be removed from mature argingipain, because the NH_2-terminal amino acid sequence of the purified mature enzyme starts with Tyr^{228}, and because the most probable site of the propeptide cleavage (Arg^{227}-Tyr^{228}) is present in the homologous sequence (from residues Asn^{222} to Thr^{229}) highly susceptible to autoproteolysis by the enzyme itself. At this time, the precise COOH-terminal end of the mature enzyme remains to be determined. However, on the basis of the observations that the purified enzyme is composed of a single polypeptide chain of *Mr.* ~50,000, we assume that the most possible COOH-terminal cleavage site to generate the mature enzyme is between Arg^{719} and Ser^{720}, which is also involved in the common sequence highly susceptible to autoproteolytic cleavage. The cleavage event thus appears to occur by intracellular autocatalysis. This is supported by the fact that the putative NH_2-terminal sequence of the COOH-terminal prosequence of

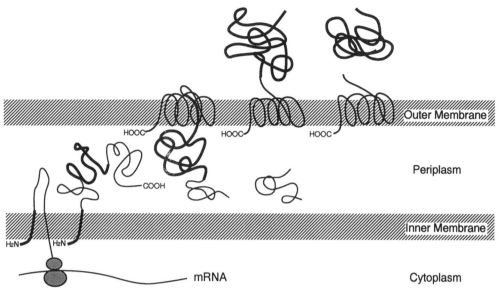

Figure 3. A schematic representation of the processing and translocation of argingipain in *P. gingivalis*.

argingipain is identical with that of hemagglutinin from *P. gingivalis* (24-26). Since the extracellular enzyme does not exhibit the hemagglutinin activity, the mature proteinase domain is more likely to be released from the COOH-terminal hemagglutinin domain during translocation of the proenzyme across the outer membrane. Based on the structural features of argingipain, we propose a model for the processing and extracellular location of the enzyme (Fig. 3). Argingipain is initially synthesized as a preproenzyme and translocated across the inner membrane with the aid of the signal peptide. After cleavage of the signal peptide, the proenzyme in the periplasm becomes incorporated into the outer membrane probably by amphipathic properties of the proteinase domain and/or the COOH-terminal prosequence. During excretion through the outer membrane, the NH_2- and COOH-terminal prosequences are processed by the proteolytic activity of argingipain itself.

The initially synthesized preproenzyme is translocated across the inner membrane with the aid of the signal peptide. After cleavage of the signal peptide, the proenzyme with both the NH_2- and COOH-terminal prosequences is localized in the periplasm and then becomes incorporated into the outer membrane probably by amphipathic properties of the COOH-terminal domain and/or the proteinase domain. During excretion through the outer membrane, the proenzyme acquires an active conformation to remove the NH_2- and COOH-terminal prosequences by autoproteolysis and then the active mature enzyme is released extracellularly.

FUNCTIONAL CHARACTERIZATION

In order to further clarify the importance of argingipain for virulence of the organism, we have constructed argingipain-deficient mutants via gene disruption by use of suicide plasmid systems (27). Firstly, the disruption of the argingipain gene was carried out by homologous recombination between *P. gingivalis* chromosomal

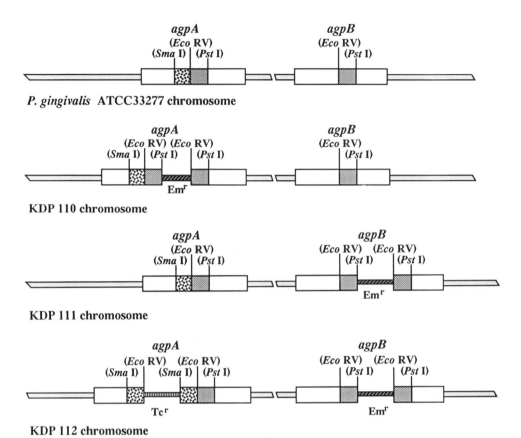

Figure 4. A schematic diagram for the structures of argingipain genes on the chromosomal DNA of *P. gingivalis* ATCC33277 and *agp*-deficient mutants.

DNA and an erythromycin-resistant suicide plasmid containing an internal DNA fragment of the argingipain gene. In the course of this study, we found that *P. gingivalis* ATCC33277 possesses two argingipain-encoding genes, designated an *agp*A and an *agp*B, on its chromosome and thus obtained two different single knockout mutants (KDP110, an *agp*A-deficient mutant; KDP111, an *agp*B-deficient mutant) (Fig. 4). Therefore, in defining the significance of the enzyme in the pathogenicity of the organism, it is necessary to construct an *agp*A *agp*B double knockout mutant (termed KDP112). For this, we have constructed the double mutant from KDP111 by use of the electroporation technique with the second suicide plasmid which has a tetracycline-resistant gene (Fig. 4). By Southern blot hybridization and Western immunoblotting analyses, the resultant transformant was found to receive complete disruption of the two *agp* genes.

Both the culture supernatant and cell extract of the double mutant are completely devoid of the hydrolytic activity with two synthetic substrates specific for argingipain (BC-Phe-Ser-Arg-MCA and Z-Phe-Arg-MCA) (Fig. 5). The *agp*A and the *agp*B single mutants also show a marked decrease of activity with these substrates but still retain significant amounts of the argingipain activity in both fractions. When general protein substrates such as casein and hemoglobin are used, only a small amount of proteolytic activity is detected in the culture supernatant of the double mutant, indicating that argingipain is a major extracellular proteinase of *P. gingivalis*.

Yoneda et al. (28) reported that *P. gingivalis* releases a potent virulence factor(s) which disrupts the functions of PMNs. The disruption of the PMN functions is of special importance in considering the initiation and development of periodontal disease, since accumulation of PMNs in the periodontal pocket and between the epithelial cells lining the pocket at sites of gingival inflammation is a characteristic feature of chronic adult periodontitis (29), since individuals with impaired PMN functions have a high prevalence of severe periodontitis (30), and since most of the patients with juvenile periodontitis exhibit defective neutrophil chemotaxis (30-32). As mentioned before, argingipain exhibits the strong suppressive activity of the CL response of PMNs. It is thus considered of interest to determine to what extent the disruption of argingipain genes affects the potency to suppress the CL response of PMNs. As shown in Fig. 5, the observed suppressive activity of the CL response of PMNs with the culture supernatant of *P. gingivalis* ATCC33277 is markedly decreased in the culture supernatant of the *agp*A *agp*B double knockout mutant. The data indicate that the disruption of PMN functions is totally due to the proteolytic activity of argingipain.

Protoheme is an absolute requirement for growth of *P. gingivalis* (33, 34) and is probably derived from erythrocytes in the natural niche of the organism. Therefore, it is particularly important for the organism to survive in vivo that it has the ability to agglutinate and lyse erythrocytes (35, 36). The association of the cysteine proteinase activity with the hemagglutinin activity has been suggested by several groups (21, 25, 26, 37, 38). However, there has been substantial controversy over the question whether these two activities result from the same molecule. Nishikata and Yoshimura (37) have found that both the proteinase and hemagglutinin activities are expressed by a single molecule. Ciborowski et al. (26) and

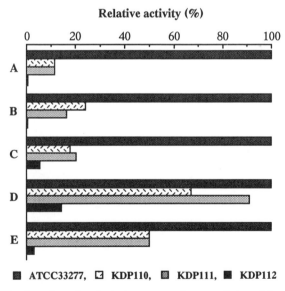

Figure 5. Effect of the disruption of argingipain genes on the proteolytic activity, the suppressive activity toward the CL response of PMNs, and the hemagglutinating activity found in a wild-type of *P. gingivalis* ATCC33277. Activities are expressed as percentages of the respective activities with the strain ATCC33277. A; the argingipain activity against Boc-Phe-Ser-Arg-MCA of the culture supernatants of the wild-type strain and the mutants, B; the argingipain activity of their cell extracts against Boc-Phe-Ser-Arg-MCA, C; the caseinolytic activity of the culture supernatants, D; the suppressive activity toward the CL response of PMNs, E; the hemaggulutinating activity of the whole cells on sheep erythrocytes.

Okamoto et al. (21) have also suggested that these activities are derived from the primary product of the same gene. On the other hand, Pike et al (16) and Shah et al. (29) reported that the two distinct activities are from two separate molecules, although they are noncovalently bound to each other. It is apparent from Fig. 5 that *P. gingivalis* ATCC33277 has the eminent hemagglutinating activity, whereas the *agp*A *agp*B double mutant is totally devoid of this activity. Taken together, the finding that the primary translation product of the argingipain genes possesses the hemagglutinin domain in the COOH-terminal prosequence indicates that the argingipain gene is involved in hemagglutination of *P. gingivalis*.

The disruption of *agp*A, *agp*B, and both genes was carried out by the use of suicide plasmid systems and the resultant mutants were termed KDP110, KDP111, and KDP112, respectively. KDP110 and KDP111 were constructed by use of an erythromycin (Em')-resistant suicide plasmid containing an internal DNA fragment of the argingipain gene, whereas KDP112 was obtained from KDP111by use of electrotransformation with a second tetracycline (Tc)-resistant suicide plasmid.

CONCLUDING REMARKS

Enzymatic and molecular genetic experiments have provided useful information on the nature, structure, and functions of the arginine-specific cysteine proteinase (termed "argingipain") of *P. gingivalis* as a major periodontal pathogenic factor. Argingipain exhibits several unusual catalytic features, as demonstrated by susceptibility to various activators and inhibitors and by the most restricted substrate cleavage specificity. The structure of the enzyme also is unique. The precursor of argingipain comprises four functional domains: signal peptide, NH_2-terminal prosequence, proteinase domain, and COOH-terminal hemagglutinin domain. It is now clear that argingipain plays a key role as a major virulence factor of *P. gingivalis* in the development of periodontal disease via the direct destruction of periodontal tissues and the disruption of normal host defense mechanisms. The pathogenicity of argingipain is further substantiated by analysis of argingipain-deficient mutants. Also, it is evident that the argingipain genes are involved in the hemagglutination of *P. gingivalis*.

REFERENCES

1. White, D., and Mayrand, D., 1981, Association of oral *Bacteroides* with gingivitis and adult periodontitis. *J. Periodont. Res.* 16: 259-265.
2. Holt, S.C., and Brumante, T.E., 1991, Factors in virulence expression and their role in periodontal disease pathogenesis. Crit. Rev. *Oral Biol. Med.* 2: 177-281.
3. Slots, J., and Rams, T.E., 1993, Pathogenicity. *In Biology of the Species Porphyromonas gingivalis* (Shah, H.N., Mayrand, D., and Genco, R.I., eds.) pp. 127-138, CRC Press Inc., Boca Raton, Ann Arbor, London, Tokyo.
4. Mayrand, D., Mouton, C., and Grenier, D., 1991, Chemical and biological properties of cell-surface components of oral black-pigmented *Bacteroides* species. *In* Periodontal Disease: Pathogens and Host Immune Responses (Hamada, S., Holt, S.C., and McGhee, J.R., eds.) pp. 99-115, Quintessence Publishing Co., Tokyo.
5. Grenier, D., and Mayrand, D., 1993, Proteinases. *In Biology of the Species Porphyromonas gingivalis* (Shah, H.N., Mayrand, D., and Genco, R.I., eds.) pp. 227-243, CRC Press Inc., Boca Raton, Ann Arbor, London, Tokyo.
6. Arnott, M.A., Rigg, G., Shah, H., Williams, D., Wallace, A., and Roberts, I.S., 1990, Cloning and expression of a *Porphyromonas* (*Bacteroides*) *gingivalis* protease gene in *Escherichia coli. Arch. Oral Biol.* 35: 97S-99S.
7. Park, Y., and McBride, B.C., 1992, Cloning od a *Porphyromonas* (*Bacteroides*) *gingivalis* protease gene and characterization of its product. *FEMS Microbiol. Lett.* 92: 273-278.

8. Nakamura, S., Takeuchi, A., Masamoto, Y., Abiko, Y., Hayakawa, M., and Takiguchi, H., 1992, Cloning of the gene encoding a glycylprolyl aminopeptidase from *Porphyromonas gingivalis*. *Arch. Oral Biol.* 37: 807-812.

9. Bourgeau, G., Lapointe, H., Peloquin, P., and Mayrand, D., 1992, Cloning, expression, and sequencing of a protease gene (tpr) from *Porphyromonas gingivalis* W83 in *Escherichia coli*. *Infect. Immun.* 60: 3186-3192.

10. Kato, T., Takahashi, N., and Kuramitsu, H.K., 1992, Sequence analysis and characterization of the *Porphyromonas gingivalis prt C* gene, which expresses a novel collagenase activity. *J. Bacteriol.* 174: 3889-3895.

11. Otogoto, J., and Kuramitsu, H.K., 1993, Isolation and characterization of the *Porphyromonas gingivalis prtT* gene, coding for protease activity. *Infect. Immun.* 61: 117-123.

12. Bedi, G.S., and Williams, T., 1994, Purification and characterization of a collagen-degrading protease from *Porphyromonas gingivalis*. *J. Biol. Chem.* 269: 599-606.

13. Sojor, H.T., Lee, J.Y., Bedi, G.S., and Genco, R.J., 1993, Purification and characterization of a protease from Porphyromonas gingivalis capable of degrading salt-solubilized collagen. *Infect. Immun.* 61: 2369-2376.

14. Lawson, D.A., and Meyer, T.F., 1992, Biochemical characterization of *Porphyromonas* (*Bacteroides*) *gingivalis* collagenase. *Infect. Immun.* 60: 1524-1529.

15. Takahashi, N., Kato, T., and Kuramitsu, H.K., 1991, Isolation and preliminary characterization of the *Porphyromonas gingivalis prtC* gene expressing collagenase activity. *FEMS Microbiol. Lett.* 84: 135-138.

16. Birkedal-Hansen, H., Taylor, R.E., Zambon, J.J., Barwa, P.K., and Nelders, M.E., 1988, Characterization of collagenolytic activity from strains of *Bacteroides gingivalis*. *J. Periodont. Res.* 23: 258-264.

17. Sundqvist, G., Carlsson, S.G., and Hanstrom, L., 1987, Collagenolytic activity of black-pigmented *Bacteroides* species. *J. Periodont. Res.* 22: 300-306.

18. Frandsen, E.V.G., Reinholdt, J.R., and Kilian, M., 1987, Enzymatic and antigenic characterization of immunoglobulin A1 proteases from *Bacteroides* and *Capnocytophaga* spp. *Infect. Immun.* 55: 631-638.

19. Mortensen, S.B., and Kilian, M., 1984, Purification and characterization of immunoglobulin A1 protease from *Bacteroides melaninogenicus*. *Infect. Immun.* 45: 550-557.

20. Kilian, M., 1981, Degradation of immunoglobulin A1, A2, and G by suspected principal periodontal pathogens. *Infect. Immun.* 34: 757-765.

21. Okamoto, K., Misumi, Y., Kadowaki, T., Yoneda, M., Yamamoto, K., and Ikehara, Y., 1995, Structural characterization of argingipain, a novel arginine-specific cysteine proteinase as a major periodontal pathogenic factor from *Porphyromonas gingivalis*. *Arch. Biochem. Biophys.* 316, in press.

22. Madden, T.E., Clark, V.L., and Kuramitsu, H.K., 1995, Revised sequence of the *Porphyromonas gingivalis* Prt cysteine protease/hemagglutinin gene: homology with Streptococcal pyrogenic exotoxin B/Streptococcal proteinase. *Infect. Immun.* 63: 238-247.

23. Kadowaki, T., Yoneda, M., Okamoto, K., Maeda, K., and Yamamoto, K., 1994, Purification and characterization of a novel arginine-specific cysteine proteinase (argingipain) involved in the pathogenesis of periodontal disease from the culture supernatant of *Porphyromonas* gingivalis. *J. Biol. Chem.* 269: 21371-21378.

24. Ikeda, T., Yoshimura, F., and Nishikata, M., 1993, Gene cloning of *Porphyromonas gingivalis* hemagglutinin. *J. Dent. Res.* 72: 119 (Abstr.).

25. Pike, R., McGraw, W., Potempa, J., and Travis, J., 1993, Lysine- and arginine-specific proteinases from *Porphyromonas gingivalis*: isolation, characterization, and evidence for the existence of complexes with hemagglutinins. *J. Biol. Chem.* 269: 406-411.

26. Ciborowski, P., Nishikata, M., Allen, R.D., and Lantz, M.S., 1994, Purification and characterization of two forms of a high-molecular-weight cysteine proteinase (porphypain) from *Porphyromonas gingivalis*. *J. Bacteriol.* 176: 4549-4557.

27. Nakayama, K., Kadowaki, T., Okamoto, K., and Yamamoto, K., 1995, Construction and characterization of arginine-specific cysteine proteinase (argingipain)-deficient mutants of *Porphyromonas gingivalis*: evidence for significant contribution of argingipain to virulence. submitted for publication.

28. Yoneda, M., Maeda, K., and Aono, M., 1990, Suppression of bactericidal activity of human polymorphonuclear leukocytes by *Bacteroides gingivalis*. *Infect. Immun.* 58: 406-411.

29. Attstrom, R., 1970, Presence of leukocytes in crevices of healthy and chronically inflamed gingivae. *J. Periodont. Res.* 5: 42-47.

30. Van Dyke, T.E., Levine, M.J., and Genco, R.J., 1985, Neutrophil function and oral disease. *J. Oral Pathol.* 14: 95-120.

31. Van Dyke, T.E., Horoszewicz, H.U., and Genco, R.J., 1982, The polymorphonuclear leukocyte (PMNL) locomotor defect in juvenile periodontitis. *J. Periodontol.* 53: 682-687.

32. Singh, S., Golub, L.M., Iacono, V.J., Ramamurthy, N.S., and Kaslick, R., 1984, In vivo crevicular leukocyte response in humans to a chemotactic challenge. *J. Periodontol.* 55: 1-8.

33. Gibbons, R.J., and MacDonald, J.B., 1960, Hemin and vitamin K compounds as required factors for the cultivation of certain strains of Bacteroides melaninogenicus. *J. Bacteriol.* 80: 164-170.

34. Shah, H.N., Bonnett, R., Mateen, B., and Williams, R.A.D., 1979, The porphyrin pigmentation of subspecies of *Bacteroides melaninogenicus. Biochem. J.* 180: 45-50.

35. Chu, L., Bramanti, T.E., Ebersole, J.L., and Holt, S.C., 1991, Hemolytic activity in the periodontpathogen *Porphyromonas gingivalis*: kinetics of enzyme release and localization. *Infect. Immun.* 59: 1932-1940.

36. Shah, H.N., and Gharbia, S.E., 1989, Lysis of erythrocytes by the secreted cysteine proteinase of Porphyromonas gingivalis W83. *FEMS Microbiol. Lett.* 61: 213-218.

37. Nishikata, M., and Yoshimura, F., 1991, Characterization of *Porphyromonas* (*Bacteroides*) *gingivalis* hemagglutinin as a protease. *Biochem. Biophys. Res. Common.* 178: 336-342.

38. Shah, H.N., Gharbia, S.E., Progulske-Fox, A., and Brockelhurst, K., 1992, Evidence for independent molecular identity and functional interaction of hemagglutination and cysteine proteinase (gingivain) of *Porphyromonas gingivalis. J. Med. Microbiol.* 36: 239-244.

5

DISCOVERY OF A NEW TYPE OF PROTEINASE INHIBITOR FAMILY WHOSE MEMBERS HAVE AN ANCHORING SEQUENCE

S. Hirose,[1] M. Furukawa,[1] I. Tamechika,[1] M. Itakura,[1] A. Kato,[1] Y. Suzuki,[1] J. Kuroki,[2] and S. Tachibana[2]

[1] Department of Biological Sciences
Tokyo Institute of Technology
4259 Nagatsuta-cho
Midori-ku, Yokohama 226, Japan
[2] Tsukuba Research Laboratories
Eisai Co., Ltd
Tsukuba 300-26, Japan

INTRODUCTION

A large number of proteolytic systems are assembled in our body and their activities are strictly regulated by a delicate balance between the production, activation, and inhibition of the proteinases. Since any disturbance in the finely tuned balance of the systems may trigger aggressive destruction of the tissue integrity and normal physiology, a variety of endogenous, highly efficient and specific proteinase inhibitors exist that include the serine proteinase inhibitors serpins, the cysteine proteinase inhibitors cystatins, and tissue inhibitors of metalloproteinases. These inhibitors are usually present either as a free, soluble form in the plasma or extracellular matrix space or as a membrane-bound form on the cell surface. In the present communication, we report covalent localization of a family of proteinase inhibitors through specially designed anchoring sequences at their amino termini, a new type of inhibitor targeting to the site of action.

TRANSGLUTAMINASE SUBSTRATE DOMAIN IN THE ELAFIN PRECURSOR

Structure of the Elafin Precursor

Elafin is a specific inhibitor for elastase and proteinase 3 and essential for the maintenance of connective tissue integrity. When we determined the complete primary

Intracellular Protein Catabolism, Edited by Koichi Suzuki and Judith Bond
Plenum Press, New York, 1996

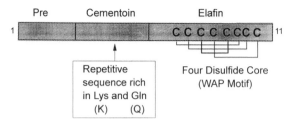

Figure 1. Structure of the elafin precursor.

structure of the elafin precursor, we noticed the presence of very interesting repetitive sequences at its N-terminus, which are rich in lysine (K) and glutamine (Q) (Fig. 1) (Saheki *et al.*, 1992). The amino acid sequence of elafin itself was determined by Wiedow *et al.* (1990) and shown to contain the WAP motif or 4 disulfide core (Fig. 1). Elafin therefore belongs to the WAP protein superfamily whose members have the characteristic compact structure held by four disulfide bonds; certain members including mucous proteinase inhibitor (HUSI-I) (Seemüller *et al.*, 1986), whose 3D structure is resolved by X-ray diffraction (Grütter *et al.*, 1988), have two such domains. Most of the WAP family members, however, lack the N-terminal extension of the repetitive sequence. Elafin and its close relatives to be described below, therefore, constitute a subfamily. The 3D structure of the elastase-elafin complex is being determined (Tsunemi *et al.*, 1993). Human elafin gene is located on the q12-q13 locus of chromosome 20 (Molhuizen *et al.*, 1994). Tissue distribution of elafin was determined by immunohistochemistry (Nara *et al.*, 1994; Schalkwijk *et al.*, 1993), RNase protection analysis (Nara *et al.*, 1994), and *in situ* hybridization histochemistry (Nonomura *et al.*, 1994).

Structural Features of Transglutaminase Substrates

For simplicity, we termed the repetitive sequence found in the elafin precursor "cementoin". A search for homology with other proteins revealed an interesting similarity between the repetitive sequence (cementoin) and the amino acid sequence of the guinea pig seminal vesicle clotting protein (Fig. 2). The cementoin moiety consists of 5 semiconserved tandem repeats of 6 amino acids. The seminal vesicle clotting protein SVP-1 also consists of repeats (Moore *et al.*, 1987). The two sequences are very similar in the aligned region KGQD. SVP is a good substrate of transglutaminase and readily cross-linked by the enzyme through the underlined residues Q and K. Involucrin (Eckert & Green, 1986) and cornifin (Marvin *et al.*, 1992) are also known to be cross-linked by keratinocyte transglutaminase and help to protect the skin. Trichohyalin, a highly expressed protein within the inner root sheath of hair follicles, also consists of full- or partial-length tandem repeats of a 23-amino acid sequence in which glutamic acids are positioned in charged environments (Fietz *et al.*, 1993); similar positioning of the amine-accepting glutamine residue is seen in other substrate proteins such as the γ-chain of fibrin (-EGQXXHL-). The transglutaminase substrates shown in Fig. 2 are predicted to form α-helical structures.

The repetitive nature and presence of K and Q, which are the properties common to a certain class of transglutaminase substrates, led us to hypothesize that the cementoin moiety is also a good substrate for transglutaminase and serves as molecular glue for covalently anchoring the elastase inhibitor elafin at its site of action.

Transglutaminase cross-links proteins by catalyzing the formation of isopeptide bonds between Lys and Gln side chains (Fig. 2B). The above mentioned substrates can serve

A

Cementoin: (VKGQDP)$_5$

SVP-1: (VTGQDSVKGRLQMKGQDSLAERFS)$_8$

Involucrin: (QEGQLKHLEQ)$_{39}$

Cornifin: (EPCQPKVP)$_{13}$

Trichohyalin: (DRKFREEEQLLQEREEQLRRQER)$_n$

B

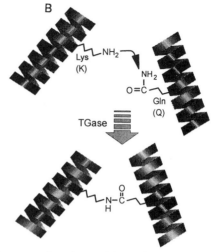

Figure 2. Similarity between the amino acid sequence of the cementoin moiety and those of transglutaminase substrate proteins (A) and the action of transglutaminase (B).

as both the donor and acceptor substrates. There are other types of transglutaminase substrates that are rich in Lys and serve as the Lys donor and that are rich in Gln and serve as the Lys acceptor. A recent example is the cross-linking between the cysteine proteinase inhibitor cystatin α, which is rich in Lys, and filaggrin, which is rich in Qln, by epidermal transglutaminase (Takahashi *et al.,* 1994). Filaggrins are, as the name (filament + aggregation) implies, proteins associated with intermediate filaments and involved in the organization of keratin filaments in the terminal stages of epidermal differentiation (Rothnagel & Steinert, 1990).

Demonstration of Cross-Linking Using Recombinant Elafin Precursor (Cementoin-Elafin)

To determine whether transglutaminase-mediated cross-linking occurs in the case of the cementoin moiety present in the elafin precursor, we produced recombinant human cementoin-elafin in *E. coli* using the pMAL-p expression vector, purified it by affinity chromatography on amylose resin, and examined its covalent cross-linking by light scattering and SDS-polyacrylamide gel electrophoresis (PAGE).

Following the addition of transglutaminase to a solution of recombinant cementoin-elafin, light-scattering of the solution increased and SDS-PAGE analysis of samples taken at various time intervals indicated that cementoin-elafin was indeed oligomerized by transglutaminase as evidenced by the appearance of higher molecular weight species. In the control in which EDTA was added to chelate Ca^{2+} which is essential for the transglutaminase activity, no cross-linking occurred. We next examined whether similar cross-linking occurs *in vivo* by measuring the size of cementoin-elafin in tissue extracts by Western blotting.

Covalent Anchoring of Cementoin-Elafin *In Vivo*

Trachea extracts were used to demonstrate covalent association of cementoin-elafin with other proteins because trachea was found to be the richest source of cementoin-elafin by an RNase protection assay (Nara *et al.,* 1994). Western blot analysis of the trachea extract using anti-cementoin antiserum and anti-elafin antiserum indicated that cementoin-elafin exists as higher molecular weight species (> 50 kDa), much higher than its monomeric form (10 kDa). This result and the fact that cementoin-elafin is a secreted protein strongly suggest that elafin is covalently cross-linked to extracellular matrix proteins through its cementoin moiety. Molhuizen *et al.* (1993) have also demonstrated the covalent association of elafin to certain extracellular matrix proteins.

NEW FAMILY MEMBERS THAT HAVE CEMENTOIN-LIKE ANCHORING SEQUENCES AND THE WAP MOTIF DOMAINS

Identification by Gene Analysis

Recently we have found, by screening a porcine genomic DNA library, the presence of a family of proteins which have cementoin-like transglutaminase substrate domains and the WAP motif domains (Fig. 3). The members are tentatively named WAP-1 (elafin), WAP-2, and WAP-3. These family proteins are very similar in their gene and protein structures. The genes consist of three exons: In each case, exon 1 contains the 5'-noncoding region and the sequence encoding the presequence or signal sequence; exon 2 corresponds to the secreted form of the proteins, namely the cementoin-like domain and WAP motif domain; and exon 3, the 3'-noncoding region.

Tissue Distribution

The family members, however, exhibited quite distinct tissue distributions; for example, the WAP-1 message is abundantly expressed in the trachea and large intestine while that of WAP-2 appears to be confined in the small intestine. These sites of localization are consistent with the locations where elastic fibers are abundant. Tissue localization of WAP-3 remains to be established; we examined the following tissues for the messages of the WAP family members by an RNase protection assay but could not detect substantial amounts of WAP-3 mRNA in any tissues: cerebrum, cerebellum, trachea, lung, atrium, stomach, duodenum, small intestine, and large intestine.

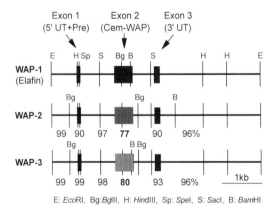

Figure 3. Gene structures of porcine WAP family members. Exon 2 of WAP-2 was found to encode a cementoin-like sequence and a 61-amino acid WAP sequence known as SPAI (Araki *et al.*, 1989).

Cross-Linking by Transglutaminase and Possible Role

Cross-linking experiments using the new members expressed in *E. coli* indicated that they can also be readily cross-linked by transglutaminase. The materials used for the cross-linking experiments were purified from cold-osmotic-shock-fluid of *E. coli* by affinity chromatography on amylose resin as a fusion protein with maltose binding protein, cleaved with Factor Xa, and separated from the maltose binding protein by hydroxyapatite column chromatography. The cross-linking was very rapid and efficient; for example, WAP-2 (100 μg in 100 μl) was completely polymerized within 5 min by 1 μg of transglutaminase (guinea pig liver, Sigma).

These findings suggest that the cementoin-like moieties of the new family members also serve as molecular glue to anchor the WAP domains to extracellular matrix proteins in order to protect elastic tissues. The proteins (WAP-1, -2, and -3) may therefore constitute a new class of proteinase inhibitors.

ACCELERATED EVOLUTION IN THE INHIBITOR DOMAINS

As mentioned above, gene cloning using porcine genomic DNA indicated the presence of three related genes, including one for cementoin-elafin, that consist of 3 exons and 2 introns (Fig. 3). They are very similar not only in the exon sequences but also in the intron sequences, suggesting that they were generated by gene duplications. Surprisingly, the similarities in the intron sequences (95 - 99%) are higher than those in the exon sequences (77 - 80%). Furthermore, within the exon sequences, the regions coding for the WAP motif are much more variable than those regions encoding the signal and cementoin-like sequences (Fig. 3). This type of mutational burst specifically affecting the active site is called "accelerated evolution".

Similar cases of accelerated evolution have been reported for the serine proteinase inhibitors serpins (Hill & Hastie, 1987; Borriello & Krauter, 1991; Rheaume *et al.*, 1994),

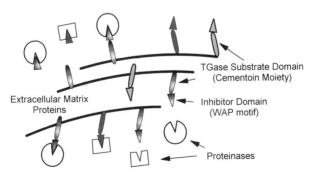

TGase Substrate Domain (Cementoin Moiety)

Extracellular Matrix Proteins

Inhibitor Domain (WAP motif)

Proteinases

Figure 4. Transglutaminase-mediated covalent anchoring of elafin (WAP-1) and its family members (WAP-2 and WAP-3) to extracellular matrix proteins (current working hypothesis). Triangles represent elafin and its newly found relatives with the WAP motif or 4 disulfide core. Rods indicate the cementoin moiety and cementoin-like transglutaminase substrate domains. Our working hypothesis is that the elastase inhibitor elafin is anchored to the extracellular matrix proteins through the cementoin moiety by the action of transglutaminase, and protects the extracellular matrix proteins from the attack of proteinases such as elastase and proteinase 3. Although the inhibitor activities of the new members remain to be demonstrated, they may play similar protective roles against tissue damages.

wheat cysteine-rich protein thionins (Castagnaro *et al.*, 1992), MHC molecules (Hughes & Nei, 1989), mouse submaxillary gland protein MSGs (Tronik-Le-Roux *et al.*, 1994), and snake venom gland phospholipase A_2 isozymes (Nakashima *et al.*, 1993).

CONCLUSION

In summary, we found the presence of a unique family of proteinase inhibitors that have anchoring sequences at their amino termini. These new type of inhibitors may have arisen by gene duplication followed by accelerated evolution in the reactive center regions.

ACKNOWLEDGEMENT

This work was supported by a Grant-in-Aid for Developmental Scientific Research from the Ministry of Education, Science and Culture of Japan.

REFERENCES

Araki, K., Kuroki, J., Ito, O., Kuwada, M. and Tachibana, S., 1989, Novel peptide inhibitor (SPAI) of Na,K-ATPase from porcine intestine, *Biochem. Biophys. Res. Commun.* 164:496-502.

Borriello, F., and Krauter, K.S., 1991, Multiple murine alpha 1-protease inhibitor genes show unusual evolutionary divergence, *Proc. Natl. Acad. Sci. U. S. A.* 88:9417-9421.

Castagnaro, A., Marana, C., Carbonero, P. and Garcia-Olmedo, F., 1992, Extreme divergence of a novel wheat thionin generated by a mutational burst specifically affecting the mature protein domain of the precursor, *J. Mol. Biol.* 224:1003-1009.

Eckert, R.L., and Green, H., 1986, Structure and evolution of the human involucrin gene, *Cell* 46:583-589.

Fietz, M.J., McLaughlan, C.J., Campbell, M.T. and Rogers, G.E., 1993, Analysis of the sheep trichohyalin gene: potential structural and calcium-binding roles of trichohyalin in the hair follicle, *J. Cell Biol.* 121:855-865.

Grütter, M.G., Fendrich, G., Huber, R. and Bode, W., 1988, The 2.5 Å X-ray crystal structure of the acid-stable proteinase inhibitor from human mucous secretions analysed in its complex with bovine α-chymotrypsin, *EMBO J.* 7:345-351.

Hill, R.E., and Hastie, N.D., 1987, Accelerated evolution in the reactive centre regions of serine protease inhibitors, *Nature* 326:96-99.

Hughes, A.L., and Nei, M., 1989, Nucleotide substitution at major histocompatibility complex class II loci: evidence for overdominant selection, *Proc. Natl. Acad. Sci. U. S. A.* 86:958-962.

Marvin, K.W., George, M.D., Fujimoto, W., Saunders, N.A., Bernacki, S.H. and Jetten, A.M., 1992, Cornifin, a cross-linked envelope precursor in keratinocytes that is down-regulated by retinoids, *Proc. Natl. Acad. Sci. U. S. A.* 89:11026-11030.

Molhuizen, H.O., Alkemade, H.A.C., Zeeuwen, P.L.J.M., de Jongh, G.J., Wieringa, B. and Schalkwijk, J., 1993, SKALP/elafin: an elastase inhibitor from cultured human keratinocytes. Purification, cDNA sequence, and evidence for transglutaminase cross-linking, *J. Biol. Chem.* 268:12028-12032.

Molhuizen, H.O., Zeeuwen, P.L., Olde-Weghuis, D., Geurts-van-Kessel, A. and Schalkwijk, J., 1994, Assignment of the human gene encoding the epidermal serine proteinase inhibitor SKALP (PI3) to chromosome region 20q12—>q13, *Cytogenet. Cell Genet.* 66:129-131.

Moore, J.T., Hagstrom, J., McCormick, D.J., Harvey, S., Madden, B., Holicky, E., Stanford, D.R. and Wieben, E.D., 1987, The major clotting protein from guinea pig seminal vesicle contains eight repeats of a 24-amino acid domain, *Proc. Natl. Acad. Sci. USA* 84:6712-6714.

Nakashima, K., Ogawa, T., Oda, N., Hattori, M., Sakaki, Y., Kihara, H. and Ohno, M., 1993, Accelerated evolution of *Trimeresurus flavoviridis* venom gland phospholipase A_2 isozymes, *Proc. Natl. Acad. Sci. U. S. A.* 90:5964-5968.

Nara, K., Ito, S., Ito, T., Suzuki, Y., Ghoneim, M.A., Tachibana, S. and Hirose, S., 1994, Elastase inhibitor elafin is a new type of proteinase inhibitor which has a transglutaminase-mediated anchoring sequence termed "cementoin", *J. Biochem.* 115:441-448.

Nonomura, K., Yamanishi, K., Yasuno, H., Nara, K. and Hirose, S., 1994, Up-regulation of elafin/SKALP gene expression in psoriatic epidermis, *J. Invest. Dermatol.* 103:88-91.

Rheaume, C., Goodwin, R.L., Latimer, J.J., Baumann, H. and Berger, F.G., 1994, Evolution of murine alpha 1-proteinase inhibitors: gene amplification and reactive center divergence, *J. Mol. Evol.* 38:121-131.

Rothnagel, J.A., and Steinert, P.M., 1990, The structure of the gene for mouse filaggrin and a comparison of the repeating units, *J. Biol. Chem.* 265:1862-1865.

Saheki, T., Ito, F., Hagiwara, H., Saito, Y., Kuroki, J., Tachibana, S. and Hirose, S., 1992, Primary structure of the human elafin precursor preproelafin deduced from the nucleotide sequence of its gene and the presence of unique repetitive sequences in the prosegment, *Biochem. Biophys. Res. Commun.* 185:240-245.

Schalkwijk, J., van-Vlijmen, I.M., Alkemade, J.A. and de-Jongh, G.J., 1993, Immunohistochemical localization of SKALP/elafin in psoriatic epidermis, *J. Invest. Dermatol.* 100:390-393.

Seemüller, U., Arnhold, M., Fritz, H., Wiedenmann, K., Machleidt, W., Heinzel, R., Appelhans, H., Gassen, H. and Lottspeich, F., 1986, The acid-stable proteinase inhibitor of human mucous secretions (HUSI-I, antileukoprotease) (complete amino acid sequence as revealed by protein and cDNA sequencing and structural homology to whey proteins and red sea turtle proteinase inhibitor), *FEBS Lett.* 199:43-48.

Takahashi, M., Tezuka, T., Kakegawa, H. and Katunuma, N., 1994, Linkage between phosphorylated cystatin alpha and filaggrin by epidermal transglutaminase as a model of cornified envelope and inhibition of cathepsin L activity by cornified envelope and the conjugated cystatin alpha, *FEBS Lett.* 340:173-176.

Tronik-Le-Roux, D., Senorale-Pose, M. and Rougeon, F., 1994, Three novel SMR1-related cDNAs characterized in the submaxillary gland of mice show extensive evolutionary divergence in the protein coding region, *Gene* 142:175-182.

Tsunemi, M., Matsuura, Y., Sakakibara, S. and Katsube, Y., 1993, Crystallization of a complex between an elastase-specific inhibitor elafin and porcine pancreatic elastase, *J. Mol. Biol.* 232:310-311.

Wiedow, O., Schröder, J., Gregory, H., Young, J.A. and Christophers, E., 1990, Elafin: an elastase-specific inhibitor of human skin, *J. Biol. Chem.* 265:14791-14795.

6

PROPOSED ROLE OF A γ-INTERFERON INDUCIBLE PROTEASOME-REGULATOR IN ANTIGEN PRESENTATION

C. A. Realini and M. C. Rechsteiner

University of Utah Medical Center
Department of Biochemistry
50 North Medical Drive
Salt Lake City, Utah 84132

INTRODUCTION

The 26S complex and the proteasome (also called 20S protease or multicatalytic protease, MCP) are major extralysosomal proteases involved in the regulation of cytosolic protein levels in eukaryotic cells (Tanaka *et al.*, 1992 for review see Rechsteiner *et al.*, 1993; Peters, 1994). The proteasome (Fig. 1 A) is a very abundant cellular component representing up to 1% of the cytoplasmic protein. It has been found in all eukaryotes and also in the archaebacterium *Thermoplasma acidophilum* (Zwickl *et al.*, 1991). Although proteasomes have not been reported in prokaryotes, structural studies and sequencing data suggest a possible evolutionary relationship between the prokaryotic Clp and Lon proteases (for review see Maurizi, 1992), and the eukaryotic 26S and proteasomes (Rechsteiner *et al.*, 1993; Arribas and Castano, 1993). The proteasome complex (Fig. 1 A) exists as a high molecular mass multimer (700 kDa) composed of at least 28 noncovalently associated subunits (20-32 kDa). The subunits share considerable inter-subunit homology, and have been classified into two subgroups, α and β, according to their similarity to the α- and β-subunit of the archaebacterial proteasome (Zwickl *et al.*, 1991). The subunits are arranged in four stacked heptameric rings to form a hollow cylinder of 11 x 16 nm (Peters *et al.*, 1991). Electron microscopy studies place α subunits at each end of the cylinder whereas β subunits, possibly containing the catalytic sites, form the two central rings. Proteasomes exhibit at least three separate catalytic activities, termed trypsin-like, chymotrypsin-like, and glutamic-site-like (Pereira *et al.*, 1992; Rivett, 1993). The regulation of these proteolytic sites as well as the exact subunit composition of the complexes are still unclear. Recent reports suggest however, that the subunit composition might represent one of the regulatory mechanisms affecting both the rate and specificity of proteasome activity. In the presence of ATP for instance, several proteasome subunits associate with higher molecular mass components to form the 26S protease, and thus participate in the degradation of ubiquitinated proteins. Alternatively, the proteasome can interact with a 11S particle composed of 6 or 7 identical 29 kDa subunits

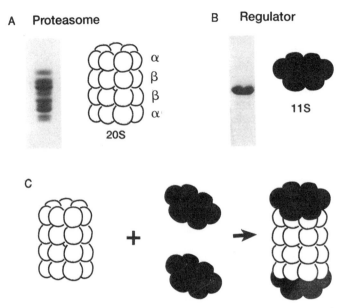

Figure 1. The proteasome (20S) and the regulator. A, The proteasome is shown as a complex composed of 28 subunits (20-32 kDa) arranged in four stacked rings. To the edge of the figure are shown the actual proteasome subunits (right), separated on a denaturing protein gel. B, The regulator is a protein of approximately 29 kDa that form hexameric rings also termed 11S due to their sedimentation on glycerol gradients. C, Interaction of two regulator hexamers with a proteasome to form an activated regulator-proteasome complex.

termed regulator or activator (Fig. 1 B). Upon association with the regulator (Fig. 1 C), proteasomes process a selected group of fluorogenic peptides up to 60 times faster, while the hydrolysis of others remains unaffected (Chu-Ping *et al.*, 1992; Dubiel *et al.*, 1992).

The functions of the proteasome are still unclear although several lines of evidence suggest an involvement in the processing of antigenic peptides or their precursors for presentation on MHC-class I molecules (see Robertson, 1991; van Bleek and Nathenson, 1992; Howard and Seelig, 1993; Michalek *et al.*, 1993; Rammensee *et al.*, 1993; Aki *et al.*, 1994 ; Engelhard, 1994, Realini *et al.*, 1994 b). First, due to its localization, the proteasome is well placed to process a variety of substrates found in the cytosol. Second, MHC heavy chains, TAP molecules, and two β–subunits of the proteasome (LMP2 and LMP7) map to the MHC gene cluster. Third, MHC-class and II MHC class I molecules, β_2-microglobulin, TAP molecules (Fruh *et al.*, 1992), LMP2 and LMP7 (Yang *et al.*, 1992) are up-regulated by γ-IFN. Here we report the cloning and expression of a human activator of the proteasome. The recombinant regulator lacks intrinsic hydrolytic activity but selectively binds and stimulates the activity of purified proteasomes. The regulator increases V_{max} and decreases the K_m of the reaction, and stimulates the cleavage of selected peptides. Immunological, physical, and biochemical tests show that the recombinant regulator and the regulator purified from human blood (red cell regulator) are very similar, if not identical. The regulator binds calcium, and proteasome-regulator complexes are reversibly inhibited by micromolar concentrations of calcium. We have identified one potential Ca^{++}-binding site in the recombinant regulator : an unusual domain composed of alternating lysine (K) and glutamic acid (E) residues (KEKE-motif) (Realini *et al.*, 1994 a, b). Similar motifs are found in α-subunits of the proteasome and subunits of 26S, and we propose that the KEKE-motif might promote

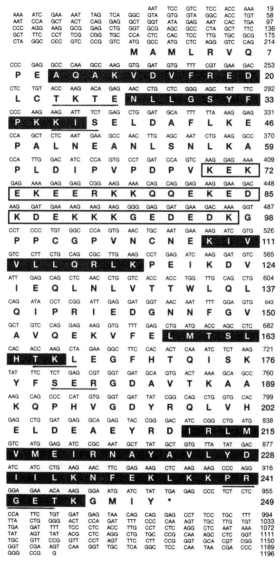

Figure 2. Nucleotide sequence of a cDNA and deduced amino acid sequence for the human proteasome regulator (from Realini *et al.*, 1994 a). The amino acids are shown in the one-letter amino acid code. V8-protease and cyanogen bromide-derived peptides obtained by direct microsequencing or putative regulator are shown white on black. The "KEKE motif" is boxed. Potential phosphorylation consensus sequences are underlined : *cAMP and cGMP-dependent kinase* phosphorylation site involving a serine (S) is found at residue 38. A serine in a *protein kinase C* consensus sequence is identified at residue 179. Threonines (T) and serines in the context of a *casein kinase II* sites are present at positions 23, 38, and 165. Finally, the threonine at position 244 is part of a possible *calmodulin-dependent phosphokinase* consensus sequence. Cysteines residues possibly involved in intra- or intermolecular disulfide bridges are found at position 101 and position 106*, stop codon TGA. Numbers indicate the nucleotide of the isolated clone or the amino acid of the open reading frame.

association between components of the proteolytic system. We further suggest that KEKE-motifs in the regulator recognize KEKE-like sequences in cellular or viral proteins directly adjacent to regions for presentation on MHC-class I receptors. A role for the proteasome and regulator in antigen presentation finds additional support in our observation that the regulator is up-regulated in HeLa cells exposed to γ–interferon (γ–IFN).

CLONING AND EXPRESSION OF THE HUMAN REGULATOR GENE

A lambda ZAP cDNA library and a lambda gt11 library were screened with synthetic non-degenerate nucleotides specific for the regulator gene. Both yielded full length clones that were isolated and sequenced as described (Realini et al., 1994 a). The sequence (Fig. 2) includes an open reading frame (ORF) of 751 nucleotides encoding a polypeptide of 249 amino acids with a calculated molecular weight of 28724 Da. The ORF is terminated by a stop codon (TGA) followed by a polyadenylation site (AATAAA) at nucleotides 1067-1072 (Fig. 2), and by a poly(A) tail (not shown). All five peptide sequences obtained by partial sequence analysis of partially purified regulator are found in the ORF accounting for about 31% of the total amino acids. The isolated gene shares no apparent homology with other known components of the proteolytic system. However, we have isolated a cDNA from a HeLa cell library, encoding a 235 residue protein 48% identical to the regulator. Furthermore, the recently cloned Ki antigen (Nikaido et al., 1990), a highly conserved nuclear protein that acts as a major autoantigen in Lupus erythematosus patients, also shares extensive homology with the regulator. Thus, most likely human cells can express at least three distinct, though closely related proteins that presumably affect the activity of the proteasome.

Both rabbit reticulocyte lysate and wheat germ extract supported the in vitro translation of capped mRNA transcribed from the full length clone. In both cases the translation product was a single ^{35}S-methionine labeled polypeptide comigrating with red cell regulator (not shown). The regulator gene was subcloned into the NdeI and BamHI sites of the expression vector pAED4 containing the potent Ø-10 promotor for T7 RNA polymerase. Analysis of the soluble protein fraction obtained from re-combinant cells treated with IPTG, revealed a major protein band migrating with the red cell regulator at approximately 29 kDa (Fig. 3 A), and specifically detected by antibodies raised against the purified human species (Fig. 3 B). Sequencing of the amino terminus region of the recombinant polypeptide produced a sequence identical to the amino terminal of the red cell regulator indicating the "in frame" expression of the gene. Upon two-dimensional analysis the recombinant regulator resolves into three distinct species with pI between 5.1-5.6, in excellent agreement with the distri-bution obtained with the red cell regulator (Dubiel et al., 1992). Although this het-erogeneity could be likely explained by posttranslational phosphorylation at serine, tyrosine, or threonine in consensus sequences found in the regulator (see legend to Fig. 2), preliminary experiments performed with the red cell and recombinant molecules failed to prove phosphate incorporation into the regulator. In particular, in vitro ex-periments performed with purified casein kinase II demonstrate phosphorylation of a previously described 30 kDa band in purified proteasomes (Ludemann et al, 1993), while recombinant regulator was not labeled under identical conditions. Recombinant regulator exhibits an apparent native molecular mass of approximately 180 kDa as determined by sizing chromatography and cosediments with red cell regulator on glycerol gradients (Johnston and Realini, unpublished observations). These observa-tions suggested that the recombinant species forms hexamers as observed for the purified red cell molecule (Chu-Ping et al.,1992; Dubiel et al., 1992). In a fluorimetric assay, recombinant regulator stimulated the hydrolytic activity of purified human and

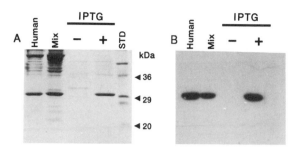

Figure 3. *In vivo* expression of the regulator gene and analysis of the gene product (from Realini *et al.*, 1994 a). A, Relative migration of recombinant and human red cell regulator separated on a 12% SDS-polyacrylamide gel. Soluble protein fractions (5 μg) obtained from either induced or non-induced recombinant *E. coli* were electrophoresed on a 12% SDS-polyacrylamide gel and stained with Coomassie Brilliant Blue R 250. B, western blots analysis of an equivalent gel using mouse antibodies against purified human red cell regulator. (IPTG, +), soluble protein fraction from recombinant cells induced with IPTG. (IPTG, -), protein fraction derived from non-induced recombinant cells. Human, partially purified human red blood cell regulator. Mix, recombinant and purified human regulator preparations were mixed prior to electrophoresis. STD, molecular weight standards from 20 -36 kDa.

rabbit proteasomes (Fig. 4 B). The stimulation was dependent on the amount of regulator added, and saturating amounts produced a 25-fold increase in peptide hydrolysis of LLVY (Fig. 4 B, insert). When increasing amounts of extract from induced *E. coli* cells were mixed with proteasomes, progressive activation of LLVY cleavage was readily detected upon peptide overlay of native gels (Fig. 4 A). Staining of the gel shown in Fig. 4 A, demonstrated a progressive retardation of the proteasomes (see also fluorescence pattern in Fig. 4 A), indicative of the formation of proteasomes-regulator complexes. The complex was confirmed by glycerol gradient analysis (not shown) and by two-dimensional electrophoresis (Fig. 4 C). The proteasome (20S, bracket) affects the migration of the regulator but not of other proteins (arrows). Kinetic analyses shown in Fig. 5 A reveal that both the recombinant and red cell regulators increase V_{max} (by approximately 50-fold at 200 μM LLVY) and decrease the K_m for hydrolysis of the same peptide (from 60 μM to ~4 μM). Moreover, re-combinant regulator stimulates LLVY-hydrolysis more than LLE- and PFR-cleavage (Fig. 5 B). The proteasome hydrolyzed LLVY 12 times faster than GPLGP in the absence of regulator, while proteasome-regulator complexes cleaved LLVY 200 times faster than GPLGP under identical conditions. This suggests that the regulator affects both the extent and specificity of peptide cleavage. The recombinant regulator is not a substrate of the proteasome, since prolonged incubation of ^{35}S-methionine labeled or ^{125}I-labeled regulator with the proteasome followed by electrophoretic separation and Phosphorimager quantification, produced no evidence for cleavage of the regulator (not shown). Furthermore, lysates obtained from non-recombinant or non-induced cells had no effect on the activity of the proteasome (see figures 4 A, B) , and no intrinsic hydrolytic activity of recombinant or purified regulators could be detected (not shown). We found that the proteasome-regulator complexes are reversibly inhibited by micro-molar concentrations of calcium (Realini and Rechsteiner, submitted). Interestingly, equimolar concentrations of calcium were ineffective in the presence of EGTA and affected the proteasome only minimally in the absence of the regulator. Similarly, proteasome-regulator complexes inhibited in the presence of 300 μM calcium, recovered their activity as increasing amounts of EGTA were added to the reaction mixture.

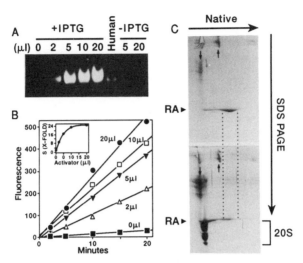

Figure 4. Stimulation of peptide hydrolysis by recombinant regulator (from Realini *et al.*, 1994 a). A, Native gel assays of purified human proteasomes in the presence of recombinant and purified human regulator. Proteasomes (300 ng), and various amounts of recombinant regulator (0-20 µl), or human activator, were mixed, separated on a 4.5% native gel and overlayed with 200 µM LLVY. Peptide cleavage was detected by translumination with long wave UV-light. (+ IPTG), soluble protein from IPTG-induced recombinant cells. (- IPTG), soluble protein from non-induced recombinant cells. human, purified human regulator. B, Fluorimetric assays. Various amounts (see figure) of extract containing recombinant regulator were added to 400 ng of purified rabbit proteasome, and the samples were incubated at 37°C in 100 µM LLVY, and fluorescence (Ex = 380 nm, Em = 440 nm) determined at the indicated times. The sample designated 0 µl contains the proteasome plus 20 µl of extract from uninduced recombinant cells. Fluorescence is graphed as a function of incubation time in the presence of increasing amounts of induced extract. The insert shows the stimulation of LLVY hydrolysis as a function of added recombinant regulator for a 5 min incubation at 37°C. Stimulation, $S = F_{Prot.+Act}/F_{Prot.}$, where $F_{Prot.+Act}$ is the rate of change in the fluorescence F in the presence of a given amount of regulator, and $F_{Prot.}$ the fluorescence measured in the absence of regulator. C, 2D-PAGE analysis of regulator-proteasome association. Purified human proteasome (400 ng) and partially purified recombinant regulator (10 µl) were mixed and electrophoresed for 6 hours at 4°C on a 8% native polyacrylamide gel. After electrophoresis, an individual lane from the gel was incubated for 10 min in 30 mM Tris-HCl, pH 6.8, 1% SDS, 5% glycerol and 5 mM β-mercaptoethanol, and then loaded on a 10% SDS-polyacrylamide gel. Proteins were stained with Coomassie Brilliant Blue R 250. The small vertical arrows indicate the relative position of proteins whose migration was not affected by the proteasome. The dotted line compares the migration of activator in the absence or presence of proteasomes, upper vs. lower panels, respectively.

THE KEKE-MOTIF OF THE REGULATOR

A domain enriched in alternating lysine (K) and glutamic acid (E) is found in the regulator ORF between lysine 70 and lysine 97 ("KEKE-motif") (Fig. 2). Circular dichroism analysis performed on the 28 amino acids long KEKE- motif of the regulator shows that this domain forms a stable α-helix in solution (Zhang , Realini and Rechsteiner, submitted). Interestingly, a KEKE domain was shown to form an α-helix in caldesmon (Marston and Redwood, 1991), and the α-helix destabilizing prolines are absent from the regulator "KEKE-motif", while prolines are enriched in both flanking regions; e.g., prolines 60, 64, 66 and 68 occupy the N-terminal edge and prolines 99, 100 and 103 are present at the C-terminal boundary. We defined KEKE motifs as greater than 12 amino acids in length containing more than 60% K and E/D, devoid of W, Y, F or P, and lacking more than four

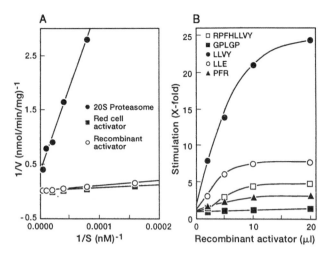

Figure 5. Comparison of recombinant and red cell regulators on kinetics and specificity of peptide hydrolysis by human red cell proteasomes (from Realini *et al.*, 1994 a). A, Lineweaver-Burke plot of LLVY hydrolysis in the presence or absence of regulator. Purified human proteasomes (400 ng) were incubate at 25°C alone, or in the presence of 30 μl of recombinant regulator, or partially purified red cell regulator with varying concentrations of LLVY. The initial velocities (nmol/min/mg proteasome) were determined for each concentration of substrate. The plot shows the reciprocal value of the initial velocity versus the reciprocal value of the substrate concentration. B, Substrate-dependent stimulation of proteasome activity. Increasing amounts of recombinant regulator (see figure) were added to 400 ng of purified human proteasome, and the mixture incubated at 37°C with 100 μM of the designated peptide. Excitation light at 380 nm and emission at 440 nm were used for the methylcoumaryl-7-amide-peptides RPFHLLVY, GPLGP, LLVY, PFR. For the carbobenzoxy-peptide LLE, 335 nm and 410 nm, respectively, were used. Stimulation is defined in Fig. 4.

consecutive positively or negatively charged residues. Based on this arbitrary definition we have performed a computer assisted search of available protein sequences. Approximately 300 proteins in the PIR library (> 100,000 entries) contain "KEKE motifs". Assuming a 5-10 fold redundancy in the library, these proteins represent approximately 3% of the submitted sequences. We have compared this distribution with the frequency of equally charged regions containing arginine (R) and aspartate (D). Among known proteins the abundance of arginine is comparable to that of lysine. Similarly aspartate and glutamate are present at equivalent frequency (Lathe, 1985). Yet we found only two RDRD proteins and more than 300 KEKE proteins, indicating that KEKE motifs are not simply statistically expected arrangements of amino acids. The reason for this biased expression of K and E in the KEKE-regions is not clear. The K/E pair however, is more frequent in α-helices than the R/D couple. A set of KEKE proteins is shown in Fig. 6. KEKE-proteins include proteasome α-subunits (e.g., subunit C9 and pros 28.1) or are found in protein complexes that bind the proteasome (e.g., subunit 12 of the 26S protease or the regulator). This led us to speculate that the KEKE-motif might serve as an association domain (Realini *et al.*, 1994 b). Interestingly, a KEKE motif was identified as the microtubulin binding site in MAP1 (Nobel *et al.*, 1989). Chaperonins, including hsp90, hsp70 and FLI G, constitute another set of proteins containing "KEKE-motifs". KEKE proteins include calcium binding proteins such as calnexin and calreticulin, an intraluminal calcium-storage protein of the endoplasmic reticulum (Milner *et al.*, 1991). A KEKE-motif in calreticulin was shown to be a regulatory calcium binding site (Baksh and Michalak, 1993). These observations prompted us to study the calcium binding of properties of regulator and its KEKE motif using ^{45}Ca overlay (Maruyama *et al.*,1984) and ruthenium

KEKE PROTEINS

Proteolytic   Components

MCP-C9	...**KKHEEEEAKAEREKKEKEQREKDK**...
MCP-PROS28	...**KIIEKEKEEELEKKKQK**...
26S-S12	...**EKKEGQEKEESKKDRKEDKEKDKDKEKSDVKKEEKK**
REG	...**KEKEKEERKKQQEKEDKDEKKGEDEDK**...
Tripept. Pept.	...**KDKEKDSEKEKDLEE**...
Clp C	...**KELEQKLDEVRKEKD**...
Calpastatin	...**EDKVKEKAKEEDREKLGEKEE**...

HSPs/Chaperonins

HSP 90	...**EEKEDKEEEKEKEEKESEDK**...
Fli L	...**EKKEEKKKEKKKEEKGDKKDAEK**...
HSP 70	...**EKLAAQRKAEAEKKEEKKDTE**
TCP 1 Cochap.	...**EKEAKQQEEKIEKMKAE**...
[a]cdc 37*	...**KKDGDEEALKKELEKIEAEGKELDRIESEMIKKEKK**...

Calcium   Binding   Proteins

Calnexin	...**EEEEKEEEKDKGDEEE**...
Calreticulin	...**KDKQDEEQRLKEEEE**...
Ca[++]/CAM Kinase	...**EDEDTKVRKQEIIK**...
Ca[++] ATPase	...**EEKKDEKKKEKK**...

Self Antigens/ Non Self Antigens

HSP90	...**EEEKKKMEESKAK**...
JAK1	...**KEKEKNKLKRKKLENKDKKDEEKNKIREE**...
eEF2	...**EKLDIKLDSEDKDKEGK**...
BBC1	...**KKEKARVITEEEK**...
Spectrin	...**ETEDNKEKKSAKD**...
P. knowlesi	...**KEGADKEKKKEKGKEKEEE**....
P. berghei	...**EGKKNEKKNEKIERNNK**...

*Some evidence for interaction with Hsp 83.
[a]*Cell* 77:1027 (1994)

Figure 6. The KEKE-motif of the regulator and other proteins (from Realini *et al.*, 1994 b). The KEKE-motifs of selected proteins are shown in the one-letter amino acid code. The motifs were found by searching updated gene banks using the "KEKE-search" algorythm that complies with a stringent definition of the a KEKE-motif.

red staining (Charuk *et al.*, 1990). We found that the purified recombinant regulator as well as a ubiquitin fusion protein containing the KEKE-motif bind ^{45}Ca and ruthenium red, while the negative control ubiquitin does not (Realini and Rechsteiner, 1994). These data support the hypothesis that the regulator is a calcium-binding and calcium-regulated protein (see above). A further class of KEKE-proteins includes precursors of antigenic peptides presented in association with class I major histocompatibility complexes (MHC-class I). Based on these observations, we proposed a role for regulator and KEKE-motifs in the selection of peptide ligands for MHC-class I molecules (Realini *et al.*, 1994 b).

ROLE OF REGULATOR AND KEKE-MOTIFS IN ANTIGEN PRESENTATION

The regulator binds and stimulates the proteasome biasing the production of peptides with positively charged C-terminals. These are common features of MHC-ligands characterized so far. Thus, proteasome-regulator complexes might preferentially generate peptides suitable for interaction with MHC-class I molecules. It has to be considered however, that the nature of the MHC-ligands identified to date is likely biased towards most frequent or easily purified ligands, and this C-terminal rule for MHC-ligands might be provisional. Furthermore, the effect of residues other than P1 on the cleavage by proteasome or proteasome-regulator complexes has still to be evaluated.

In addition to the KEKE-motifs of the regulator and proteasome subunits, we found KEKE domains in proteins processed for MHC-class I - restricted antigen presentation such as JAK1 a protein tyrosine kinase strongly activated in cells exposed to γ–IFN. While most peptides are presented at 100-500 copies, the natural K^d-ligands SYFPEITHI derived from JAK-1 is present in 10,000 copies on P815 mastocytoma cells (Falk *et al.*, 1991), and is located directly downstream from the KEKE-motif of JAK1. This finding led us to analyze the sequences of the precursors for known antigenic peptides. In approximately 20% of 54 eukaryotic, plasmodial or viral proteins that provide peptides for presentation on MHC-class I molecules, we found a KEKE-motif directly upstream of the natural ligands of MHC-class I molecules. These include a tandemly repeated highly immunogenic dodecapeptide in the circumsporozoite protein of *Plasmodium knowlesi*. This enrichment of KEKE-motifs among antigenic precursors generated our hypothesis that KEKE motifs facilitate the processing of precursors for antigen presentation. As an extension of the hypothesis presented above, we propose that the KEKE-motif promotes association between antigenic precursors and components of the proteolytic system. This hypothesis was tested by studying the interaction between regulator and a ubiquitin fusion protein containing a KEKE motif (Ub-KEKE) on glycerol gradients, and the effect of this fusion protein on the activity of proteasome-regulator complexes (Realini and Rechsteiner, in preparation). We found that Ub-KEKE binds the regulator and stimulates proteasome-regulator complexes in a dose-dependent manner. Proteasomes in the absence of regulator are not affected. Ubiquitin had no effect on the regulator-proteasome complexes, while other fusion proteins stimuled to a lesser extent. The distribution of regulator and Ub-KEKE after sedimentation revealed radiolabeled Ub-KEKE is associated with the regulator. By contrast, equivalent amounts of Ub do not sediment in this region. It is possible that the regulator binds KEKE-containing substrates and thus, due to its association with the proteasome, facilitates processing of these precursors by a strongly activated proteolytic complex. It is not clear however, whether these interactions involve assembly of KEKE-motifs with KEKE-sequences or KEKE binding sites on the regulator. To further support the role of proteasome-regulator complexes in antigen presentation we have tested the effect of γ–IFN on the intracellular concentration of regulator. γ–IFN is a pleiotropic cytokine that up-regulates several proteins involved in the presentation of antigens (Sen and Lengyel, 1992). We found that HeLa cells exposed to recombinant human γ-IFN show 6-fold higher levels of regulator than untreated cells (Realini *et al.*, 1994 a). These results were confirmed by a recent data bank submission reporting the discovery of a γ-IFN inducible protein with sequence identical to the regulator gene.

In conclusion, a cell exposed to γ–IFN might potentiate its antigen processing apparatus at different levels. The proteasomes, affected by factors such as regulator or LMPs, would generate higher amounts of peptides or peptide precursors particularly suitable for binding to MHC-class I heavy chains. The other components of the system such as the peptide transporters, class I, and class II molecules would ensure the efficient translocation and presentation of these peptides at the cell surface.

CONCLUSION

Antigenic determinants derived from intracellular proteins are exposed on the cell surface in association with MHC-class I molecules (see Yewdell and Bennink, 1992; Germain, 1994). Recent developments implicate the proteasome in the processing of cytosolic proteins for the production of antigenic peptides. Thus, understanding of proteasome structure and function as well as the identification of regulatory factors might be crucial for clarifying various aspects of antigen presentation. We have analyzed the gene for the regulator, a potent stimulator of the proteasome, and studied the recombinant molecule expressed *in vivo*. We have discovered that the regulator belongs to a class of proteins up-regulated by γ-IFN. We have also found evidence that the regulator, in concert with other elements such as the proteasome LMPs subunits, might be one of the factors enhancing the role of the proteasome in antigen processing. In fact, the regulator strongly stimulates the activity of the proteasome and favors the production of peptides suitable for interaction with MHC-class I molecules. An important number of proteins that provide antigenic peptides for MHC-class I presentation contain KEKE-motifs variably spaced from their natural MHC-ligands. Approximately 20% of the presented peptides that we have identified in the literature, exhibit a strict KEKE motif variably spaced from the natural MHC-ligand. These include circumsporozoite proteins from *Plasmodium*, various viral coat proteins as well as the γ-IFN-dependent tyrosine kinase JAK1. On the other hand, only 3% of the PIR library contains the same motifs. Thus, KEKE-motifs are enriched in regions adjacent to MHC-ligands. Furthermore, we found that regulator can bind an ubiquitin fusion protein containing a KEKE motif. These preliminary results add further evidence for a role of the regulator and the proteasome in antigen processing. One might in fact speculate that, the KEKE-motif of the regulator would selectively bind KEKE motifs in antigenic precursors and facilitate the interaction of peptides containing KEKE-motifs with the proteasome's active sites. This would channel these peptides into the processing pathway and finally promote their presentation on the cell surface.

REFERENCES

Aki, M., Shimbara, N., Takashina, M., Akiyama, K., Kagawa, S., Tamura, T., Tanahashi, N., Yoshimura, T., Tanaka, K., and Ichihara, A., 1994, Interferon-γ induces different subunit organization and functional diversity of proteasomes, *J. Biochem.* 115:257-269.

Arribas, J., and Castano, J. G., 1993, A comparative study of the chymotrypsin-like activity of the rat liver multicatalytic proteinase and the ClpP from *Escherichia coli*, *J. Biol. Chem.* 268:21165-21171.

Baksh, S. and Michalak, M., 1991, Expression of calreticulin in *Escherichia coli* and identification of its Ca^{2+}-binding domains, *J. Biol. Chem.* 266:21458-21465.

van Bleek, G M., and Nathenson, S. G., 1992, Presentation of antigenic peptides by MHC class I molecules, *Trends Cell Biol.* 2:202-207.

von Boehmer, H., 1994, Positive selection of lymphocytes, *Cell* 76:219-228.

Boes, B., Hengel, H., Ruppert, T., Multhaup, G., Koszinowski, U. H., and Kloetzel, P. -M., 1994, Interferon gamma stimulation modulates the proteolytic activity and cleavage site preference of 20S mouse proteasomes, *J. Exp. Med.* 179:901-909.

Charuk, J. H. M., Pirraglia, C. A., and Reithmeier, R. A. F., 1990, Interaction of ruthenium Red with Ca^{2+}-binding proteins, *Anal. Biochem.* 188:123-131.

Chu-Ping, M., Slaughter, C. A., and DeMartino, G. N., 1992, Identification, purification, and characterization of a protein activator (PA28) of the 20S proteasome (macropain), *J. Biol. Chem.* 267:10515-10523.

Dubiel, W., Pratt, G., Ferrell, K., and Rechsteiner, M., 1992, Subunit 4 of the 26S protease is a member of a novel eukaryotic ATPase family, *J. Biol. Chem.* 267:22369-22377.

Engelhard, V. H., 1994, Structure of peptides associated with class I and class II MHC molecules, *Annu. Rev. Immunol.* 12:181-207.

Falk, K., Rötzschke, O., Stevanovic, S., Jung, G., and Rammensee, H. -G., 1991, Allele-specific motifs revealed by sequencing of self-peptides eluted from MHC molecules, *Nature* 351:290-296.

Fruh, K., Yang, Y., Arnold, D., Chambers, J., Wu, L., Waters, J. B., Spies, T., and Peterson, P. A., 1992, Displacement of housekeeping proteasome subunits by MHC-encoded LMPs : a newly discovered mechanism for modulating the multicatalytic proteinase complex, *J. Biol. Chem.* 267:22131-22140.

Gaczynska, M., Rock, K. L., and Goldberg, A. L., 1993, Gamma-interferon and expression of MHC genes regulate peptide hydrolysis by proteasomes, *Nature* 365:264-267.

Germain, R. N., 1994, MHC-dependent antigen processing and peptide presentation: providing ligands for the T lymphocytes activation, *Cell* 76:287-299.

Howard, J. C., and Seelig, A., 1993, Peptides and the proteasome, *Nature* 365:211-212.

Ikai, A., Hishigai, M., Tanaka, K., and Ichihara, A., 1991, Electron microscopy of 26 S complex containing 20 S proteasome, *FEBS Lett.* 292:21-24.

Lathe, R., 1985, Synthetic oligonucleotide probes deduced from amino acid sequence data. Theoretical and practical considerations, *Biochemistry* 183:1-12.

Ludemann, R., Lerea, K.M., and Etlinger, J. D., 1993, Copurification of casein kinase II with 20S proteasomes and phosphorylation of a 30kDa proteasome subunit, *J. Biol. Chem.* 268:17413-17417.

Marston, B., and Redwood, C. S., 1991, The molecular anatomy of caldesmon, *Biochem. J.* 279:1-16.

Maruyama, K., Mikawa, T., and Ebashi, S., 1984, Detection of calcium-binding proteins by ^{45}Ca autoradiography on nitrocellulose membranes after sodium dodecyl sulfate gel electrophoresis, *J. Biochem.* 95:511-519.

Maurizi, M. R., 1992, Proteases and protein degradation in *Escherichia coli*, *Experentia* 48:178-201.

Michalek, M. T., Grant, E. P., Gramm, C., Goldberg, A. L. and Rock, K. L., 1993, A role for the ubiquitin-dependent proteolytic pathway in MHC-restricted antigen presentation, *Nature* 363:552-554.

Milner, R. E., Baksh, S., Shemanko, C., Carpenter, M. R., Smilie, L., Vance, J. E., Opas, M., and Michalak, M., 1991, Calreticulin and not calsequestrin, is the major calcium binding protein of smooth muscle sarcoplasmic reticulum and liver endoplasmic reticulum, *J. Biol. Chem.* 266:7155-7165.

Nikaido, T., Shimada, K., Shibata, M., Hata, M., Sakamoto, M., Takasaki, Y., Sato, C., Takahashi, T, and Nishida, Y., 1990, Cloning and nucleotide sequence of cDNA for Ki antigen, a highly conserved nuclear protein detected with sera from patients with systemic lupis erythematosus, *Clin. Exp. Immunol.* 79:209-214.

Nobel, M., Lewis, S.A. and Cowan, N.J., 1989, The microtubule domain of microtubule-associated protein MAP1B contains a repeated sequence motif unrelated to that of MAP2 and Tau, *J. Cell Biol.* 109:3367-3376.

Pereira, M., Nguyen, T., Wagner, B. J., Margolis, J. W., Yu, B., and Wilk, S., 1992, 3,4-dichloroisocoumarin-induced activation of the degradation of β-casein by the bovine pituitary multicatalytic proteinase complex, *J. Biol. Chem.* 267:7949-7955.

Peters, J. -M., Harris, J. R., and Kleinschmidt, J. A., 1991, Ultrastructure of the 26S complex containing the 20S cylinder particle (multicatalytic proteinase/proteasome), *Eur. J. Biochem.* 56:422-432.

Peters, J.-M., 1994, Proteasomes: protein degradation machines of the cell, *Trends Biochem. Sci.* 19:377—382.

Rammensee, H. -G., Falk, K., and Rötschke, O., 1993, Peptides naturally presented by MHC class I molecules, *A. Rev. Immun.* 11:213-244.

Realini, C. A., Dubiel, W., Pratt, G., Ferrel, K., and Rechsteiner, M., 1994 a, Molecular cloning and expression of a γ-interferon-inducible activator of the multicatalytic protease, *J. Biol. Chem.* 269:20727-20732

Realini, C. A., Rogers, S., and Rechsteiner, M., 1994 b, Proposed roles in protein-protein association and presentation of peptides by MHC class I receptors, *FEBS Lett.* 348:109-113.

Rechsteiner, M., Hoffman, L., and Dubiel, W., 1993, The multicatalytic and 26S proteinase complexes, *J. Biol. Chem.* 268:6065-6068.

Rivett, A. J., 1993, Proteasomes : multicatalytic proteinase complexes, *Biochem. J.* 291:1-10.

Robertson, M., 1991, Proteasomes in the pathway, *Nature* 353:300-301.

Sen, G. C., Lengyel, P., 1992, The interferon system, *J. Biol. Chem.* 267:5017-5020.

Tanaka, K., Tamura, T., Yoshimura, T., and Ichihara, A., 1992, Proteasomes : protein and gene structures, *New Biologist* 4:173-187.

Yang, Y., Waters, J. B., Früh, K., and Peterson, P. A., 1992, Proteasomes are regulated by interferon-γ: implications for antigen processing, *Proc. Natl. Acad. Sci. U.S.A.* 89:4928-4932.

Yewdell, J. W., and Bennink, J. R., 1992, Cell biology of antigen processing and presentation to major histocompatibility complex class I molecule-restricted T lymphocytes, *Adv. Immunol.* 52:1-123.

Zwickl P., Lottspeich, F., Dahlmann, B., and Baumeister, W., 1991, Cloning and sequencing of the gene encoding the large (α)-subunit of the proteasome from *Thermoplasma acidophilum*, *FEBS Lett.* 278:217-221.

7

NOVEL MEMBERS OF MAMMALIAN KEXIN FAMILY PROTEASES, PACE 4C, PACE 4D, PC 7A AND PC 7B

Y. Matsuda, A. Tsuji, H. Nagamune, T. Akamatsu, C. Hine,
K. Muramatsu, K. Mori, Y. Tamai, and K. Yonemoto

The University of Tokushima
Department of Biological Science and Technology
Faculty of Engineering
Minamijosanjima 2, Tokushima 770, Japan

INTRODUCTION

Biologically active peptides, such as peptide hormones, neuropeptides and growth factors, are known to be synthesized as larger precursor proteins and converted to active peptides by proteolytic cleavage at the carboxyl side of paired basic amino acids[1,2,3]. This process is found also in the activation of membrane-bound receptor proteins such as insulin receptor and many secreted proteins like serum albumin. Kexin family proteases were identified as processing proteases for these proteins; they share a subtilisin-like catalytic domain and require calcium ion for their activities[2,3]. Kexin family proteases are widely distributed in nature, from yeast, to mammals[4,5,6]. Currently seven mammalian kexins have been reported: furin[4], PC1/3[7], PC2,[8] PACE4[5,6], PC4[9], PC5/6[10], and PC7 [11].

We have isolated four novel cDNA s encoding new kexin family proteases. Two of them were PACE 4 isoforms[12,13], PACE 4C and PACE 4D, and the other two were PC 7A and PC 7B[11]. This paper presents the domain structures of mammalian kexin family proteases, the nucleotide sequences of newly found kexins and their cell specific expression by the methods of *in situ* hybridization and immunohistochemistry, and discusses the physiological function of kexin family proteases.

STRUCTURES OF KEXIN FAMILY PROTEASES

Domain Structure of Kexin Family Proteases

The structures of kexin family proteases are highly conserved among mammalian species. In general, all members have a subtilisin-like catalytic domain and a homo-B region, and these two domains are known to be essential for their enzymatic activities.

Schematic representation of Kexin family proteases

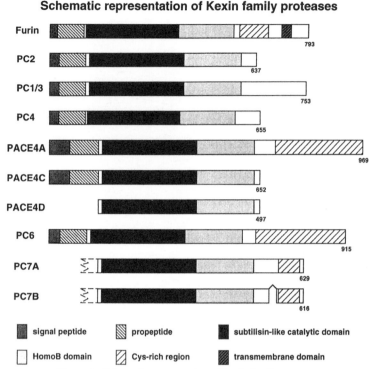

Figure 1. Schematic representaion of kexin family proteases.

Furin is composed of a signal peptide, propeptide, subtilisin-like catalytic domain, homo-B region, cysteine-rich region, and transmembrane region; it is localized in the trans Golgi apparatus. By contrast, other kexin family proteases lack the transmembrane domain and seem to be soluble enzymes. PACE 4A, PC 6 and PC 7 contain the cysteine-rich region at the C-terminus but, PC 1/3 , PC 2, PC 4, PACE 4C and PACE 4D lacked both the cysteine- rich and transmembrane domains. There has been little data to explain the role of the cysteine- rich region of kexins; one possibility is that the cysteine-rich region may participate as a signal for targetting the enzyme.

The alignment analysis of the sequences of the subtilisin-like catalytic domain of kexins is shown in Fig.2. Active sites amino acids, aspartic acid, histidine and serine, are well conserved among kexins, and are shown by asterisks. Amino acids sequences of the catalytic domains of PACE 4A and PC 7 are highly homologous; only six amino acids were different out of 288 amino acids as shown by closed dots in Fig. 2.

Sequences of PACE 4C and PACE 4D

We have isolated two novel cDNAs encoding PACE 4 isoforms (PACE 4C and PACE 4D) from both human placenta and rat pituitary cDNA libraries. Sequences of h (human)- PACE 4C and h-PACE 4D are shown in Fig. 3 and Fig. 4, respectively. The deduced amino acid sequence of PACE 4C was 652 amino acids, and the estimated molecular mass was 72kDa . The signal peptide, propeptide, subtilisin-like catalytic domain, and homo-B domain of PACE 4C were exactly identical to those of PACE 4A; the C-terminal 32 amino acid peptide and nucleotide 1861 to 2968

Amino acid sequences of catalytic domain among kexin family protease

```
                                                        *
Furin    114:TDPKFPQQWY------LSGVTQRDLNVKEAWAQGFTGRGIVVSILDDGIEKNHPDLAGNY
PC1      122:NDPMWNQQWYLQDTRMTASLPKLDLHVIPVWQKGITGKGVVITVLDDGLEWNHPDIYANY
PC2      121:NDPLFTKQWYLFNTGQADGTPGLDLNVAEAWELGYTGKGVTIGIMDDGIDYLHPDLAYNY
PC4      116:TDPWFSKQWY------MNKEIEQDLNILKVWNQGLTGRGVVVSILDDGIEKDHPDLWANY
PC6      129:NDPKWPSMWYMHC-SDNTHPCQSDMNIEGAWKRGYTGKNIVVTILDDGIERTHPDLMQNY
PACE4    161:NDPIWSNMWYLHC-GDKNSRCRSEMNVQAAWKRGYTGKNVVVTILDDGIERNHPDLAPNY
PC7       10:NDPIWSNMWYMHC-ADKNSRCRSEMNVQAAWKRGYTGKNVVVTILDDGIERNHPDLAPNY
                      .    .
                                         *
Furin    168:DPGASFDVNDQDPDPQPRYTQMNDNRHGTRCAGEVAAVANNGVCGVGVAYNARIGGVRML
PC1      182:DPEASYDFNDNDHDPFPRYDPTNENKHGTRCAGEIAMQANNHKCGVGVAYNSKVGGIRML
PC2      181:NSDASYDFSSNDPYPYPRVTDDWFNSHGTRCAGEVSAAASNNICGVGVAYNSKVAGIRML
PC4      170:DPLASYDFNDYDPDPQPRYTPNDENRHGTRCAGEVSATANNGFCGAGVAFNARIGGVRML
PC6      188:DALASCDVNGNDLDPMPRYDASNENKHGTRCAGEVAATANNSHCTVGIAFNAKIGGVRML
PACE4    220:DSYASYDVNGNDVDPSPRYDASNENKHGTRCAGEVAASANNSYCIVGIAYNAKIGGIRML
PC7       69:DSYASYDVNGNDYDPSPRYDASNENKHGTRCAGEVAASANNSYCIVGIAYNARIGGIRML
                                                                        .

Furin    228:D-GEVTDAVEARSLGLNPNHIHIYSASWGPEDDGKTVDGPARLAEEAFFRGVSQGRGGLG
PC1      242:D-GIVTDAIEASSIGFNPGHVDIYSASWGPNDDGKTVEGPGRLAQKAFEYGVKQGRQGKG
PC2      241:DQPFMTDIIEASSISHMPQLIDIYSASWGPTDNGKTVDGPRELTLQAMADGVNKGRGGKG
PC4      230:D-GAITDIVEAQSLSLQPQHIHIYSASWGPEDDGKTVDGPGLLTQEAFRRGVTKGRQGLG
PC6      248:D-GDVTDMVEAKSVSYNPQHVHIYSASWGPDDDGKTVDGPAPLTRQAFENGVRMGRRGLG
PACE4    280:D-GDVTDVVEAKSLGIRPNYIDIYSASWGPDDDGKTVDGPGRLAKQAFEYGIKKGRQGLG
PC7      129:D-GDVTDVVEAKSLGIRPNYIDIYSASWGPDDDGKTVDGPGRLAKQAFEYGIKKGRQGLG

Furin    287:SIFVWASGNGGREHDSCNCDGYTNSIYTLSISSATQFGNVPWYSEACSSTLATTYSSG--
PC1      301:SIFVWASGNGGRQGDNCDCDGYTDSIYTISISSASQQGLSPWYAEKCSSTLATSYSSG-D
PC2      301:SIYVWASGDGGSY-DDCNCDGYASSMWTISINSATNDGRTALYDESCSSTLASTFSNGRK
PC4      289:TLFIWASGNGGLHYDNCNCDGYTNSIHTLSVGSTTRQGRVPWYSEACSSTFTTTFSSG--
PC6      307:SVFVWASGNGGRSKDHCSCDGYTNSIYTISISSTAESGKKPWYLEECSSTLATTYSSG--
PACE4    339:SIFVWASGNGGREGDYCSCDGYTNSIYTISVSSATENGYKPWYLEECSSTLATTYSSG--
PC7      188:SIFVWASGNGGREGDHCSCDGYTNSIYTISVSSTTENGHKPWYLEECSSTLATTYSSG--
                         .               .
                                    *
Furin    345:NQNEKQIVTTDLRQKCTESHTGTSASAPLAAGIIALTLEANKNLTWRDMQHL
PC1      360:-YTDQRITSADLHNDCTETHTGTSASAPLAAGIFALALEANPNLTWRDMQHL
PC2      360:RNPEAGVATTDLYGNCTLRHSGTSAAAPEAAGVFALALEANVDLTWRDMQHL
PC4      347:VVTDPQIVTTDLHHQCTDKHTGTSASAPLAAGMIALALEANPLLTWRDLQHL
PC6      365:ESYDKKIITTDLRQRCTDNHTGTSASAPMAAGIIALALEANPFLTWRDVQHV
PACE4    397:AFYERKIVTTDLRQRCTDGHTGTSVSAPMVAGIIALALEANSQLTWRDVQHL
PC7      246:AFYERKIVTTDLRQRCTDGHTGTSVSAPMVAGIIALALEANNQLTWRDVQHL
                                                            .
```

Figure 2. Amino acids sequences of the catalytic domain of kexins.

were unique to PACE 4C. In addition, the sequence of PACE 4C was highly conserved between human and rat, only two nucleotides out of 2940 bases were different.

On the other hand, the predicted amino acid sequence of PACE 4D was 497 amino acids; it had an identical catalytic domain and homo-B domain to PACE 4A and 4C, but lacked the signal peptide, propeptide and cysteine-rich region. In addition, 5' - and 3'-untranslated region of the cDNA sequence of PACE 4D were completely different from either PACE 4A or 4C.

These results suggested that PACE 4A, PACE 4C and PACE 4D were probably the product of alternatively spliced mRNAs.

Nucleotide and deduced amino acid sequence of rat-PACE4C

```
-27                                                                    CGGGCGCCGCGCGAGCCTGTCGCCGCT

   1  ATG CCT CCG CGC GCG CCG CCT GCG CCC GGG CCC CGG CCG CCG CCC CGG GCC GCC GCC GCC ACC GAC ACC GCC GCG GGC GCG GGG GGC GCG
   1  Met Pro Pro Arg Ala Pro Pro Ala Pro Gly Pro Arg Pro Pro Pro Arg Ala Ala Ala Ala Thr Asp Thr Ala Ala Gly Ala Gly Gly Ala

  91  GGG GGC GCG GGG GGC GCC GGC GGG CCC GGG TTC CGG CCG CTC GCG CCG CGG CGT CCC TGG CGC TGG CTG CTG CTG CTG GCG CTG CCT GCC GCC
  31  Gly Gly Ala Gly Gly Ala Gly Gly Pro Gly Phe Arg Pro Leu Ala Pro Arg Pro Trp Arg Trp Leu Leu Leu Leu Ala Leu Pro Ala Ala

 181  TGC TCC GCG CCC CCG CCG CGC CCC GTC TAC ACC AAC CAC TGG GCG GTG CAA GTG CTG GGC GGC CCG GCC GAG GCG GAC CGC GTG GCG GCG
  61  Cys Ser Ala Pro Pro Pro Arg Pro Val Tyr Thr Asn His Trp Ala Val Gln Val Leu Gly Gly Pro Ala Glu Ala Asp Arg Val Ala Ala

 271  GCG CAC GGC TAC CTC AAC TTG GGC CAG ATT GGA AAC CTG GAA GAT TAC TAC CAT TTT TAT CAC AGC AAA ACC TTT AAA AGA TCA ACC TTG
  91  Ala His Gly Tyr Leu Asn Leu Gly Gln Ile Gly Asn Leu Glu Asp Tyr Tyr His Phe Tyr His Ser Lys Thr Phe Lys Arg Ser Thr Leu
                                                                                                                           ↓
 361  AGT AGC AGA GGC CCT CAC ACC TTC CTC AGA ATG GAC CCC CAG GTG AAA TGG CTC CAG CAA CAG GAA GTG AAA CGA AGG GTG AAG AGA CAG
 121  Ser Ser Arg Gly Pro His Thr Phe Leu Arg Met Asp Pro Gln Val Lys Trp Leu Gln Gln Gln Glu Val Lys Arg Arg Val Lys Arg Gln

 451  GTG CGA AGT GAC CCG CAG GCC CTT TAC TTC AAC GAC CCC ATT TGG TCC AAC ATG TGG TAC CTG CAT TGT GGC GAC AAG AAC AGT CGC TGC
 151  Val Arg Ser Asp Pro Gln Ala Leu Tyr Phe Asn Asp Pro Ile Trp Ser Asn Met Trp Tyr Leu His Cys Gly Asp Lys Asn Ser Arg Cys

 541  CGG TCG GAA ATG AAT GTC CAG GCA GCG TGG AAG AGG GGC TAC ACA GGA AAA AAC GTG GTG GTC ACC ATC CTT GAT GAT GGC ATA GAG AGA
 181  Arg Ser Glu Met Asn Val Gln Ala Ala Trp Lys Arg Gly Tyr Thr Gly Lys Asn Val Val Val Thr Ile Leu [Asp] Asp Gly Ile Glu Arg

 631  AAT CAC CCT GAC CTG GCC CCA AAT TAT GAT TCC TAC GCC AGC GTG AAC GGC AAT GAT TAT GAC CCA TCT CCA CGA TAT GAT GCC
 211  Asn His Pro Asp Leu Ala Pro Asn Tyr Asp Ser Tyr Ala Ser Val Asn Gly Asn Asp Tyr Asp Pro Ser Pro Arg Tyr Asp Ala

 721  AGC AAT GAA AAT AAA [CAC] GGC ACT CGT TGT GCG GGA GAA GTT GCT GCT TCA GCA/AAC\AAT TCC TAC TGC ATC GTG GGC ATA GCG TAC AAT
 241  Ser Asn Glu Asn Lys [His] Gly Thr Arg Cys Ala Gly Glu Val Ala Ala Ser Ala\Asn/Asn Ser Tyr Cys Ile Val Gly Ile Ala Tyr Asn

 811  GCC AAA ATA GGA GGC ATC CGC ATG CTG GAC GGC GAT GTC ACA GAT GTG GTC GAG GCA AAG TCG CTG GGC ATC AGA CCC AAC TAC ATC GAC
 271  Ala Lys Ile Gly Gly Ile Arg Met Leu Asp Gly Asp Val Thr Asp Val Val Glu Ala Lys Ser Leu Gly Ile Arg Pro Asn Tyr Ile Asp

 901  ATT TAC AGT GCC AGC TGG GGG CCG GAC GAC GAC GGC AAG ACG GTG GAC GGG CCC GGC CGA CTG GCT AAG CAG GCT TTC GAG TAT GGC ATT
 301  Ile Tyr Ser Ala Ser Trp Gly Pro Asp Asp Asp Gly Lys Thr Val Asp Gly Pro Gly Arg Leu Ala Lys Gln Ala Phe Glu Tyr Gly Ile

 991  AAA AAG GGC CGG CAG GGC CTG GGC TCC ATT TTC GTC TGG GCA TCT GGG AAT GGC GGG AGA GAG GGG GAC TAC TGC TCG TGC GAT GGC TAC
 331  Lys Lys Gly Arg Gln Gly Leu Gly Ser Ile Phe Val Trp Ala Ser Gly Asn Gly Gly Arg Glu Gly Asp Tyr Cys Ser Cys Asp Gly Tyr

1081  ACC AAC AGC ATC TAC ACC ATC TCC GTC AGC AGC GCC ACC GAG AAT GGC TAC AAG CCC TGG TAC CTG GAA GAG TGT GCC TCC ACC CTG GCC
 361  Thr Asn Ser Ile Tyr Thr Ile Ser Val Ser Ser Ala Thr Glu Asn Gly Tyr Lys Pro Trp Tyr Leu Glu Glu Cys Ala Ser Thr Leu Ala

1171  ACC ACC TAC AGC AGT GGG GCC TTT TAT GAG CGA AAA ATC GTC ACC ACG GAT CTG CGT CAG CGC TGT ACC GAT GGC CAC ACT GGG ACC [TCA]
 391  Thr Thr Tyr Ser Ser Gly Ala Phe Tyr Glu Arg Lys Ile Val Thr Thr Asp Leu Arg Gln Arg Cys Thr Asp Gly His Thr Gly Thr [Ser]

1261  GTC TCT GCC CCC ATG GTG GCG GGC ATC ATC GCC TTG GCT CTA GAA GCA AAC AGC CAG TTA ACC TGG AGG GAC GTC CAG CAC CTG CTA GTG
 421  Val Ser Ala Pro Met Val Ala Gly Ile Ile Ala Leu Ala Leu Glu Ala Asn Ser Gln Leu Thr Trp Arg Asp Val Gln His Leu Leu Val
                                                                                   *
1351  AAG ACA TCC CGG CCG GCC CAC CTG AAA GCG AGC GAC TGG AAA GTG AAC GGC GCG GGT CAT AAA GTT AGC CAT TTC TAT GGA TTT GGT TTG
 451  Lys Thr Ser Arg Pro Ala His Leu Lys Ala Ser Asp Trp Lys Val Asn Gly Ala Gly His Lys Val Ser His Phe Tyr Gly Phe Gly Leu
                                                                                              *
1441  GTG GAC GCA GAA GCT CTC GTT GTG GAG GCA AAG AAG TGG ACA GCA GTG CCA TCG CAG CAC ATG TGC GTG GCC GCC TCG GCA AAG AGA CCC
 481  Val Asp Ala Glu Ala Leu Val Val Glu Ala Lys Lys Trp Thr Ala Val Pro Ser Gln His Met Cys Val Ala Ala Ser Ala Lys Arg Pro

1531  AGG AGC ATC CCC TTA GTG CAG GTG CTG CGG ACT ACG ACC CTG ACC AGC ACC TGC GCG GAG CAC TCG GAC CAG CGG GTG GTC TAC TTG GAG
 511  Arg Ser Ile Pro Leu Val Gln Val Leu Arg Thr Thr Ala Leu Thr Ser Ala Cys Ala Glu His Ser Asp Gln Arg Val Val Tyr Leu Glu

1621  CAC GTG GTG GTT CGC ACC TCC ATC TCA CAC CCA CGC CGA GGA GAC CTC CAG ATC TAC CTG GTT TCT CCC TCG GGA ACC AAG TCT CAA CTT
 541  His Val Val Val Arg Thr Ser Ile Ser His Pro Arg Arg Gly Asp Leu Gln Ile Tyr Leu Val Ser Pro Ser Gly Thr Lys Ser Gln Leu

1711  TTG GCA AAG AGG TTG CTG GAT CTT TCC AAT GAA GGG TTT ACA AAC TGG GAA TTC ATG ACT GTC CAC TGC TGG GGA GAA AAG GCT GAA GGG
 571  Leu Ala Lys Arg Leu Leu Asp Leu Ser Asn Glu Gly Phe Thr Asn Trp Glu Phe Met Thr Val His Cys Trp Gly Glu Lys Ala Glu Gly

1801  CAG TGG ACC TTG GAA ATC CAA GAT CTG CCA TCC CAG GTC CGC AAC CCG GAG AAG CAA GGT GAT CTT GAG ACT CCT GTT GCA AAT CAA CTG
 601  Gln Trp Thr Leu Glu Ile Gln Asp Leu Pro Ser Gln Val Arg Asn Pro Glu Lys Gln Gly Asp Leu Glu Thr Pro Val Ala Asn Gln Leu

1891  ACC ACA GAA GAG AGG GAA CCT GGA CTA AAA CAC GTG TTC CGG TGG CAG ATT GAA CAA GAG CTT TGG TAA TCTCTGTGAAATACTGCAGAGGGACAG
 631  Thr Thr Glu Glu Arg Glu Pro Gly Leu Lys His Val Phe Arg Trp Gln Ile Glu Gln Glu Leu Trp ***

1987  ATGACCAAAGCCACCACATTTAGAACTTTGGCTGCCTTTGGAAGTCCAGAGCTGGATCTCTCAGCTCCCGCCCCCAGAGGGTCAGCACTTTGGACATGGCTCACAAGCAGTTTTTGATT
2106  GACTGCATGAATGCATGTGCGTGCAAGCATGAACCTTGTTTAAATCAAGAGGCTTACATAATTTTAACCAGTTCTGTCTTCAGCTGTACATACTCAGTAAAATGTTTAATGAAGGGGAA
2225  GAGATTAGTCTCTTCTGTGTGACCATGTTTTCCCTTTATTCATCCTAAAAAGTTCCATGAATTCTTGATTTCCTTTCAGTGGCCCTTTCAACAATGTCTTTTTTCCCAAGAGCATAACT
2344  GTTCTCATTTTATTGCTAGCCATCTTGATCTGTGTTTTATTGACATCTCTTTTGAGCTAATCTTCATTTCTAAGATAAGAGTTGAGATTTTGCAATCTGTGTTCGATGGCTCAATCTAT
2463  CCTGTGCTTGATGCTAGAAAGGAAGACAGATTTAAAGCACATGCCTTCTGTGCCGGCTTTCAAGTTTGTCACTAAACTCTCATTTCTGGAAAGTGCAATTATAGAGTATCACTCCCACT
2582  TCCTTGGAAACAGAGCTGAAGAACTTGGCACACTCTCCAAACAGTCACCATACACACTGTTGTCAAAAAGTTCCATTTTTAACCCCATTTGCATTAATATTGCAGTCAATCTCTTTACC
2701  TCGTTTTCTCTTTCACGGGGCCGTGACAGTGACGCCTTTCCCCAAAACTCTCCTCGTTTGAGAAAAAAGAAGTATGTATCCCCACTTATCTCGGGGAGAAATGCAACCAACTGCTGCT
2820  GTGCACATTTATGAATCACAGTATTGTTTAGTCGGTTCTGTATCTCCAGTAGAAAGCATAACAAAAAGATGACCTTTGTCTCACCTCATAGCTAATTTTTGCAAATAAAATCCTAAACA
2939  TTG(Aln
```

Figure 3. Nucleotide and predicted amino acid sequences of PACE 4C.

Nucleotide and deduced amino acid sequence of h-PACE4D

```
-546                                              GTCGACTGCTACCTTTTTTTTTTTCAAACTCCAATTGTTCACTGCTGGCATATAATCGATTTTTATATATT
-476 AACTTTGTACCCTGCAACCTTGCTAAACTCACTGATTGGTCCGAGGAGTTTTTTTTATAATTTCTTTACACTTTTCTACATGAAAGAATCTATCTCCTATGAATAGAGGAAGTTTAATT
-357 TCTCTCTTCCCGATATATATATGCATTTTATTCATTTATTTTTGCCTAACTGTACTGGCTAGGACTTTTACTACCATGGTAAATATGACTGGTGAGAAAAAAAAATATCATTACCTAGCTC
-238 CAAATGCTAGAAGACTGAATGCTTTCCCCTTATGAAGTATTTTAGCTGTATATTTTTTGTAGATGCTCTTTATTAAGTTCCTTCTATTACTAGTTCATTGAGATTTTTTTTTTTAAAT
-119 GATGAATAGATATTCCCCAGGTGAAATGGCTCCAGCAACAGGAAGTGAAACGAAGGGTGAAGAGACAGGTGCGAAGTGACCCGCAGGCCCTTTACTTCAACGACCCCATTTGGTCCAAC

  1 ATG TGG TAC CTG CAT TGT GGC GAC AAG AAC AGT CGC TGC CGG TCG GAA ATG AAT GTC CAG GCA GCG TGG AAG AGG GGC TAC ACA GGA AAA
  1 Met Trp Tyr Leu His Cys Gly Asp Lys Asn Ser Arg Cys Arg Ser Glu Met Asn Val Gln Ala Ala Trp Lys Arg Gly Tyr Thr Gly Lys

 91 AAC GTG GTG GTC ACC ATC CTT GAT GAT GGC ATA GAG AGA AAT CAC CCT GAC CTG GCC CCA AAT TAT GAT TCC TAC GCC AGC TAC GAC GTG
 31 Asn Val Val Val Thr Ile Leu Asp Asp Gly Ile Glu Arg Asn His Pro Asp Leu Ala Pro Asn Tyr Asp Ser Tyr Ala Ser Tyr Asp Val

181 AAC GGC AAT GAT TAT GAC CCA TCT CCA CGA TAT GAT GCC AGC AAT GAA AAT AAA CAC GGC ACT CGT TGT GCG GGA GAA GTT GCT GCT TCA
 61 Asn Gly Asn Asp Tyr Asp Pro Ser Pro Arg Tyr Asp Ala Ser Asn Glu Asn Lys His Gly Thr Arg Cys Ala Gly Glu Val Ala Ala Ser

271 GCA AAC AAT TCC TAC TGC ATC GTG GGC ATA GCG TAC AAT GCC AAA ATA GGA GGC ATC CGC ATG CTG GAC GGC GAT GTC ACA GAT GTG GTC
 91 Ala Asn Asn Ser Tyr Cys Ile Val Gly Ile Ala Tyr Asn Ala Lys Ile Gly Gly Ile Arg Met Leu Asp Gly Asp Val Thr Asp Val Val

361 GAG GCA AAG TCG CTG GGC ATC AGA CCC AAC TAC ATC GAC ATT TAC AGT GCC AGC TGG GGG CCG GAC GAC GAC GGC AAG ACG GTG GAC GGG
121 Glu Ala Lys Ser Leu Gly Ile Arg Pro Asn Tyr Ile Asp Ile Tyr Ser Ala Ser Trp Gly Pro Asp Asp Asp Gly Lys Thr Val Asp Gly

451 CCC GGC CGA CTG GCT AAG CAG GCT TTC GAG TAT GGC ATT AAA AAG GGC CGG CAG GGC CTG GGC TCC ATT TTC GTC TGG GCA TCT GGG AAT
151 Pro Gly Arg Leu Ala Lys Gln Ala Phe Glu Tyr Gly Ile Lys Lys Gly Arg Gln Gly Leu Gly Ser Ile Phe Val Trp Ala Ser Gly Asn

541 GGC GGG AGA GAG GGG GAC TAC TGC TCG TGC GAT GGC TAC ACC AAC AGC ATC TAC ACC ATC TCC GTC AGC AGC GCC ACC GAG AAT GGC TAC
181 Gly Gly Arg Glu Gly Asp Tyr Cys Ser Cys Asp Gly Tyr Thr Asn Ser Ile Tyr Thr Ile Ser Val Ser Ser Ala Thr Glu Asn Gly Tyr

631 AAG CCC TGG TAC CTG GAA GAG TGT GCC TCC ACC CTG GCC ACC ACC TAC AGC AGT GGG GCC TTT TAT GAG CGA AAA ATC GTC ACC ACG GAT
211 Lys Pro Trp Tyr Leu Glu Glu Cys Ala Ser Thr Leu Ala Thr Thr Tyr Ser Ser Gly Ala Phe Tyr Glu Arg Lys Ile Val Thr Thr Asp

721 CTG CGT CAG CGC TGT ACC GAT GGC CAC ACT GGG ACC TCA GTC TCT GCC CCC ATG GTG GCG GGC ATC ATC GCC TTG GCT CTA GAA GCA AAC
241 Leu Arg Gln Arg Cys Thr Asp Gly His Thr Gly Thr Ser Val Ser Ala Pro Met Val Ala Gly Ile Ile Ala Leu Ala Leu Glu Ala Asn

811 AGC CAG TTA ACC TGG AGG GAC GTC CAG CAC CTG CTA GTG AAG ACA TCC CGG CCG GCC CAC CTG AAA GCG AGC GAC TGG AAA GTA AAC GGC
271 Ser Gln Leu Thr Trp Arg Asp Val Gln His Leu Leu Val Lys Thr Ser Arg Pro Ala His Leu Lys Ala Ser Asp Trp Lys Val Asn Gly

901 GCG GGT CAT AAA GTT AGC CAT TTC TAT GGA TTT GGT TTG GTG GAC GCA GAA GCT CTC GTT GTG GAG GCA AAG AAG TGG ACA GCA GTG CCA
301 Ala Gly His Lys Val Ser His Phe Tyr Gly Phe Gly Leu Val Asp Ala Glu Ala Leu Val Val Glu Ala Lys Lys Trp Thr Ala Val Pro

991 TCG CAG CAC ATG TGT GTG GCC GCC TCG GAC AAG AGA CCC AGG AGC ATC CCC TTA GTG CAG GTG CTG CGG ACT ACG GCC CTG ACC AGC GCC
331 Ser Gln His Met Cys Val Ala Ala Ser Asp Lys Arg Pro Arg Ser Ile Pro Leu Val Gln Val Leu Arg Thr Thr Ala Leu Thr Ser Ala

1081 TGC GCG GAG CAC TCG GAC CAG CGG GTG GTC TAC TTG GAG CAC GTG GTG GTT CGC ACC TCC ATC TCA CAC CCA CGC CGA GGA GAC CTC CAG
361 Cys Ala Glu His Ser Asp Gln Arg Val Val Tyr Leu Glu His Val Val Val Arg Thr Ser Ile Ser His Pro Arg Arg Gly Asp Leu Gln

1171 ATC TAC CTG GTT TCT CCC TCG GGA ACC AAG TCT CAA CTT TTG GCA AAG AGG TTG CTG GAT CTT TCC AAT GAA GGG TTT ACA AAC TGG GAA
391 Ile Tyr Leu Val Ser Pro Ser Gly Thr Lys Ser Gln Leu Leu Ala Lys Arg Leu Leu Asp Leu Ser Asn Glu Gly Phe Thr Asn Trp Glu

1261 TTC ATG ACT GTC CAC TGC TGG GGA GAA AAG GCT GAA GGG CAG TGG ACC TTG GAA ATC CAA GAT CTG CCA TCC CAG GTC CGC AAC CCG GAG
421 Phe Met Thr Val His Cys Trp Gly Glu Lys Ala Glu Gly Gln Trp Thr Leu Glu Ile Gln Asp Leu Pro Ser Gln Val Arg Asn Pro Glu

1351 AAG CAA GGT GAT CTT GAG ACT CCT GTT GCA AAT CAA CTG ACC ACA GAA GAG AGG TTC GTT TCC ACA CCC TCG ATT CTG TTC CAT TGG TCT
451 Lys Gln Gly Asp Leu Glu Thr Pro Val Ala Asn Gln Leu Thr Thr Glu Glu Arg Phe Val Ser Thr Pro Ser Ile Leu Phe His Trp Ser

1441 GTA TAT CTA TCT TGG AGT CAG TAC CAT ATT GTT TTG ATC ACT GTA GCT TTG TAG TAAGTTTTGAAATAAGAAAGTGCGAGTCCTC(A)n
481 Val Tyr Leu Ser Trp Ser Gln Tyr His Ile Val Leu Ile Thr Val Ala Leu ***
```

Figure 4. Nucleotide and predicted amono acid sequences of PACE 4D.

Sequence of PC 7

Another new member of the kexin family protease cDNA was cloned from a rat anterior pituitary gland cDNA library and named PC 7. As shown in Fig.6, the nucleotide sequence of rat PC 7 was very similar to human PACE 4A; there is 86% identity between them. However, rat PC 7 is not the rat counterpart of human PACE 4A. Because the human counterpart of rat PC 7 was identified from mRNA prepared from a human cell line by the method of RT-PCR, and nucleotide sequences of this region between rat PC 7 and human PC 7 was the same. Although, Johnson et al.[14] reported the nucleotide sequence of rat PACE 4A, it might be one of rat PC 7 isoforms but not PACE 4A. These results indicated that PC 7 was a different gene product than that from the PACE 4 gene. PC 7 cDNA was not

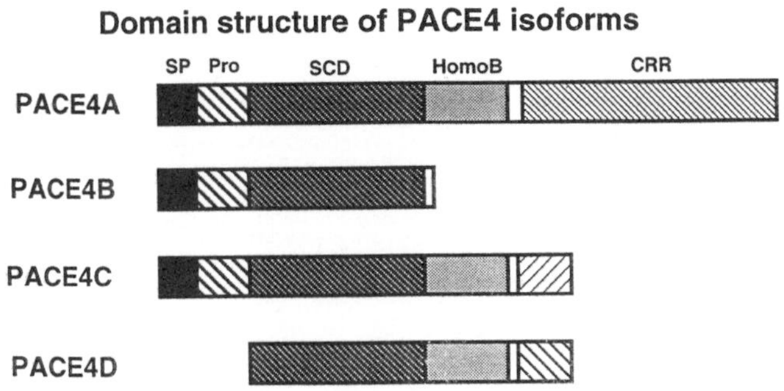

Domain structure of PACE4 isoforms

Figure 5. Schematic representation of the structures of PACE 4 isoforms, PACE 4A, 4C and 4D cDNAs. SP, signal peptide; Pro, propeptide; SCD, subtilisin like catalytic domain; Open boxes indicate region of complete identity among the cDNAs. Closed boxes indicate the unique 3'-end of PACE 4A, PACE 4C and PACE 4D cDNAs and the unique 5'-end of PACE 4D cDNA.

full-length, but it coded for the complete sequence of the mature PC 7. The predicted molecular mass of the mature form of PC 7 was 67 kDa; it contained subtilisin-like catalytic domain, homo-B domain, and cysteine-rich region , but lacked a transmembrane domain. In addition, we have identified PC 7 isoforms with an extra 39 bp nucleotide, named PC 7A ; the former isoform was named PC 7B.

TISSUE DISTRIBUTION OF KEXIN FAMILY PROTEASES

Kexin family proteases are expressed in a tissue-specific manner. Among PACE 4 isoforms, PACE 4A was not highly expressed in rat tissues but, PACE 4C was expressed in B cells of pancreatic islets, cardiac musle, liver and kidney. On the other hand, PACE 4D was expressed abundantly in pituitary and some cultured cell lines. We have analysed mRNA expression by the method of the reverse transcriptase-polymerase chain reaction (RT-PCR). In addition, specific antibody was prepared against each kexin family protease and immunochemical studies were carried out . Pancreatic islets are composed of insulin- producing B cells, glucagon-producing A cells, somatostatin-producing D cells and pancreatic polypeptide-producing P cells. PC 2 was expressed highly in A and D cells, to a lesser extant in B cells, but not in P cells. PC 1 was expressed only in B cells, and furin and PC 6 were weakly expressed in all islet cells. PACE 4C was detected only in B cells of pancreatic islets as shown in Fig.7. These results suggested that PACE 4C and PC 1 might participate in processing bioactive peptides secreted from B cells and other kexins were less important in B cells in the rat pancreas[15]. On the other hand, PC 7 was expressed in rat central nervous system and adrenal medulla but not much in human tissues. We have detected PC 7 mRNA in human neuroblastoma cell line. By contrast, rat heart contained PACE 4A and PACE 4C. These results suggested that hPACE 4A was expressed in various tissues and the expression of PC 7 was very low in human tissues. On the contrary, PC 7 was expressed at high levels in rat or mouse brain and neural cells, but the expression of PACE 4A was very low in these tissues. However, mRNAs of both PACE 4A and PC 7 were detected in rat and human tissues respectively. Therefore expression of PACE 4A and PC 7 were regulated in a species-specific manner and may contribute to processing of the precursors of biologically active peptides.

Nucleotide and deduced amino acid sequence of PC7B

↓

```
   1  CGG CTC GGG CTC CAG CAA CAG GAA GTG AAA CGC AGG GTC AAG ACA CAG GCG CGA AGC GAC TCT CTT TAT TTC AAT GAT CCC ATT TGG TCC
   1  Arg Leu Gly Leu Gln Gln Gln Glu Val Lys Arg Arg Val Lys Arg Gln Ala Arg Ser Asp Ser Leu Tyr Phe Asn Asp Pro Ile Trp Ser

  91  AAC ATG TGG TAT ATG CAT TGT GCT GAT AAG AAC AGT CGC TGT CGG TCA GAG ATG AAC GTC CAG GCG GCA TGG AAG CGC GGC TAC ACA GGA
  31  Asn Met Trp Tyr Met His Cys Ala Asp Lys Asn Ser Arg Cys Arg Ser Glu Met Asn Val Gln Ala Ala Trp Lys Arg Gly Tyr Thr Gly

 181  AAG AAC GTG GTT GTC ACC ATC CTC GAT GAC GGC ATA GAA AGG AAT CAC CCA GAC CTG GCC CCC AAC TAC GAT TCC TAT GCA AGC TAC GAT
  61  Lys Asn Val Val Val Thr Ile Leu Asp Asp Gly Ile Glu Arg Asn His Pro Asp Leu Ala Pro Asn Tyr Asp Ser Tyr Ala Ser Tyr Asp

 271  GTC AAC GGA AAC GAT TAT GAC CCA TCA CCG AGA TAT GAT GCC AGC AAC GAG AAC AAA CAT GGT ACT CGC TGT GCG GGA GAA GTC GCT GCT
  91  Val Asn Gly Asn Asp Tyr Asp Pro Ser Pro Arg Tyr Asp Ala Ser Asn Glu Asn Lys His Gly Thr Arg Cys Ala Gly Glu Val Ala Ala

 361  TCA GCC AAC AAC TCC TAC TGC ATC GTG GGC ATA GCA TAT AAT GCA AGG ATA GGA GGC ATC CGG ATG CTG GAC GGT GAC GTG ACC GAC GTG
 121  Ser Ala Asn Asn Ser Tyr Cys Ile Val Gly Ile Ala Tyr Asn Ala Arg Ile Gly Gly Ile Arg Met Leu Asp Gly Asp Val Thr Asp Val

 451  GTT GAG GCC AAG TCT CTG GGC ATC AGA CCC AAC TAC ATT GAC ATT TAC AGC GCT AGT TGG GGG CCA GAT GAT GAT GGG AAG ACC GTG GAT
 151  Val Glu Ala Lys Ser Leu Gly Ile Arg Pro Asn Tyr Ile Asp Ile Tyr Ser Ala Ser Trp Gly Pro Asp Asp Asp Gly Lys Thr Val Asp

 541  GGG CCC GGC CGT CTG GCT AAA CAG GCT TTC GAG TAT GGC ATT AAA AAG GGC CGC CAA GGT CTG GGC TCC ATT TTT GTC TGG GCC TCT GGG
 181  Gly Pro Gly Arg Leu Ala Lys Gln Ala Phe Glu Tyr Gly Ile Lys Lys Gly Arg Gln Gly Leu Gly Ser Ile Phe Val Trp Ala Ser Gly

 631  AAT GGT GGG AGA GAA GGG GAC CAC TGC TCC TGT GAT GGC TAC ACC AAC AGC ATC TAC ACC ATC TCT GTG AGC AGC ACC ACT GAG AAC GGC
 211  Asn Gly Gly Arg Glu Gly Asp His Cys Ser Cys Asp Gly Tyr Thr Asn Ser Ile Tyr Thr Ile Ser Val Ser Ser Thr Thr Glu Asn Gly

 721  CAC AAA CCC TGG TAC CTG GAG GAA TGT GCT TCC ACC TTG GCT ACC ACC TAC AGC AGC GGG GCC TTC TAT GAA CGG AAG ATC GTC ACC ACG
 241  His Lys Pro Trp Tyr Leu Glu Glu Cys Ala Ser Thr Leu Ala Thr Thr Tyr Ser Ser Gly Ala Phe Tyr Glu Arg Lys Ile Val Thr Thr

 811  GAC CTG CGT CAG CGC TGC ACC GAC GGC CAC ACT GGG ACA TCT GTC TCA GCT CCC ATG GTG GCT GGC ATC ATT GCC CTG GCT CTA GAA GCA
 271  Asp Leu Arg Gln Arg Cys Thr Asp Gly His Thr Gly Thr Ser Val Ser Ala Pro Met Val Ala Gly Ile Ile Ala Leu Ala Leu Glu Ala

 901  AAC AAC CAG TTG ACC TGG AGG GAC GTG CAG CAC CTG TTA GTA AAG ACG TCA CGG CCG GCT CAT CTG AAG GCG AGT GAC TGG AAA GTC AAC
 301  Asn Asn Gln Leu Thr Trp Arg Asp Val Gln His Leu Leu Val Lys Thr Ser Arg Pro Ala His Leu Lys Ala Ser Asp Trp Lys Val Asn

 991  GGA GCT GGG CAT AAA GTT GAC CAT CTC TAT GGA TTT GGC TTG GTG GAT GCT GAA GGG CTC GTC CTA GAG GCA AGG AAG TGG ACG GCA GTG
 331  Gly Ala Gly His Lys Val Asp His Leu Tyr Gly Phe Gly Leu Val Asp Ala Glu Gly Leu Val Leu Glu Ala Arg Lys Trp Thr Ala Val

1081  CCA TCC CAG CAC ATG TGC GTG GCC ACC GCA GAC AAA AGG CCC AGG AGC ATC CCC GTA GTG CAG GTG CTG CGG ACC ACA GCC CTG ACC AAT
 361  Pro Ser Gln His Met Cys Val Ala Thr Ala Asp Lys Arg Pro Arg Ser Ile Pro Val Val Gln Val Leu Arg Thr Thr Ala Leu Thr Asn

1171  GCC TGT GCA GAC CAC TCT GAC CAG CGT GTG GTG TAC CTG GAG CAT GTG GTA GTC CGA ATC TCT ATC TCA CAT CCA CGA CGG GGT GAC CTC
 391  Ala Cys Ala Asp His Ser Asp Gln Arg Val Val Tyr Leu Glu His Val Val Val Arg Ile Ser Ile Ser His Pro Arg Arg Gly Asp Leu

1261  CAG ATC CAC CTG ATT TCT CCC TCT GGA ACC AAG TCT CAA CTT TTG GCA AAG AGA TTG CTG GAT TTT TCC AAT GAG GGG TTC ACG AAC TGG
 421  Gln Ile His Leu Ile Ser Pro Ser Gly Thr Lys Ser Gln Leu Leu Ala Lys Arg Leu Leu Asp Phe Ser Asn Glu Gly Phe Thr Asn Trp

1351  GAG TTC ATG ACT GTC CAC TGC TGG GGA GAA AAG GCT GAA GGT GAA TGG ACC CTG GAA GTC CAG GAT ATA CCA TCG CAG GTC CGC AAC CCA
 451  Glu Phe Met Thr Val His Cys Trp Gly Glu Lys Ala Glu Gly Glu Trp Thr Leu Glu Val Gln Asp Ile Pro Ser Gln Val Arg Asn Pro

1441  GAG AAA CAA GGA AAG TTG AAA GAA TGG AGC CTC ATT TTA TAT GGC ACT GCA GAG CAC CCA TAC CGC ACC TTC AGC TCC CAC CAG TCT CGC
 481  Glu Lys Gln Gly Lys Leu Lys Glu Trp Ser Leu Ile Leu Tyr Gly Thr Ala Glu His Pro Tyr Arg Thr Phe Ser Ser His Gln Ser Arg

1531  TCA CGG ATG CTG GAG CTT TCA GTC CCG GAA CAG GGA CCT CTC AAG GCT GAG GGA CCA CCA CCG CAG GCA GAG ACT CCA GAA GAA GAG GAA
 511  Ser Arg Met Leu Glu Leu Ser Val Pro Glu Gln Gly Pro Leu Lys Ala Glu Gly Pro Pro Pro Gln Ala Glu Thr Pro Glu Glu Glu Glu

1621  GAG TAC ACA GGT GTG TGC CAT CCA GAG TGT GGT GAT AAA GGC TGC GAT GGT CCC AGT GCA GAC CAG TGC TTG AAC TGC GTC CAC TTC AGC
 541  Glu Tyr Thr Gly Val Cys His Pro Glu Cys Gly Asp Lys Gly Cys Asp Gly Pro Ser Ala Asp Gln Cys Leu Asn Cys Val His Phe Ser

1711  CTG GGA AAC TCC AAG ACA AAC AGG AAG TGT GTG AGC GAG TGC CCC TTG GGC TAC TTT GGG GAC ACA GCA GCA AGA CGT TGC CGT CGA TGC
 571  Leu Gly Asn Ser Lys Thr Asn Arg Lys Cys Val Ser Glu Cys Pro Leu Gly Tyr Phe Gly Asp Thr Ala Ala Arg Arg Cys Arg Arg Cys

1801  CAT AAG GGA TGT GAG ACA TGC ACG GGC AGG AGC CAA CAC AGT GCC TGT CTT GTC GCC GTG GGT TCT ATC ACC ACC AGG AAA CGA ACA CAT
 601  His Lys Gly Cys Glu Thr Cys Thr Gly Arg Ser Gln His Ser Ala Cys Leu Val Ala Val Gly Ser Ile Thr Thr Arg Lys Arg Thr His

1891  GTG TGA CCCTGTGTCCTGCCGGACTTTATGCTGATGAAAGTCAGAGACTCTGCCTCAGGTGCCACCCGAGCTGTCAGAAGTGTGTGGATGAACCTGAGAAGTCACTGTGTGCAAGGA
 631  Val ***

2008  GGGATTCAGCCTCGCACGGGGCAGCTGCATTCCGGACTGTGAACCAGGTACCTACTTCGATTCTGAGCTCATCAGATGTGGGGAATGCCATCACACCTGCCGGACCTGCGTGGGGCCCA
2127  GCAGAGGAAGAATGTATTCACTGTGCAAAAAGCTTCCACTTCCAAGACTGGAAATGTGTGCCGGCCTGTGGTGAGGGCTTCTACCCGGAGGAGATGCCTGGCTTACCCCACAAAGTGTGT
2246  CGAAGATGTGATGAAAACTGCCTGAGCTGCGAGGGCTCCAGCAGGAACTGCAG
```

Figure 6. Nucleotide sequence and deduced amino acid sequence of rat PC 7B.

REFERENCES

1. Docherty, K. and Steiner, D.F., 1982, Post-translational proteolysis in polypeptide hormone biosynthesis, *Ann. Rev. Physiol.* 44: 625-638.
2. Fuller, R.S., Sterne, R.E. and Thorner, J., 1988, Enzymes required for yeast prohormone processing, *Ann. Rev. Physiol.* 50: 345-362.

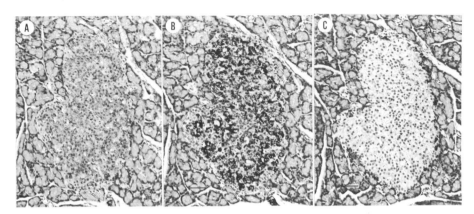

Figure 7. Immunohistochemical staining of PACE 4C in rat pancreatic islet. Adjacent sections A, B and C were immunostained, then counter-stained with Mayer's hemalum solution. A ; anti-insulin, B; anti-PACE 4C, C; anti-pancreaticpolypeptide. Scalebars are 50 μm.

3. Fuller, R. S., Brake, A.and Thorner, J., 1989, Yeast prohormone processing enzyme (KEX2 gene product) is a Ca^{2+}-dependent serine protease, *Proc. Natl. Acad. Sci U.S.A.*, 86: 1434-1438.

4. Bresnahan, P.A., Leduc, R., Thomas, L., Thorner, J., Gibson, H. L.Brake, A. J., Barr, P. J. and Thomas, G., 1990, Human fur gene encodes a yeast KEX2-like endoprotease that cleaves pro-β- NGF in vivo, *J. Cell. Biol.* 111: 2851-2859.

5. Kiefer, M.C., Tucher, J. E., Joh, R., Landsberg, K.E., Saltman, D. and Barr, P., 1991, Identification of second human subtilisin-like protease gene in the fes/fps region of chromosome 15, *DNA Cell Biol.* 10: 757-769.

6. Barr, P.J., Mason, O. B., Landsberg, K.E., Wong,P.A., Kiefer, M.C. and Brake, A.J., 1991, cDNA and gene structure for a subtilisin-like protease with cleavage specificity for paired basic amino acids residues, *DNA Cell Biol.* 10: 319-328.

7. Benjannet, S., Rondeau, N., Day, R., Chretien M. and Seidah N.G., 1991, PC 1 and PC 2 are proprotein convertases capable of cleaving propiomelanocortin at distinct pairs of basic residues, *Proc. Natl. Acad. Sci.USA*, 88: 3564-3568.

8. Bennet, D.L., Bailyes, E.M., Nielsen, E., Guest, P.C., Rutherford, N.G., Arden, S. D. and Hutton J. C., 1992, Identification of the type 2 proinsulin processing endopeptidase as PC 2, a member of the eukaryote subtilisin family, *J. Biol. Chem.* 267: 15229-15236.

9. Nakayama, K., Kim, W-S., Torii, S., Hosaka, M., Nakagawa, T. Ikemizu, J., Baba, T., and Murakami, K., 1992, Identification of the fourth member of the mammalian endopeptidase family homologous to the yeast Kex2 protease, *J. Biol. Chem.*, 267: 5897-5900.

10. Nakagawa, T., Hosaka,M., Torii, S., Watanabe, T., Murakami,K. and Nakayama, K., 1993, Identification and functional expression of a new member of the mammalian Kex2-like processing protease family: Its striking structural similarity to PACE 4, *J. Biochem.* 113: 132-135.

11. Tsuji,A.,Hine,C., Mori,K., Tamai,Y., Higashine,K., Nagamune,H. and Matsuda,Y., 1994, A novel member, PC 7, of the mammalian kexin-like protease family: Homology to PACE 4A, Its brain specific expression and identification of isoforms, *Biochem. Biophys. Res. Commun.*, 202: 1452-1459.

12. Tsuji,A., Higashine,K., Hine,C., Mori,K., Tamai,Y., Nagamune,H. and Matsuda,Y., 1994, Identification of novel cDNAs encoding human kexin-like protease, PACE 4 isoforms, *Biochem. Biophys. Res. Commun.* 200 (2): 943-950.

13. Tsuji,A., Mori,K., Hine,C., Tamai,Y., Nagamune,H. and Matsuda,Y., 1994, The tissue distribution of mRNAs for the PACE 4 isoforms, Kexin-like processing protease: PACE 4C and PACE 4D are major transcripts among PACE 4 isoforms, *Biochem., Biophys. Res. Commun.*, 202:1215-1221.

14. Johnson, R.C., Darlington, D.N., Hand,T.A., Blooquist,B.T. and Mains, R.E., 1994, A subtilisin-like endoprotease prevalent in the anterior pituitary and regulated by thyroid status, *Endocrinology*, 135: 1178-1185.

15. Nagamune,H., Muramatsu,K., Akamatsu,T., Tamai,Y. Izumi,K., Tsuji,A. and Matsuda,Y., 1995, Distribution of the kexin family proteases in pancreatic islets: PACE 4C is specifically expressed in B cells of pancreatic islets, *Endocrinology* 136: 357-360.

8

PHYTOCYSTATINS AND THEIR TARGET ENZYMES - MOLECULAR CLONING, EXPRESSION AND POSSIBLE FUNCTIONS

S. Arai, K. Abe,[1] and Y. Emori[2]

[1] Department of Applied Biological Chemistry
Division of Agriculture and Agricultural Life Sciences
[2] Department of Biology and Biophysics, Faculty of Science
The University of Tokyo
Bunkyo-ku, Tokyo ll3, Japan

INTRODUCTION

Protein inhibitors of cysteine proteinases [EC 3.4.22] are referred to as cystatins (Barrett et al., 1986). A number of cystatins originating from animal tissues and products have been well investigated and, based on their primary structures, they are unified as members belonging to the cystatin superfamily which comprises three families. The naming for each of the three families has also been proposed (Bode et al., 1988). Structures and functions of these animal cystatins have been extensively studied primarily in the field of biochemistry and physiology. Structurally, each of them is a small protein with a little more than 100 amino acid residues, characterized by having two target enzyme-binding sites in the form of QVVAG or related sequence in a central region, and PW or, in some cases XW, in a C-terminal region (Turk and Bode, 1991). As to possible functions of animal cystatins it is put forward that these act to regulate endogenous catheptic cysteine proteinases as they happen to come out of the lysosome for any reasons and also some exogenous cysteine proteinases which might be carried by infecting pests (Korant et al., 1986; Bjorck et al., 1989; Kondo et al., 1992). Animal cystatins are thus understood as substances contributing to biodefense through a non-immunological mechanism.

Cystatins of plant origin, i.e. phytocystatins, on the other hand, had long been veiled until almost 10 years ago when we disclosed the occurrence of a proteinaceous cysteine proteinase inhibitor in the seed of rice, *Oryza sativa L. japonica* (Abe et al., 1985; l987a). Subsequently, we investigated this inhibitor in detail on both protein and gene levels and named it oryzacystatin (Abe et al.,1987b).This is now renamed as oryzacystatin I (OC-I) since, shortly after, we found the second oryzacystatin (OC-II) by molecular cloning (Kondo et al., 1990a).

This article deals with genotypical and phenotypical charactaristics of OC-I and other phytocystatins, and their endogenous as well as exogenous target enzymes.

Intracellular Protein Catabolism, Edited by Koichi Suzuki and Judith Bond
Plenum Press, New York, 1996

PHYTOCYSTATINS

OC-I is a protein with a molecular weight of ca. 11,500 consisting of 102 amino acid residues (Abe *et al.*, 1985; 1987a; 1987b). Quantitative experiments using an anti-OC-I antibody showed that it occurs at an amount of approximately 3 mg/kg seed (Kondo *et al.*, 1989a; 1990b). OC-I is also characterized by its heat stability. Even treated at 100° C for 30 min, it maintains its original cystatin activity. The result indicates that those who have rice as a stable food take OC-I in an active form as they eat cooked rice day by day. Enzymologically, OC-I inhibits the proteolytic activity exlusively of cysteine proteinases such as papain, ficin, actinidin and aleurain of plant origin and cathepsin B, H and L of animal origin. Assayed based on a decrease in N-benzoyl-DL-arginine-2-naphthylamide hydrolyzing activity of papain, OC-I gave a Ki value at a 10^{-8}M which almost accorded with Ki values observed when this enzyme is inhibited by animal cystatins (Abe *et al.*, 1987a).

cDNA cloning was carried out using two oligonucleotides synthesized according to codons deduced from the chemically determined partial amino acid sequences, KPWNDF and KPVDAS, of OC-I. Screening a cDNA library constructed from two-week-ripened seeds of rice (cultivar Nakateshinsenbon), we obtained a cDNA clone encoding a 102-mer protein exactly containing these two partial sequences (Abe *et al.*, 1987b). The sequence QVVAG and PW also existed, and are known as papain-binding sites of chicken egg white cystatin (Bode *et al.*,1988).We then screened the same library using the OC-I cDNA clone as a probe and obtained another cDNA clone which encoded a 107-mer protein, i.d., OC-II, containing the target enzyme-binding sites in the form of QVVGG and AW (Kondo *et al.*, 1990a). An amino acid sequence similarity of as much as 55% was found between OC-I and OC-II.

A system was established to obtain recombinantly produced OC-I, OC-II and their N-truncated mutants by transformation of *Escherichia coli* (Abe *et al.*,1988; Arai *et al.*,1991). Kinetic studies showed that OC-I inhibited papain and cathepsin H with Ki values of 3.02 x 10^{-8} M and 1.29 x 10^{-6} M, respectively, while Ki values for OC-II against these enzymes were 0.83 x 10^{-6} M and 1.84 x 10^{-8} M, respectively (Kondo *et al.*, 1990a). A differential relationship may thus exist between OC-I and OC-II as they act on different target enzymes.

Both OC-I and OC-II, compared with animal cystatins, are structurally unique. Each lacks disulfide bonds unlike family-2 cystatins. With respect to overall primary structure, however, they resemble family-2 rather than family-1 cystatins. An unreasonable problem then arises when these phytocystatins are forced to be included in either of the current classification families based exclusively on cystatins of animal origin. The unreasonableness is enlarged when, as discussed below, the genotypes of OC-I and OC-II are compared with those of animal cystatins. We constructed a genomic DNA library from rice seeds, screened it with OC-I and OC-II cDNA clones as probes, and obtained respective genes spanning ca. 2.5 kb each (Kondo *et al.*,1989b; 1991). Nucleotide sequence analysis demonstrated that each gene was organized with three exons and two introns. The first intron, spanning 336 bp, intervened between A38 and N39 and second one, 372 bp, existed in the 3'-noncoding region, particulary at the G nucleotide residue next to the stop codon TAA. Interestingly enough, the intron positions found are quite different from those observed for animal cystatin genes. This is the reason why we have proposed that OC-I and OC-II be classified into family-4 as a new category in the cystatin superfamily (Abe *et al.*, 1991).

The reasonableness of our proposal is going to be confirmed by the cloning studies we are conducting on cystatins of corn, wheat and soybean origins. We have actually cloned two corn cystatins (Abe *et al.*, 1992; 1994), a wheat cystatin (unpublished) and soyacystatin (unpublished). These have significant degrees of structural similarity to OC-I and OC-II. Of interest is the finding that the corn cystatins have signal sequences, the soyacystatin of dicotyledon origin attaches extremely large N- and C-extention sequences which have not been found in any other cystatins known. Genotypically, corn cystatins and soyacystatin are similar to OC-I and OC-II in respect to intron positions. These results suggest that among all cystatins which are probably being cognate, phytocystatins apparently deviated at an early stage in the evolutional pedigree. Incidentally, it is noted that the term "phytocystatin" has been quoted (Turk and Bode, 1991).

ENDOGENOUS TARGET ENZYMES

Cysteine proteinases are ubiquitous in plant tissues. Some of them are secreted, popular examples of which are papain, ficin and bromelain. A gibberellin-inducible cysteine proteinase has been found in the aleurone of barley seeds and named aleurain (Rogers *et al.*, 1985). We also found, in germinating rice seeds, a cysteine proteinase whose expression was enhanced to a large extent in the presence of added gibberellic acid (Arai *et al.*, 1988). We then carried out molecular cloning using an aleurain cDNA clone as a probe, with the result that three cysteine proteinases, named oryzain a, b and g (Fig.1), are expressed in rice seeds as soon as they are made to imbibe for germination (Watanabe *et al.*, 1991). Subsequently, we demonstrated that these three oryzain species do occur in mature protein forms; the enzyme we first found (Arai *et al.*, 1988) corresponds to oryzain a. All the three species should initially exist in prepro forms to be processed into maturity. Each of oryzain a, b and g has the so-called catalytic triad constituted with C25, H159 and N175 (papain numbering). Alignment of amino acid sequences showed that oryzain a and b resembled papain, whereas oryzain g was similar to cathepsin H rather than papain (Watanabe *et al.*, 1991). This again indicates that, in germinating rice seeds, OC-I and OC-II could exhibit their inhibitory activities differentially toward these endogenous target enzymes, i.e., the former toward oryzain a and b, and the latter toward oryzain g. By the way, OC-I and OC-II genes are separately located in rice chromosomes (Kishimoto *et al.*, 1994). The expression of oryzain has also been studied (Watanabe *et al.*, 1994).

EXOGENOUS TARGET ENZYMES

Besides the regulatory function discussed above, some defensive function may exist in OC-I, OC-Il and other phytocystatins of seed origin. This function involves protection of

Figure 1

```
α  LPESVDWRIKGAVAEIKDQGGCGSCWAFSAIAAVEDINQIVTGDLISLSEQELVDCDTSY  188
β  LPESVDWREKGAVAPVKNQGQCGSCWAFSAVSTVESINQLVTGEMITLSEQELVECSTNG  199
γ  LPEITKDWREDGIVSPVKDQGHCGSCWPFSTTGSLEARYTQATGPPVSLSEQQLADCATRY  204

α  -NEGCNGGLMDYAFDFIIN-NGGIDTEDDYPYKGKDERCDVNRKNAKVVTIDSYEDVTPN  246
β  QNSGCNGGLMADAFDFIIK-NGGIDTEDDYPYKAVDGKCDINRENAKVVSIDGFEDVPQN  258
γ  NNFGCSGGLPSQAFEYLI-KYNGGLDTEEAYPYTGVNGICHYKPENAGVKVLDSV-NITLV  262

α  SETSLQKAVR-NQPVSVAIEAGGRAFQLYSSGIFTG-KCGTA-LD--HGVAAVGYGTENG  301
β  DEKSLQKAVA-HQPVSVAIEAGGREFQLYHSGVFSG-RCGTS-LD--HGVVAVGYGTDNG  313
γ  AEDELKNAVGLVRPVSVAFQVIN-GFRMYKSGVYTSDHCGTSPMDVNHAVLAVGYGVENG  321

α  KDYWIVRNSWGKSWGESGYVRMERNIKASSGK--CGIAVEPSYPLKKGENPPNPGPTPPS  359
β  KDYWIVRNSWGPKWGESGYVRMERNINVTTGK--CGIAMMASYPTTKSGANPPKPSPTPPT  371
γ  VPYWLIKNSWGADWGDNGYFTMEM------GKNMCGIATCASYPIVA               362
```

seeds and other plant tissues from invading insect pests. The crop damage inflicted by these pests is a serious problem for agricultural and food industries. A great deal of effort has been directed toward establishing a countermeasure. A number of low-molecular-weight chemicals have been developed to occupy a central position in the pesticide for agricultural use, but there is often the case that insects have developed resistance to these pesticides when they are used over a long period of time. Residual chemicals remaining in foodstuffs may also cause a safety problem.

Insects, as well as many other invaders of crops, utilize proteins storage stored therein by digestion with their intralumen proteases. This means that blocking the digestive proteases of insects could cause starvation, often with lethal results. Therefore, much attention has been paid to the use of proteinase inhibitors for this blocking. This idea has arisen from the fact that plant seeds per se contain a variety of proteinase inhibitors probably as bioprotectants for the plants themselves. A number of naturally occurring proteinase inhibitors have been isolated and characterized; most of them, however, are serine proteinase inhibitors including trypsin inhibitors. For practical application, a report has been presented describing that feeding crop insects on a diet with added serine proteinase inhibitors of plant origin can retard their growth to a significant extent (Hilder *et al.*, 1987; Johnson *et al.*, 1989). Despite that, most crops are still damaged by invading insect pests. This suggests that the target should be set preferably at phytocystatins of food seed origin which are probably safe for human consumption and yet are able to block cysteine proteinases in the gut of insect pests.

The occurrence of gut cysteine proteinases has been found in the example of a coleoptera insect which inflicts great damage on beans (Murdock *et al.*, 1987). This finding has now become a testimony to the significant involvement of this class of proteinases. Since, however, there is no phytocystatin industrially available so far, the only way has long been to use low-molecular-weight cysteine proteinase inhibitors such as E-64. We have completed a system for overexpression of OC-I and OC-II by introduction of their genes into *Escherichia coli* (Abe *et al.*, 1988), making it easy to use recombinantly produced phytocystatins for the antipest purpose. In our study, we targetted *Callosobruchus chinensis* as an invader of azuki and cowpea seeds, and *Rip tortus clavatus* as a representative soybean pest. Feeding each of these insects on their respective diets with added OC-I or OC-II, we found that it clearly showed a growth-retarding effect, even with a lethal result when fed at higher concentration (unpublished data). We also tested the inhibitory effect on the growth of *Sitophilus zeamais*, a notorious invader of rice seeds. The result was almost similar in that both OC-I and OC-II can retard the growth of this insect and even kill it at the larval stage, also indicating that rice per se has such an inhibitor as a self-protectant, though its original quantity may be insufficient for complete protection from an attack. Recently, recombinant OC-I was found to elicit a similar growth-inhibitory effect on *Tribolium castaneum*; this probably is a result of blocking its gut cysteine proteinase (Liang *et al.*, 1991).

There is thus an experimental background warranting the effectiveness of phytocystatins as potential antipest substances. The effect, however, varies depending on the quantity and quality of the phytocystatin used. Such a variation may be due to a difference in specificities of cystatins toward an insect gut cysteine proteinase, also reflecting a difference in terms of how much the cysteine proteinase is involved in the digestive process of the insect. It suggests the importance of fundamentally studying insect gut cysteine proteinases in detail. Despite that, no general information has been presented in this respect so far; no attempts have been made from the standpoint of enzymology and molecular cloning.

We started cloning a cysteine proteinase of *Drosophila melanogaster* as a model insect material with the expectation that the study would provide generally applicable information. We tried with success to isolate a *Drosophila* genomic DNA clone encoding a cysteine proteinase (Matsumoto *et al.*, 1995). The mature protein region of the encoded enzyme, termed DCPl, consisted of 218 amino acid residues, having a structural similarity of 67% to a lobster cysteine proteinase which has been reported to occur in the digestive juice (Laycock *et al.*, 1991). DCP1 also showed significant similarities to cysteine proteinases of animal origin, e.g., cathepsin H and L, as well as to those of plant origin, e.g., oryzain a and b (Watanabe *et al.*, 1991). *In situ* hybridization studies adopted for the embryo demonstrated that DCP1 mRNA was predominantly expressed in larval alimentary organs such as salivary gland. The adult midgut including the gastric caeca also expresses the mRNA to a significant extent. These observations suggest that DCP1 is an enzyme potently involved in the digestion of fed proteins and, in the light of the knowledge mentioned above, could also be used as a model target of phytocystatins.

In this regard, it is added that molecular breeding has started to produce transgenic rice with its cystatin level greatly enhanced by introduction of phytocystatin genes. A positive result has been obtained (Hosoyama *et al.*, 1994), which, we hope, will eventually lead to new strategies for the regulation of crop insect pests in practice.

REFERENCES

Abe, K. and Arai, S., 1985, Purification of a cysteine proteinase inhibitor from rice, Oryza sativa L. japonica, *Agric. Biol. Chem.* 49:3349-3350.

Abe, K., Kondo, H., and Arai, S., 1987a, Purification and characterization of a rice cysteine proteinase inhibitor, *Agric. Biol. Chem.* 51:2763-2768.

Abe, K., Emori, Y., Kondo, H., Suzuki, K., and Arai, S., 1987, Molecular cloning of a cysteine proteinase inhibitor of rice (oryzacystatin): homology with animal cystatins and transient expression in the ripening process of rice seeds, *J. Biol. Chem.* 262:16793-16797.

Abe, K., Kondo, H., Watanabe, H., Emori, Y., and Arai, S., 1991, Oryzacystatins as the first well-defined cystatins of plant origin and their target proteinases in rice seeds, *Biomed. Biochem. Acta*, 50:637-641.

Abe, K., Emori, Y., Kondo, H., Arai, S., and Suzuki, K., 1988, The NH_2-terminal 21 amino acid residues are not essential for the papain-inhibitory activity of oryzacystatin, a member of the cystatin superfamily: expression of oryzacystatin cDNA and its truncated fragments in *Escherichia coli*, *J. Biol. Chem.* 263:7655-7659.

Abe, M., Abe, K., Kuroda, M., and Arai, S., 1992, Corn kernel cysteine proteinase inhibitor as a novel cystatin superfamily member of plant origin, *Eur. J. Biochem.* 209:933-937.

Abe, M., Abe, K., Iwabuchi, K., Domoto, C., and Arai, S., 1994, Corn cystatin I expressed in Escherichia coli: investigation of its inhibitory profile and occurrence in corn kernels, *J. Biochem.* 116:488-492.

Arai, S., Watanabe, H., Kondo, H., Emori, Y., and Abe, K., 1991, Papain-inhibitory activity of oryzacystatin, a rice seed cysteine proteinase inhibitor, depends on the central Gln-Val-Val-Ala-Gly region conserved among cystatin superfamily members. *J. Biochem.* 109:294-298.

Arai, S., Hosoyama, H., and Abe, K., 1988, Gibberellin-induced cysteine proteinase occurring in germinating rice seeds and its specificity for digesting insulin B-chain. *Agric. Biol. Chem.* 52:2957-2959.

Barrett, A. J., Fritz, H., Grubb, A., Isemura, S., Jarvinen, N., Katunuma.,N., Machleidt, W., MullerEsterl, W., SasaKi, M., and Turk, V., 1986, Nomenclature and classification of the proteins homologous with the cysteine proteinase inhibitor chicken cystatin, *Biochem. J.* 236:312.

Bjorck, L., Akesson, P., Bohus, M., Trojnar, J., Abrahamson, M., Olasfsson, I.,and Grubb, A., 1989, Bacterial growth blocked by a synthetic peptide based on the structure of a human proteinase inhibitor, *Nature* 337:385-386.

Bode,W., Engh, R., Musil, D., Thiele, V., Huber, R., Karshikov, A., Brzin. J., Kos, J., and Turk, V., 1988, The 2.0∂ X-ray crystal structure of chicken egg white cystatin and its possible mode of interaction with cysteine proteinases. *EMBO J.* 7:2593-2599.

Hilder, V. A., Gatehouse, A. M. R., Sheerman, S. E., Barker, R'. F., and Boulter, D., 1987, A novel mechanism of insect resistance engineered into tobacco, *Nature* 330:160-163.

Hosoyama, H., Irie, K., Abe, K., and Arai, S., 1994, Oryzacystatin exogenously introduced into protoplasts and regeneration of transgenic rice, *Biosci. Biotech. Biochem.* 58:1500-1505.

Johnson, R., Narvaez, J., An, G., and Ryan, C., 1989, Expression of proteinase inhibitors I and II in transgenic tobacco plants: effects on natural defense against Manduca sexta larvae, *Proc. Natl. Acad. Sci. U.S.A.* 86:9871-9875.

Kishimoto, N., Higo, H., Abe, K., and Arai, S., 1994, Identification of duplicated segments in rice chromosomes 1 and 5 by linkage analysis of cDNA markers of known functions, *Theor. Appl. Genet.* 88:722-726.

Kondo, H., Abe, K., and Arai, S., 1989a, Immunoassay of oryzacystatin occurring in rice seeds during maturation and germination, *Agric. Biol. Chem.* 53:2949-2954.

Kondo, H., Emori, Y., Abe, K., Suzuki. K., and Arai., S., 1989b, Cloning and sequence analysis of the genomic DNA fragment encoding oryzacystatin, *Gene* 81:259-265.

Kondo, H., Abe, K., Nishimura, I., Watanabe, H., Emori. Y., and Arai, S., 1990a, Two distinct cystatin species in rice seeds with different specificities against cysteine proteinases, *J. Biol. Chem.* 265:15832-15837.

Kondo, H., Abe, K., Fujisawa, I., Tada, T., and Arai, S., 1990b, Establishment of monoclonal antibodies and their application to the purification of oryzacystatin, *Agric. Biol. Chem.* 54:2153-2155.

Kondo, H., Abe, K., Emori, Y., and Arai, S., 1991, Gene organization of oryzacystatin II, a new cystatin superfamily member of plant origin is closely related to that of oryzacystatin I but different from those of animal cystatins, *FEBS Lett.* 278:87-90.

Kondo, H., Ijiri, S., Abe, K., Maeda, H., and Arai, S., 1992, Inhibitory effect of oryzacystatins and a truncation mutant on the replication of poliovirus in infected Vero cells, *FEBS Lett.* 299:48-50.

Korant, B, D., Towatari, T., Ivanoff, L., Petteway, S. Jr., Brzin, J., Lenarcic, B., and Turk, V., 1986, Viral therapy: prospects for protease inhibitors, *J. Cell. Biochem.* 32:91-95.

Laycock, M. V., Mackay, R. H., Fruscio, M. D., and Gallant, J. W., 1991, Molecular cloning of three cDNAs that encode cysteine proteinases in the digestive gland of the American lobster (Homarus americanus), *FEBS Lett.* 292:ll5-120.

Liang, C., Brookhart, G., Feng, G. H., Reeck, G. R., and Kramer, K. J., 1991, Inhibition of digestive proteinases of stored grain coleoptera by oryzacystatin, a cysteine proteinase inhibitor from rice seed, *FEBS Lett.* 278:139-142.

Matsumoto, I., Watanabe, H., Abe, K., Arai, S., and Emori, Y., 1995, A putative digestive cysteine proteinase of an insect, Drosophila melanogaster, predominantly expressed in embryonic and larval midgut, Eur. *J. Biochem.* 227:582-587.

Murdock, L. L., Shade, R. E., and Pomeroy, M. A., 1988, Effects of E-64, a cysteine proteinase inhibitor, on cowpea weevil growth, development, and fecundity, *Environ. Entomol.* 17:467-469.

Rogers, J. C., Dean. D., and Heck, G. R., 1985, Aleurain: a barley thiol protease closely related to mammalian cathepsin H, *Proc. Natl. Acad. Sci. U.S.A.* 82:6512-6516.

Turk, V. and Bode. W., 1991, The cystatins: protein inhibitors of cysteine proteinases, *FEBS Lett.* 285:213-219.

Watanabe, H., Abe, K., Emori, Y., Hosoyama, H., and Arai, S., 1991, Molecular cloning and gibberellin-induced expression of multiple cysteine proteinases of rice seeds (oryzains) *J. Biol Chem.* 266:16897-16902.

Watanabe, H., Abe, K., and Arai, S.,1992, Gibberellin-responsive gene expression taking place with oryzains a and g as cysteine proteinases of rice seeds, *Biosci. Biotech. Biochem.* 56:1154-1155.

9

HORSESHOE CRAB FACTOR G: A NEW HETERODIMERIC SERINE PROTEASE ZYMOGEN SENSITIVE TO (1→3)-β-D-GLUCAN

Tatsushi Muta,[1,2] Noriaki Seki,[2] Yoshie Takaki,[1] Ryuji Hashimoto,[1] Toshio Oda,[1] Atsufumi Iwanaga,[1] Fuminori Tokunaga,[2] Daisuke Iwaki,[1] and Sadaaki Iwanaga[1,2]

[1] The Department of Biology, Faculty of Science
[2] The Department of Molecular Biology
Graduate School of Medical Science
Kyushu University 33
Fukuoka 812-81, Japan

INTRODUCTION

Horseshoe crabs (or limulus), as well as other invertebrate animals, do not have an ordinary immune system like mammals. Although they prefer to live in clean oceans, there are still a number of bacteria in their environment. In order to defend themselves from such microorganisms, they have developed a unique and sophisticated defense system in their hemolymph.

Hemolymph of horseshoe crabs contains only a single type of hemocytes, which plays a major role in the defense system of this animal. The hemocyte contains two types of secretory granules, large granules and small but dense granules (1). This cell is extremely sensitive to bacterial endotoxin, lipopolysaccharide (LPS) (2). When Gram-negative bacteria invade their hemolymph, the hemocyte detects LPS molecules on the bacteria. Then, the cells quickly release the contents of the granules by exocytosis (3). LPS-sensitive coagulation factors released from the larger granules are activated in the hemolymph to make clots on the surface of the bacteria. This coagulation reaction proceeds by a cascade type of reaction (4 - 6; Figure 1). In the presence of LPS, an LPS-sensitive serine protease zymogen, factor C, is autocatalytically activated to factor $\bar{c}$ (7 - 9). Then, two other serine protease zymogens, factor B (10) and proclotting enzyme (11) are successively activated by limited proteolysis. Clotting enzyme finally activates a clottable protein, coagulogen, to form an insoluble coagulin gel (12). The microorganisms are engulfed by the clot, and subsequently killed by antimicrobial substances, such as tachyplesins, anti-LPS factor, and big defensin, which are also released from the granules (13).

Intracellular Protein Catabolism, Edited by Koichi Suzuki and Judith Bond
Plenum Press, New York, 1996

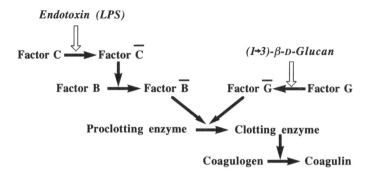

Figure 1. Horseshoe crab coagulation cascade.

This is the principal of the so-called limulus test or Limulus amebocyte lysate (LAL) test. This test is now widely being utilized to detect and quantitate a trace amount of contaminating pyrogen or LPS in the clinical field because of its high sensitivity and convenience (14). However, during the diagnostic application of the limulus test, it was pointed out that positive reactions were observed even in the absence of LPS with some patients' plasma (15). Since some of those patients were suffering from fungus infection or undergoing hemodialysis, this pseudo-positive reaction had been suspected to be at least in part due to glucans. In 1981, we and others obtained evidence that β-glucan-sensitive protease zymogen is present in the hemocyte lysate (16, 17). We named the zymogen factor G. Since then, the purification of this protein was hampered by its instability. However, we recently succeeded in purification, characterization, and cDNA cloning of the protein (18, 19).

PURIFICATION

Factor G was purified from the hemocyte lysate by three successive chromatographies, namely dextran sulphate-Sepharose CL-6B, ConA Sepharose, and Sephacryl S-200 HR chromatographies (18). Potential factor G-activating factors such as Sephadex-based matrix or dialysis tubing were avoided during the purification process, which was performed under sterile conditions. Factor G activity was detected as an amidase activity hydrolyzing Boc-E(OBzl)GR-MCA after incubation with β-glucan.

With Sephacryl S-200 chromatography, by reducing the pH of the buffer down to 6.5, factor G, unexpectedly, eluted in a very late fraction and was completely separated from other proteins. This observation indicates a weak interaction between factor G and the resin, which is not evident at neutral pH.

PRIMARY STRUCTURE

The purified factor G gave two bands of 72 kDa and 37 kDa on SDS-PAGE under reducing conditions (18). Since the two bands are apparent even under non-reducing conditions, no covalent bonds are present between the two polypeptides. The native molecular mass of factor G was estimated to be 110 kDa by gel filtration on Superose 6, which is very close to the summation of 72 kDa and 37 kDa. Thus, we concluded that factor G is a 1:1 heterodimer of these two polypeptides.

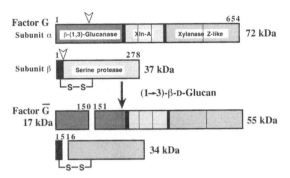

Figure 2. Gross structure of factor G and its activation.

The amino acid sequences deduced from their cDNAs are consistent with our conclusion (19; Figure 2.). We obtained two independent cDNAs and identified independent mRNAs for each subunit, indicating that these two subunits are derived from separate genes and assembled in the cell. We named the large subunit α and smaller one β.

Subunit β is a serine protease zymogen with a short, 15 amino acid-amino-terminal extension. The serine protease domain is most homologous to horseshoe crab factor B (40.5%

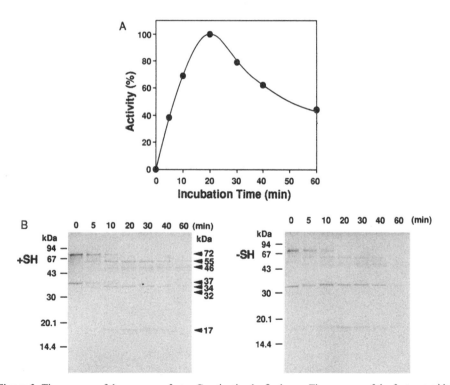

Figure 3. Time course of the zymogen factor G activation by β-glucan. Time course of the factor -amidolytic activity to hydrolyze Boc-E(OBzl)GR-MCA (A) and the structural changes (B) of factor G after the incubation with curdlan are shown. (B) SDS-polyacrylamide gel electrophoresis on 12.5% gel in the presence (+SH) or absence (-SH) of 2-mercaptoethanol.

Table 1. Activation of horseshoe crab factor G by various glucans

Saccharides	Type of linkages	Activation	ED_{50} [a]
D-Glucans			*g/ml*
Curdlan	$(1\rightarrow3)$-β-D	+	4×10^{-9}
Paramylon	$(1\rightarrow3)$-β-D	+	4×10^{-8}
Krestin	$(1\rightarrow3),(1\rightarrow4),(1\rightarrow6)$-$\beta$-D	+	4×10^{-8}
Baker's yeast β-D-glucan	$(1\rightarrow6),(1\rightarrow3)$-$\beta$-D	+	4×10^{-8}
Barley β-glucan	$(1\rightarrow3),(1\rightarrow4)$-$\beta$-D	+	5×10^{-8}
Lichenan	$(1\rightarrow3),(1\rightarrow4)$-$\beta$-D	+	1×10^{-7}
Lentinan	$(1\rightarrow6),(1\rightarrow3)$-$\beta$-D	+	3×10^{-6}
Laminaran			
oligosaccharides (d.p.[b] 2-7)	$(1\rightarrow3)$-β-D	–	
$(1\rightarrow3)$-β-D-Galactan	$(1\rightarrow3)$-β-D	+	2×10^{-4}
α-D-Mannan	$(1\rightarrow2),(1\rightarrow3),(1\rightarrow6)$-$\alpha$-D	+	4×10^{-4}
$(1\rightarrow4)$-β-D-xylan	$(1\rightarrow4)$-β-D	+	4×10^{-5}

[a] Minimal concentration (ED_{50}) of each saccharide required to activate 50 % of the purified factor G (0.32 µg/ml). [b] Degree of polymerization.

identity) or proclotting enzyme (37.7%). On the other hand, subunit α has a unique mosaic structure. The amino-terminal region contains a bacterial β-(1,3) glucanase-like sequence (20). The carboxyl-terminal region has two tandem repeats, each of which has similarity with that found in xylanase Z (21, 22), but this region is not a catalytic domain and its function is not yet known even in the xylanase Z. In the middle of the molecule, there are three tandem repeat structures. This type of tandem-repeat structure is found in xylanase A (23), *Ralobacter* protease I (24), β–(1,3)-glucanase (25), and ricin B chain (26). In all of these proteins, this structure is responsible for the carbohydrate binding. Thus, such region in factor G is most likely to be the recognition site for β-glucan, although there is no direct evidence for that.

ACTIVATION OF THE ZYMOGEN FACTOR G

In the presence of β-glucan, factor G follows a complex process including autoactivation and autoinactivation (18; Figure 3.). Although factor G has no amidase activity itself, soon after the incubation with β-glucan, the amidase activity appears and increases for 20 min. Then, the activity gradually decreases. During the activation process, an Arg^{15}-Ile^{16} bond in subunit β and an Arg^{150}-Glu^{151} in β-(1,3) glucanase-like domain in subunit α are cleaved, resulting in the conversion of the 37-kDa subunit β to a 34-kDa fragment and the 72-kDa subunit α to a 55-kDa and a 17-kDa fragments. These chain conversions are complete by 20 min, and then the 55 kDa fragment is further degraded to 46 kDa, which seems to be responsible for the decrease in activity. This suggests also that the activity of subunit β is tightly regulated by subunit α.

Although the activation of factor G is accompanied by limited proteolyses of Arg-X bonds, horseshoe crab coagulation factors or other digestive proteases, such as trypsin, did not activate factor G at all. This fact implies that the cleavage site in subunit β is somehow masked by subunit α and some conformational changes induced by β-glucan binding to subunit α may be necessary to expose the activation site.

TRIGGERS FOR ACTIVATION OF FACTOR G

Table I shows the amount of various glucans to activate 50% of factor G as ED_{50} (18). The zymogen factor G is activated by most of the β-glucans with β-(1,3) linkages from

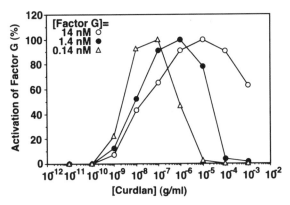

Figure 4. Dose dependent curve of curdlan for the activation of factor G. Three different concentrations of factor G were incubated with various concentrations of curdlan and the amidolytic activity was measured with Boc-E(OBzl)GR-MCA as a substrate.

various sources. The most potent glucan examined is curdlan or paramylon, linear $(1,3)$-β-glucans. Branching with β-$(1,4)$ or β-$(1,6)$ linkages tends to decrease the activity. Shorter β-glucans with 2 to 7 glucose residues do not activate factor G at all. In this experiment, xylan, mannan, and galactan also activated factor G, but we can not rule out the possibility of the trace amount of the contamination of β-glucan because their ED_{50} is very high. A 0.01% contamination of β-glucan in the preparations can explain this result. Factor G is insensitive to smaller oligosaccharides as well as LPS, sulfatide, and steroid sulfate.

GLUCAN DOSE-DEPENDENCY

When we looked at the dose dependency of curdlan for the activation of factor G, it showed a good bell-shaped curve (18). As low as 1 ng/ml of curdlan significantly activates factor G and an optimal concentration of curdlan is evident for the activation. Furthermore, the optimum concentration shifts when a different amount of factor G is used. At the optimum, the molar ratio of curdlan to factor G is almost constant (approximately 10:1). This type of bell-shaped kinetics strongly suggests that factor G is activated by intermolecular interaction with β-glucan. Similar phenomena have been observed in factor C activation by LPS (27) or factor XII activation by a negatively-charged surface (28).

ROLE OF FACTOR G IN THE COAGULATION CASCADE

Using highly purified proteins, we examined whether active factor G activates some of the coagulation factors (18). In the presence of factor G, only proclotting enzyme was activated to the clotting enzyme and it did not activate the zymogen factor C and coagulogen involved in the clotting cascade. We further succeeded in the reconstitution of the β-glucan-mediated coagulation reaction. The three proteins, namely, factor G, proclotting enzyme, and coagulogen are minimum and sufficient components for the reaction. Factor G is autocatalytically activated in the presence of β-glucan, then active factor activates proclotting enzyme to clotting enzyme, which resulted in the conversion of coagulogen into coagulin gel (Figure 1).

SUMMARY

Factor G, a new heterodimeric serine protease zymogen sensitive to β-glucan is composed of two subunits α (72 kDa) and β (37 kDa). Subunit α contains a mosaic structure including glucanase like- and xylanase like-domains. On the other hand, subunit β is a serine protease zymogen homologous to the horseshoe crab factor B or proclotting enzyme. It is autocatalytically activated in the presence of long linear (1,3)-β-glucans. Upon activation, both subunits α and β are cleaved at a specific Arg-X bond. The activated factor induces the clot formation through the activation of proclotting enzyme. Since factor G is co-localized with the other coagulation factors in the large granules (unpublished results), it would be released into the hemolymph when the cells are stimulated. The released factor G is activated by β-glucans on fungi or other possible pathogens, resulting in the engulfment of the foreign materials by the coagulin clot.

The biochemical evidence described above suggests the following hypothesis on the activation of factor G: In the zymogen form, subunit α sterically hinders the activation site of subunit β. β-Glucan-binding to subunit α then exposes the active site of subunit β, allowing autocatalytic activation through bimolecular interaction between subunit β's. The active subunit ß then quickly hydrolyzes the Arg^{150}-Glu^{151} bond in subunit α. Then, another site in the same subunit is cleaved, which is followed by the inactivation of the protease activity. Since the cDNAs for both subunits α and β are now available, the above hypotheses could be examined in expression experiments of the native and mutated forms of factor G.

ACKNOWLEDGMENTS

We would like to express our thanks to Dr. T. Nakamura (Tokushima University) for valuable advice and helpful discussion, and to Dr. J. Aketagawa (Seikagaku Corp.) for providing glucans. We are also grateful to S. Watanabe, H. Hashimoto, and C. Yano for technical assistance on amino acid analyses. This work was supported by a Grant-in-Aid for Scientific Research from the Ministry of Education, Science and Culture of Japan, and by a research grant from Japan Foundation for Applied Enzymology.

REFERENCES

1. Toh, Y., Mizutani, A., Tokunaga, F., Muta, T., and Iwanaga, S., 1991, Morphology of the granular hemocytes of the Japanese horseshoe crab *Tachypleus tridentatus* and immunocytochemical localization of clotting factors and antimicrobial substances, *Cell Tissue Res.* 266:137-147.
2. Levin, J., and Bang, F. B., 1964, The role of endotoxin in the extracellular coagulation of Limulus blood., *Bull. Johns Hopkins Hosp.* 115:265-274.
3. Ornberg, R. L., and Reese, T. S., 1979, Secretion in *Limulus* amebocytes is by exocytosis, *Prog. Clin. Biol. Res.* 29:125-130.
4. Iwanaga, S., 1993, The limulus clotting reaction, *Curr. Opin. Immunol.* 5:74-82.
5. Iwanaga, S., 1993, Primitive coagulation systems and their message to modern biology, *Thrombos. Haemostas.* 70:48-55.
6. Iwanaga, S., Muta, T., Shigenaga, T., Miura, Y., Seki, N., Saito, T., and Kawabata, S., 1994, Role of hemocyte-derived granular components in invertebrate defense, *Ann. N. Y. Acad. Sci.* 712:102-116.
7. Nakamura, T., Morita, T., and Iwanaga, S., 1986, Lipopolysaccharide-sensitive serine-protease zymogen (factor C) found in Limulus hemocytes. Isolation and characterization, *Eur. J. Biochem.* 154:511-521.
8. Tokunaga, F., Nakamura, T., Muta, T., Miyata, T., and Iwanaga, S., 1989, Intracellular serine protease zymogen, factor C, from Limulus hemocytes: Its structure and function, In *Intracellular Proteolysis. Mechanisms and Regulations. Proceedings of the 7th ICOP meeting, Shimoda, Japan* (Katunuma, N. and Kominami, E. eds.), pp.120-128, Japan Scientific Society Press, Tokyo.

9. Muta, T., Miyata, T., Misumi, Y., Tokunaga, F., Nakamura, T., Toh, Y., Ikehara, Y., and Iwanaga, S., 1991, Limulus factor C. An endotoxin-sensitive serine protease zymogen with a mosaic structure of complement-like, epidermal growth factor-like, and lectin-like domains, *J. Biol. Chem.* 266:6554-6561.

10. Muta, T., Oda, T., and Iwanaga, S., 1993, Horseshoe crab coagulation factor B. A unique serine protease zymogen activated by cleavage of an Ile-Ile bond, *J. Biol. Chem.* 268:21384-21388.

11. Muta, T., Hashimoto, R., Miyata, T., Nishimura, H., Toh, Y., and Iwanaga, S., 1990, Proclotting enzyme from horseshoe crab hemocytes. cdNA cloning, disulfide locations, and subcellular localization, *J. Biol. Chem.* 265:22426-33.

12. Miyata, T., Hiranaga, M., Umezu, M., and Iwanaga, S., 1984, Amino acid sequence of the coagulogen from *Limulus polyphemus* hemocytes, *J. Biol. Chem.* 259:8924-8933.

13. Iwanaga, S., Muta, T., Shigenaga, T., Seki, N., Kawano, K., Katsu, T., and Kawabata, S., 1994, Structure-function relationships of tachyplesins and their analogues, In *Ciba Foundation Symposium 186 Antimicrobial peptides* eds.), pp.160-175, John Wiley & Sons, Chichester, England.

14. Tanaka, S., and Iwanaga, S., 1993, Limulus test for detecting bacterial endotoxins, *Methods Enzymol.* 223:358-364.

15. Pearson, F. C., Bohon, J., Lee, W., Bruszer, G., Sagona, M., Dawe, R., Jakubowski, G., Morrison, D., and Dinarello, C., 1984, Comparison of chemical analyses of hollow-fiber dialyzer extracts, *Artif. Organs* 8:291-298.

16. Morita, T., Tanaka, S., Nakamura, T., and Iwanaga, S., 1981, A new $(1{\to}3)$-β-D-glucan-mediated coagulation pathway found in *Limulus* amebocytes, *FEBS Lett.* 129:318-321.

17. Kakinuma, A., Asano, T., Torii, H., and Sugino, Y., 1981, Gelation of *Limulus* amoebocyte lysate by an antitumor $(1{\to}3)$-β-D-glucan, *Biochem. Biophys. Res. Commun.* 101:434-439.

18. Muta, T., Seki, N., Takaki, Y., Hashimoto, R., Oda, T., Iwanaga, A., Tokunaga, F., and Iwanaga, S., 1995, Purified horseshoe crab factor G: Reconstitution and characterization of the $(1{\to}3)$-β-d-glucan-sensitive serine protease cascade, *J. Biol. Chem.* 270:892-897.

19. Seki, N., Muta, T., Oda, T., Iwaki, D., Kuma, K., Miyata, T., and Iwanaga, S., 1994, Horseshoe crab $(1,3)$-β-D-glucan-sensitive coagulation factor G. A serine protease zymogen heterodimer with similarities to β-glucan-binding proteins, *J. Biol. Chem.* 269:1370-1374.

20. Yahata, N., Watanabe, T., Nakamura, Y., Yamamoto, Y., Kamimiya, S., and Tanaka, H., 1990, Structure of the gene encoding β-1,3-glucanase A1 of *Bacillus circulans* WL-12, *Gene* 86:113-117.

21. Grépinet, Chebrou, M. C., and Béguin, P., 1988, Nucleotide sequence and deletion analysis of the xylanase gene (*xyn* Z) of *Clostridium thermocellum*, *J. Bacteriol.* 170:4582-4588.

22. Gosalbes, M. J., A., P.-G. J., González, R., and A., N., 1991, Two beta-glycanase genes are clustered in *Bacillus polymyxa*: molecular cloning, expression, and sequence analysis of genes encoding a xylanase and an endo-beta-(1,3)-(1,4)-glucanase, *J. Bacterol.* 173:7705-7710.

23. Shareck, F., Roy, C., Yaguchi, M., Morosoli, R., and Kluepfel, d., 1991, Sequences of three genes specifying xylanases in *Streptomyces lividans*, *Gene* 107:75-82.

24. Shimoi, H., Iimura, Y., Obata, T., and Tadenuma, M., 1992, Molecular structure of *Rarobacter faecitabidus* protease I. A yeast-lytic serine protease having mannose-binding activity, *J. Biol. Chem.* 267:25189-25195.

25. Shen, S. H., Chrétien, P., Bastien, L., and Slilaty, S. N., 1991, Primary sequence of the glucanase gene from *Oerskovia xanthineolytica*. Expression and purification of the enzyme from *Escherichia coli*, *J. Biol. Chem.* 266:1058-1063.

26. Lamb, F. I., Roberts, L. M., and Lord, J. M., 1985, Nucleotide sequence of cloned cDNA coding for preproricin, *Eur. J. Biochem.* 148:265-270.

27. Nakamura, T., Tokunaga, F., Morita, T., Iwanaga, S., Kusumoto, S., Shiba, T., Kobayashi, T., and Inoue, K., 1988, Intracellular serine-protease zymogen factor C, from horseshoe crab hemocytes. Its activation by synthetic lipid A analogues and acidic phospholipids, *Eur. J. Biochem.* 176:89-94.

28. Sugo, T., Kato, H., Iwanaga, S., Takada, K., and Sakakibara, S., 1985, Kinetic studies on surface-mediated activation of bovine factor XII and prekallikrein. Effect of kaolin and high-Mr kininogen on the activation reactions, *Eur. J. Biochem.* 146:43-50.

10

THE 43 kDa PAPAIN-INHIBITING PROTEIN IN PSORIATIC EPIDERMIS IS IDENTICAL TO SQUAMOUS CELL CARCINOMA ANTIGEN (SCC-ANTIGEN)

M. Järvinen[1], N. Kalkkinen[2], A. Rinne[3], and V. K. Hopsu-Havu[4]

[1] Department of Pathology
University of Oulu
Oulu, Finland
[2] Institute of Biotechnology
University of Helsinki
Helsinki, Finland
[3] Department of Pathology
University of Tromso
Tromso, Norway
[4] Department of Dermatolgy
University of Turku
Turku, Finland

INTRODUCTION

The scales of patients suffering psoriasis as a rich source of cysteine proteinase inhibitors. The major inhibitor in psoriatic epidermis is cystatin A (1), which also is highly expressed in normal epidermis (2). In addition to cystatin A (12.6 kDa), psoriatic epidermis contains a high amount of another papain-inhibiting protein (43 kDa) which can easily be separated from cystatin A by gel chromatography and then purified by ion exchange chromatography (3). By isoelectric focusing, three major activity peaks with pI's of 7.3, 6.9 and 6.5 are separated. By immunohistochemical techniques, using antibodies against the isoelectric variant pI 6.9, the inhibitor is located in the cytoplasm of suprabasal layers of the psoriatic epidermis. In the immature psoriatic stratum corneum the staining for the 43 kDa inhibitor is very variable and uneven: sometimes a strong uniform staining is seen and sometimes the stained and unstained areas vary both vertically and horizontally in a single section (3). In addition to psoriasis, the expression of the inhibitor is elevated in several skin disorders where proliferation of the epidermal cell is accelerated (4). In normal epidermis the inhibitor is seen only in granular layer and in stratum corneum. The inhibitor seems to be a constant property of various squamous epithelia of skin, mouth, oesophagus, uterine

Intracellular Protein Catabolism, Edited by Koichi Suzuki and Judith Bond
Plenum Press, New York, 1996

portio and thymus and it is also expressed in squamous cell carcinomas and in the so-called basal cells of bronchial epithelium (5).

In order to further characterize the papain-inhibiting protein, we isolated it from psoriatic epidermis and subjected eight of is tryptic peptides to sequence analysis. All sequenced peptides gave a 100 % identity to a protein previously known under names tumor antigen 4 or squamous cell carcinoma antigen (6). The partial sequence together with comparison of other properties of the psoriatic epidermis inhibitor and SCC antigen strongly suggest that they represent the same protein.

MATERIAL AND METHODS

Psoriatic scales were collected from untreated voluntary patients. The scales were pooled and stored frozen at -20 °C. Human liver cathepsin B and cathepsin L were purified as previously described (7).

Assay Methods

Inhibition of papain, ficin, trypsin, cathepsin B and cathepsin H was assayed as described earlier (3). Double radial immunodiffusion and specific antibodies were used to recognize kininogen (8) cystain A (2) and psoriatic inhibitor (3) protein in chromatography fractions. Protein concentrations were assayed by Bio-Rad Protein Assay using bovine albumin as a standard. The proteins in the chromatographic fractions were estimated by absorption at 280 nm.

Purification of the Psoriatic Inhibitor

The method described by Järvinen et al. (3) was used with small modifications. Briefly, the psoriatic scale extract was fractionated on Sephadex G-75 (Pharmacia) and the 43 kDa papain inhibitor peak was subjected to anion exchange chromatography on DEAE Sephacel (Pharmacia) using linear NaCl gradient in 10 mM Tris-HCl buffer, pH 8,0. The pooled material from DEAE Sephacel was fractionated into two peaks by a Mono-Q (Pharmacia) anion exchange chromatography using a linear NaCl gradient in 10 mM Tris-HCl buffer, pH 8.0. The two peaks were further purified on reversed phase chromatography and dried under vacuum.

Enzyme Digestions

The inhibitor was digested with 10%(w/w) human liver cathepsin B in 50 mM phosphate buffer, 1 mM EDTA, 2 mM DTT at 37 °C for 0, 30 and 60 min. The digestion products were separated by SDS-polyacrylamide gel electrophoresis under reducing conditions (9).

For sequencing, the inhibitor peaks from reversed phase chromatography were digested with 2 % (w/w) modified trypsin (Promega, sequenal grade, V5111) at 37 °C for 6 h. The resulting peptides were isolated by reversed phase chromatography and concentrated to a volume of 50 µl for sequencing.

Protein and Peptide Sequencing

Protein and peptide sequencing was performed using a gas-pulsed liquid phase sequencer with on-line PTH-amino acid analysis (10).

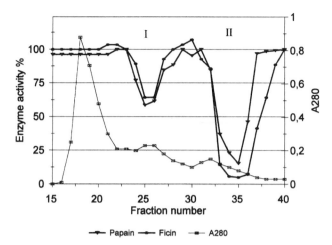

Figure 1. Sephadex G-75 gel chromatogram of psoriatic scale extract. Two papain- inhibiting peaks I (43 kDa) and II (12.6 kDa) were separated.

RESULTS AND DISCUSSION

The 43 kDa inhibitor was separated from cystatin A by gel chromatography (Fig 1). Both cystatin A (Peak II) and the 43 kDa inhibitor (Peak I) were demonstrated using papain or ficin as the test enzyme and α-N-Bz-DL-Arg-2-NNap or azocasein as a substrate, and also by immunodiffusion against the specific antisera for the inhibitors. No kininogen immunore-activity was found in the chromatography fractions by immunodiffusion. The pooled peak I was further purified by DEAE-Sephacel. Papain-inhibiting activity was eluted as a partially separated double peak, which was precipitated with the antibody against the 43 kDa inhibitor. The peak was pooled, dialyzed and rechromatographed on Mono Q column (Fig 2).

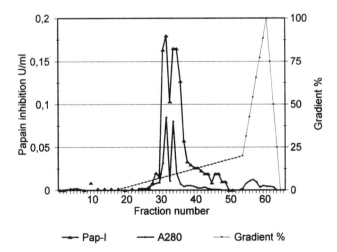

Figure 2. Anion exchange chromatography on a Mono-Q column. Two papain- inhibiting protein peaks A and B were resolved. In immunodiffusion the polyclonal antibody of the psoriatic inhibitor precipitated a protein in both peaks.

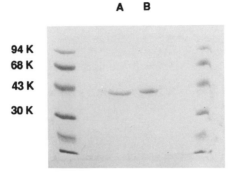

Figure 3. SDS-polyacrylamide gel electrophoresis of anion exchange chromatography peaks A and B after reversed phase chromatography. Pharmacia Phast gel (12.5 %) was used.

Linear NaCl-gradient resolved the inhibitor into 2 distinct peaks, which inhibited papain and contained immunoreactive 43 kDa protein. The pooled peaks 1 and 2 contained 135 and 130 μg of protein, respectively.

Pooled peaks 1 and 2 from Mono-Q chromatography were further purified by reversed phase chromatography in which both samples gave essentially a single peak with identical retention times. The peaks from reversed phase chromatography also gave in SDS-PAGE a single band corresponding an apparent molecular mass of 43 kDa (Fig 3).

Attempts to obtain N-terminal sequence from either of the intact proteins failed indicating that they both have a blocked amino terminus. In order to further characterize the purified proteins, they were cleaved with trypsin and the resulting peptides were separated by reversed phase chromatography. The peptide maps obtained for the purified proteins were very similar suggesting that the two proteins actually represent different forms of the same protein. This was further supported by peptide sequencing. No differences in the primary structure was found when corresponding peptides from the two protein forms were sequenced. The sequencing results for eight analyzed peptides are shown in Table 1.

The sequence homology search among the existing protein databases gave a 100% match of all determined sequences to a protein known under names tumor antigen 4 and squamous cell carcinoma antigen (Table 2), published by Suminami et al. (6).

The amino acid sequence of SCC as deduced from the cDNA sequence by Suminami et al. (6) is 40-50 % homologous with antithrombin III, PAI-II, ovalbumin, or chicken gene Y protein, which are known members of serpin family of proteins. In that respect, it is interesting that the psoriatic 43 kDa inhibitor does not inhibit the serine proteinases bovine

Table 1. Sequences of the tryptic peptides of the psoriatic epidermis papain inhibitor.

Peptide #	Sequence
1	AATYHVDR
2	SGNVHQFQK
3	LLTEFNK
4	INSWVESQGTNE
5	GQWEK
6	SIQMMR
7	QYTSFHFASLED
8	FSSP

Table 2. The amino acid sequence of Squamous Carcinoma Cell Antigen (6). The locations of the eight tryptic peptides of the psoriatic epidermis papain inhibitor are indicated by lines and numbers above the sequence.

```
001    MNSLSEANTK FMFDLFQQFR KSKENNIFYS PISITSALGM VLLGAKDNTA

                 ┌─────1────┐ ┌──────2─────┐ ┌──3───┐
051    QQIKKVLHFD QVTENTTGKA ATYHVDRSGN VHHQFQKLLT EFNKSTDAYE

                                                 ┌──
101    LKIANKLFGE KTYLFLQEYL DAIKKFYQTS VESVDFANAP EESRKKINSW

       ──4──────┐                       ┌─5─┐
151    VESQTNEKIK NLIPEGNIGS NTTLVLVNAI YFKGQWEKKF NKEDTKEEKF

                 ┌──6──┐ ┌──────7──────┐
201    WPNKNTYKSI QMMRQYTSFH FASLEDVQAK VLEIPYKGKD LSMIVLLPNE

251    IDGLQKLEEK LTAEKLMEWT SLQNMRETRV DLHLPRFKVE ESYDLKDTLR

301    TMGMVDIFNG DADLSGMTGS RGLVLSGVLH KAFVEVTEEG AEAAAATAVV

                                        ┌─8─┐
351    GFGSSPASTN EEFHCNHPFL FFIRQNKTNS ILFYGRFSSP
```

trypsin or chymotrypsin or the human cysteine proteinases cathepsin B or cathepsin H (3), while the plant cysteine proteinases papain and ficin are inhibited. In fact, cathepsin B is capable of cleaving the psoriatic inhibitor (Fig 4). When a 10-fold excess of isolated psoriatic inhibitor is incubated with cathepsin B , the protein is rapidly cleaved and a product with molecular mass of about 40 kDa appears.

The blood plasma concentration of SCC-antigen has been shown to correlate with the malignant growth of a squamocellular carcinoma (11,12,13) and the cytoplasmic expression of the antigen is most prominent in non-keratinizing large-cell carcinomas (12,13). However, recent reports indicate that the concentration of SCC-antigen (14) and its mRNA (6, 15) in normal squamous epithelia is higher than in carcinoma cells but the inhibitor leaks from carcinoma cells into circulation. Our findings that the concentration of the epidermal 43 kDa inhibitor is present in all squamous epithelia tested (5) and its expression is greatly elevated in psoriasis and other eczematous skin diseases (4) support the hypothesis that SCC-antigen, alias 43 kDa inhibitor, is related to differentiation and proliferation of squamous epithelial cells (6,12,14).

The papain inhibitor from psoriatic epidermis and SCC-antigen from squamous cell carcinomas seem to be identical molecules. The properties on which this identity is based include (i) partial amino acid sequences, (ii) molecular size, (iii) their presence in all squamous epithelia studied and (iiii) expression in squamous cell carcinoma. Even the purification procedures used by us and by professor Kato's group (11) resemble each other. Paradoxically, the molecule is cleaved by cathepsin B, it inhibits papain and ficin but is homologous to serpins.

The names used to identify the molecule - Psoriatic Epidermis 43 kDa Papain Inhibitor - Tumor-Antigen 4 - Squamous Cell Carcinoma antigen - are inadequate and misleading. The molecule seems to be a constant property of normal, and to a lesser extent,

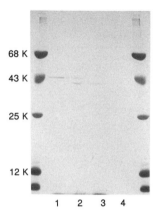

Figure 4. Cleavage of epidermal papain inhibitor by cathepsin B. 60 μg of purified inhibitor was incubated with 6 μg of human liver cathepsin β. Incubation times: 0 min (lane 1), 30 min (lane 2), and 60 min (lane 3). Lane 4: no inhibitor, 30 min incubation.

malignant squamous epithelia, and related to the differentiation of the epithelium. We suggest the name squamin (an abbreviation of Squamocellular Inhibitor) for it.

ACKNOWLEDGEMENTS

This work was supported by Arvo and Inkeri Suominen Foundation, Salo, Finland

REFERENCES

1. Hopsu-Havu, V.K., Joronen, I.A., Järvinen, M., and Rinne, A., 1983, Cysteine proteinase inhibitors in psoriatic epidermis. *Arch. Dermatol. Res.* 275:305-309.
2. Järvinen, M., 1978, Purification and some characteristics of the human epidermal SH-protease inhibitor. *J. Invest. Dermatol.* 71:114-118.
3. Järvinen, M., Rinne, A., and Hopsu-Havu, V.K., 1984, Partial purification and some properties of a new papain inhibitor from psoriatic scales. *J. Invest. Dermatol.* 82:471-476.
4. Aho, H., Joronen, I., Järvinen, M., Rinne, A., and Hopsu-Havu, V.K., 1988, The 43 kDa papain inhibitor in normal and diseased skin. *J. Cutan. Pathol.* 15:367-370.
5. Hopsu-Havu, V.K., Rinne, A., and Järvinen, M., 1986, The 43 kDa papain inhibitor in human tissues. In: *Cysteine proteinases and their inhibitors* (ed. Turk, V.), pp. 535 - 544. Walter de Gruyter & Co., Berlin, New York.
6. Suminami, Y., Kishi, F., Sekugchi, K., and Kato, H., 1991, Squamous cell carcinoma antigen is a new member of the serine protease inhibitors *Biochem. Biophys. Res. Commun.* 181:51-58.
7. Schwartz, W.N., and Barrett, A.J., 1980, Human Cathepsin H. *Biochem. J.* 191:487-491.
8. Järvinen, M., 1979, Purification and some characteristics of two human serum proteins inhibiting papain and other thiol proteinase. *FEBS Letters* 108:461-464.
9. Laemmli, U.K., 1970, Cleavage of structural proteins during the assembly of the head of bacteriophage T4. *Nature* 227:680-685.
10. .Kalkkinen, N., Tilgman, C., 1988, A gas-pulsed-liquid-phase sequencer constructed from Beckman 890D instrument by using Applied Biosystems delivery and cartridge blocks. *J. Prot. Chem.* 7:242-243.
11. Kato, H., and Torigoe, T., 1977, Radioimmunoassay for tumor antigen of human cervical squamous cell carcinoma. *Cancer* 40, 1621-1628.

12. Maruo, T., Shibata, K., Kimura, A., Hoshina, M., and Mochizuki, M., 1985, Tumor-associated antigen, TA-4, in the monitoring of the effects of therapy for squamous cell carcinoma of the uterine cervix. Serial deteminations and tissue localization. *Cancer* 56:302-308.
13. Mino, N., Iio, A., and Hamamoto, K, 1988, Availability of Tumor-Antigen 4 as a marker of squamous cell carcinoma in the lung and other organs. *Cancer* 62:730-734.
14. Crombach, G., Scharl, A., Vierbuchen, M., Würtz, H., and Bolte, A., 1989, Detection of squamous cell carcinoma antigen in normal squamous epithelia and in squamous cell carcinomas of the uterine cervix. *Cancer* 63:1337-1342.
15. Takeshima, N., Suminami, Y., Takeda, O., Abe, H., Okuno, N., and Kato, H., 1992, Expression of mRNA of SCC antigen in squamous cells. *Tumor Biol.* 13:338-342.

11

ALPHA-MERCAPTOACRYLIC ACID DERIVATIVES AS NOVEL SELECTIVE CALPAIN INHIBITORS

Kevin K. W. Wang,[1] Avigail Posner,[1] Kadee J. Raser,[1]
Michelle Buroker-Kilgore,[1] Rathna Nath,[1] Iradj Hajimohammadreza,[1]
Albert W. Probert,[1] Frank W. Marcoux,[1] Elizabeth A. Lunney,[2]
Sheryl J. Hays,[3] Po-wai Yuen[3]

[1] Departments of Neuroscience Therapeutics
[2] Biological and Structural Drug Design
[3] Neuroscience Chemistry
   Parke-Davis Pharmaceutical Research
   Division of Warner-Lambert Company
   2800 Plymouth Road
   Ann Arbor, Michigan 48105

INTRODUCTION

Calpain is a family of cytosolic cysteine proteases activated by calcium (1-5). There are two major isoforms, μ- and m-calpain, requiring low and high micromolar calcium for *in vitro* activity, respectively. Both μ- and m-calpain have a large catalytic subunit (80 kDa) and a small regulatory subunit (29 kDa). Both subunits contain EF-hand calcium-binding structures which control the proteolytic activity of the enzyme. Calpains are implicated in a number of pathological conditions, including stroke, myocardial ischemia and cataract, where intracellular calcium levels are elevated markedly (6). In cerebral ischemia, synaptic glutamate (excitotoxin) buildup leads to excessive activation of ionotropic N-methyl-D-aspartate NMDA- and AMPA/kainate-receptors (7-8). The resultant calcium influx triggers calpain activation. Calpain then degrades cytoskeletal proteins which leads to the loss of cell integrity and cell death (9). Several peptide calpain inhibitors (e.g., calpain inhibitor I (acetyl-Leu-Leu-Nle-H) and MDL28170 (carbobenzoxy-Val-Phe-H)) were found to have neuroprotective effects in various models of excitotoxicity and/or ischemia (10-12). However, these agents are nonselective and cross-inhibit other cysteine proteases. Thus, it is highly desirable to develop selective and potent calpain inhibitors.

Intracellular Protein Catabolism, Edited by Koichi Suzuki and Judith Bond
Plenum Press, New York, 1996

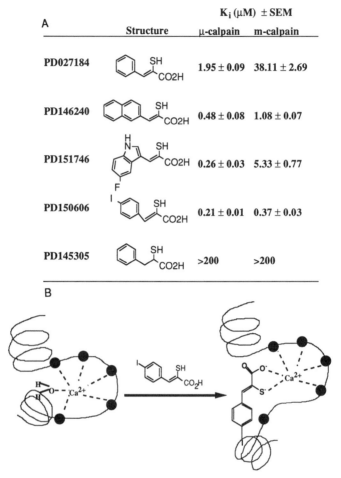

Figure 1. Chemical structure of mercatoacrylates as calpain inhibitors (A) and proposed interaction of PD150606 with the helix-loop-helix calcium-binding site(s) of calpain (B). In (A) the calpain microplate assay was carried out as previously described (13). Briefly, a mixture containing 0.5 mg/ml casein, 20 mM dithiothreitol (DTT), 50 mM Tris-HCl (pH 7.4), 0.01 unit μ-calpain (human erythrocytes) or II (rabbit skeletal muscle) (about 0.08-0.12 μg) (14), calpain inhibitor and 4 mM CaCl$_2$ was added to a microtiter plate (250 μL) and incubated for 60 min at 25°C The samples were then processed for colorimetric development and read at 595 nm. In (B), an EF-hand (helix-loop-helix) calcium-binding domain of calpain is depicted (15). The solid circles represent residues coordinating the calcium ion. The proposed binding of PD150606 to this site provides two coordinates in its S- and COO- groups and thus perturbes the conformation of the calcium-binding site.

ALPHA-MERCATOACRYLATES AS SELECTIVE CALPAIN INHIBITORS

We recently discovered that several mercaptoacrylic acid derivatives are potent inhibitors of μ- and m-calpain (Fig.1A). For example, using casein as a substrate, PD150606 has an inhibition constant (K$_i$) of 0.21 μM and 0.37 μM against μ- and m-calpain, respectively (Fig. 1A). The saturated analog (PD145305) was inactive up to 200 μM (Fig. 1A). Unmodified sulfhydryl and carboxylic acid groups were found to be necessary for activity

Table 1. Selectivity of alpha-mercaptoacrylate PD150606 and as a calpain inhibitor. The calpain microplate assay was carried out as previously described (17). Papain (papaya latex), thermolysin (Bacillus thermoproteolyticus) and trypsin (bovine pancreas) were assayed similarly to calpain except that the reaction mixture contained 7-10 ng papain or 5 μg thermolysin or 1.25 μg trypsin instead of calpain. Non-protease samples were also run as controls. Cathepsin B (0.002 unit, bovine spleen) was assayed with 50 mM MES (pH 5.5), 100 μM carbobenzoxy-Arg-Arg-4-methoxy-beta-naphthlamide (cbz-Arg-Arg-MNA), inhibitor, and 2 mM DTT in 200 μL for 60 min at room temperature. Fifty-μL aliquots were transferred to a fluorescence-compatible plate and then read with a Perkin-Elmer fluorometer LS-50B (excitation 340 nm and emission 425 nm). Calcineurin (bovine brain, Sigma) (5 μg, 0.8 unit) was assayed with 10 mM p-nitrophenyl phosphate in 1 mM DTT, 1 mM CaCl2, 1 mM MnCl2, 200 μM calmodulin (CaM, if added), and 50 mM Tris-HCl (pH 7.4) at 25°C for 60 min or more. Production of p-nitrophenol was monitored at 405 nm. Basal phosphatase activity and calmodulin-activated component were calculated separately. Data were means ± SEM.

Ki (μM) ± SEM	PD150606	Calpain inhibitor I
μ-Calpain	0.21 ± 0.01	0.086 ± 0.023
m-Calpain	0.37 ±0.03	0.192 ± 0.023
Cathepsin B	127.8 ± 9.2	0.022 ± 0.009
Papain	>500	2.24 ± 0.22
Trypsin	>500	>500
Thermolysin	204.1 ± 23.3	>500
Calcineurin-Basal	>200	>200
Calcineurin-CaM	12.98 ± 0.47	>200

(data not shown). We propose that these compounds interact with the EF-hand (helix-loop-helix) calcium binding site of calpain (further discussion to follow later) (Fig. 1B).

PD150606 was tested against several other proteases. In parallel, calpain inhibitor I was also examined. PD150606 appeared to be quite specific to calpains among proteases (Table I). We also tested the effect of PD150606 on the phosphatase activity of calcineurin as an assay for calmodulin, another EF-hand calcium-binding protein. While we found no inhibition of basal calcineurin, the calmodulin stimulation component was dose-dependently inhibited at higher concentrations of PD150606 (K_i 12.98 ± 0.47 μM) (Table I). In kinetic experiments using casein as a substrate, PD150606 was found to inhibit μ-calpain uncom-

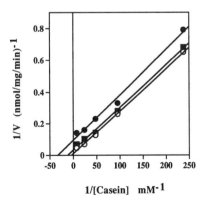

Figure 2. Inhibition kinetics of alpha-mercaptoacrylate PD146240 as a calpain inhibitor.

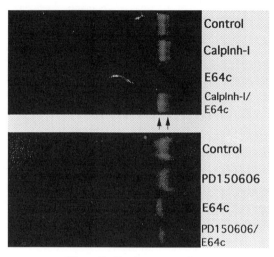

Figure 3. Casein zymography.

petitively with respect to substrate, as reflected on the parallel curves in a double recipicol plot (Fig. 2).

Kinetic experiments were performed with casein as a substrate (42-134 µM) in 50 mM Tris-HCl (pH 7.4 at 25°C), 20 mM DTT, 4 mM CaCl$_2$, 5 µg µ-calpain in the absence (open circles) or the presence of PD146240 (2.5 µM (filled squares) or 10 µM (filled circles)). Initial rate of hydrolysis (V) was then calculated. Means from two experiments were shown.

PD150606 and calpain inhibitor I were compared in casein zymography (16). Both calpain inhibitor I- and PD150606-treated µ-calpain samples were active in casein gels (Fig. 3). Presumably if an inhibitor were reversible, it would be removed from the embedded calpain during electrophoresis and the subsequent gel incubation. Therefore, both compounds appear reversible. On the other hand, irreversible inhibitor E64c-treated calpain was largely inactive in the casein zymogram. Secondly, when calpain was first exposed to calpain inhibitor I, the enzyme was protected from subsequent inactivation by E64c, since both

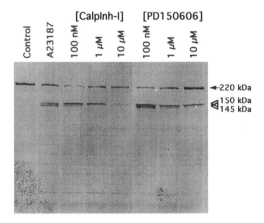

Figure 4. Inhibition of calpain activity in intact human leukemic Molt-4 cells.

compounds target the active site cysteine residue. Interestingly, PD150606 did not protect calpain from inactivation by E64c, suggesting that they bind to different sites (Fig. 3).

Casein gels were poured as a 12% (w/v) polyacrylamide and 0.2% (w/v) co-polymerized sodium casein into mini-gel casts. The casein gels were pre-run with Tris-glycine buffer with 1 mM EDTA and 1 mM DTT for 15 min (4°C). µ-calpain (4 µg) was incubated in 50 mM Tris-HCl (pH 7.4), 3 mM DTT, and 1 mM $CaCl_2$ in the presence of an inhibitor for 5 min on ice (30 µL). If desired, a second inhibitor (E64c) can be introduced and further incubated for 5 min. 10 mM EGTA was added and 5 µL of non-SDS sample buffer was added to the mixture (final volume 38 µL). The samples were then loaded onto the casein gel and electrophoresed at 125 V for 3 h. The gel was then washed twice with 20 mM Tris-HCl (pH 7.4), 10 mM DTT, and 1 mM $CaCl_2$ for 30 min and incubated overnight at ambient temperature in the same buffer with shaking. Finally, the gel was stained with Coomassie blue.

Our hypothesis that PD150606 acts on the calcium-binding domain(s) of calpain is supported by several lines of evidence: i) the uncompetitive inhibition with respect to substrate (Fig. 2), ii) the lack of crossreactivity with other cysteine protease (Table I); iii) low affinity antagonism for calmodulin (Table I); and iv) the requirement of unsubstituted sulfhydryl and carboxyl groups for activity, presumably for chelation of calcium ion (Fig. 1B).

PD150606 IS CELL PERMEABLE

Next we examined the ability of PD150606 to inhibit calpain in intact cells. Saido *et al.* previously demonstrated that A23187 treatment of human leukemic Molt-4 cells lead to calpain mediated spectrin breakdown of spectrin (220 kDa) to the 150 kDa and 145 kDa fragments (17). When Molt-4 cells were treated with 15 µM A23187 for 90 min, the 145 kDa fragment was more prominent than the 150 kDa fragment (Fig. 4). Pretreatment of cells with calpain inhibitor I diminished the 145 kDa fragment formation at 0.1-1 µM. At 10 µM, only a very faint band of 150 kDa was observed (Fig. 4). Pretreatment with 1 µM PD150606 also prevented the formation of the 145 kD fragment. At 10 µM, the 145 kDa band was further diminished but not completely blocked (Fig. 4). This is likely due to the uncompetitive kinetics of PD150606 which suggest that the inhibitor would only bind to the protease with high affinity after the protease is substrate-bound. This would mean that calpain could hydrolyse to some extent before being inhibited by PD150606.

Molt-4 cells were washed three times with serum-free RPMI-1640 medium and resuspended to 4 million/0.5 mL and transferred to a 12-well plate (0.5 mL/well). Calpain inhibitor I (CalpInh-I), PD150606 or PD145305 was added to the wells if desired and preincubated for 1 h. A23187 (15 µM) was then added and the cells were further incubated for 90 min at 37°C. The cells were then lysed with 2% (w/v) SDS, 5 mM EGTA, 5 mM EDTA, 150 mM NaCl, 0.5 mM PMSF, 10 µg/mL AEBSF, 5 µg/mL leupeptin, 10 µg/mL pepstatin, 10 µg/mL TLCK, 10 µg/mL TPCK, and 20 mM Tris-HCl (pH 7.4) at room temperature for 5-10 min. One hundred-µL of 100% (w/v) TCA was added and total protein precipitate was collected by microcentrifugation. The final sediments were neutralized with Tris base. Protein samples (50 µg) were run on SDS-PAGE with the Tris-glycine running buffer system and transferred onto a PVDF membrane. The blots were probed with an anti-spectrin (non-erythroid) antibody. The arrow indicates intact spectrin (220 kDa) while the triangles indicate the spectrin breakdown products (150 kDa and 145 kDa).

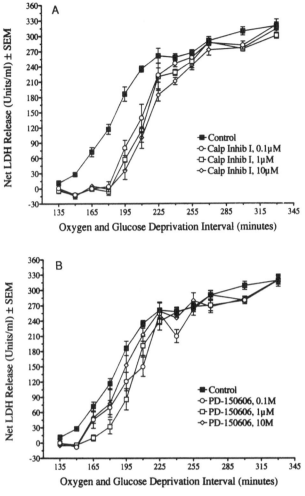

Figure 5. Protective effects of PD150606 against neuronal death induced by hypoxia/hypoglycemia challenge in cerebrocortical cultures. Cerebrocortical cells were harvested from fetal rats (Sprague-Dawley) in their 18th day of gestation and cultured with DMEM/F12 medium containing 10% horse and 6% fetal bovine serum (heat inactivated) in a 96-well poly-L-lysine-coated plates as described previously (19). Non-neuronal cell division was halted three days into culture by adding 25 µg/ml uridine and 10 µg/ml 5-flouro-2i-deoxyuridine. On the 17th daypost-plating, the cultures were washed three times with serum-free medium. Inhibitor was added at this point for a 1 h preincubation. The cultures were then challenged with hypoxia/hypoglycemia (195 min) (exposure atmosphere in gas incubator: 1% O_2, 8% CO_2, 91% N_2; exposure medium: 1.8 mM Ca^{2+}, 0.8 mM Mg^{2+}, 0.2 g/L D-glucose) in the presence of inhibitor. The plates were then returned to normal serum-free medium (with calpain inhibitor) in an oxygenated incubator (21% O_2, 8% CO_2, 71% N_2) until 24 h after the experiment initiation. A set of normoxic cultures with the same number of media changes were used as controls. Neuronal death was then assessed by measuring the cytosolic enzyme, lactate dehydrogenase (LDH) released into the medium (25 µL samples) as described before (20). Data were means ± SEM.

PD150606 IS A NEUROPROTECTANT

The role of calpains in excitotoxic and ischemic injuries to central neurons has been suggested (9-12). The neuroprotective effects of PD150606 and calpain inhibitor I were investigated. In fetal rat cerebrocortical mixed cultures, a combination of hypoxia and hypoglycemia

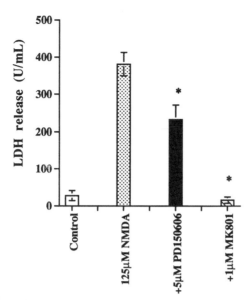

Figure 6. Protective effects of PD150606 against neuronal death induced by NMDA in cerebrocortical cultures.

produced delayed neuronal death (measured by lactate dehydrogenase (LDH) release) as a function of deprivation time (Fig. 5) (18). PD150606 (at 0.1-10 μM) and calpain inhibitor I (0.1-10 μM) protected the neurons by making them more tolerant to the insults (Fig. 5A & B). The concomitant spectrin breakdown by cellular calpains was also decreased (data not shown). PD145305 (50 μM) failed to protect the neurons (data not shown). Also, PD150606 provided partial protection to the same type of cultures challenged with 125 μM NMDA (30 min), although the protection was less effective than NMDA receptor antagonist MK-801 (Fig. 6).

Cortical hemispheres were sectioned from fetal rats (Sprague Dawley) in their 18th day of gestation and were trypsin digested and triturated into single cell suspension. Cells were pipetted into individual wells of poly-L-lysine-coated plates, yielding a final cell concentration of 200000 cells/cm^3 using Gibcoís Minimum Essential Medium (MEM; containing 10% horse and 10% fetal bovine serum (heat inactivated)). Non-neuronal cell division was halted three days into culture by adding 25 μg/ml uridine and 10 μg/ml 5-flouro-2í-deoxyuridine. Feedings were performed as necessary with MEM with 10% horse serum. Experiments were performed on 17-21 day (post-isolation) old neurons. Cultures were triple washed with MEM (serum free) with or without the test agents (inhibitors). NMDA (125 μM) was added directly to the culture media. After 30 min incubation at 37°C, NMDA was washed by triple exchange with serum-free MEM (+inhibitors). Final concentration of glucose in the MEM throughout the culture period was 30 mM. LDH release was measured as above. Data were means ± SEM. * indicates data significantly different from NMDA alone (p< 0.02, Studentís t-test).

CONCLUSION

In summary, we have discovered the first selective non-peptide calpain inhibitors. We propose that they bind to the calcium-binding domain(s) rather than the active site, thus providing superior selectivity over exisiting peptide-based calpain inhibitors. With demonstrated cell permeability, these compounds can be used in defining the roles of calpain in various physiological and pathophysiological conditions. We would caution that further improvement of their chemical properties are needed before they can be useful in *in vivo* studies.

REFERENCES

1. Murachi, T., 1989, Intracellular regulatory system involving calpain and calpastatin, *Biochem. Int.* 18: 263-294.
2. Melloni, E., and Pontremoli, S., 1989, The calpains, *Trends Neurosci.* 12: 438-444.
3. Wang, K.K.W., Villalobo, A. and Roufogalis, B.D., 1989, Calmodulin-binding proteins as calpain substrates, *Biochem. J.* 262: 693-706.
4. Croall, D.E., and Demartino, G.N., 1991, Calcium-activated neutral protease (calpain) system: structure, function, and regulation, *Physiol. Reviews* 71: 813-847.
5. Saido, T.C., Sorimachi, H., and Suzuki, K., 1994, Calpain: new perspectives in molecular diversity and physiological-pathological involvement, *FASEB J.* 8: 814-822.
6. Wang, K.K.W., and Yuen P-w., 1994, Calpain inhibition: an overview of its therapeutic potentials, *Trends Pharmacol. Sci.* 15: 412-419.
7. Lipton, S.A., and Rosenberg, P.A., 1994, Excitatory amino acids as a final common pathway for neurologic disorders, *New England J. Med.* 330: 613-622.
8. Meldrum, B., and Garthwaite, J., 1990, Excitatory amino acid neurotoxicity and neurodegenerative disease, *Trends Pharmacol. Sci.* 11: 379-387.
9. Siman, R., and Noszek, J.C., 1988, Excitatory amino acids activate calpain I and induce structural protein breakdown in vivo, *Neuron* 1: 279-287.
10. Arlinghaus, L., Mehdi, S., and Lee, K.S., 1991, Improved posthypoxic recovery with a membrane-permeable calpain inhibitor, *Europ. J. Pharmacol.* 209: 123-125.
11. Hong, S.-C., Goto, Y., Lanzino, G., Soleau, S., Kassell, N.F., and Lee, K.S., 1994, Neuroprotection with a calpain inhibitor in a model of focal cerebral ischemia, *Stroke* 25: 663-669.
12. Bartus, R.T., Baker, K.L., Heiser, A.D., Sawyer, S.D., Dean, R.L., Elliott, P.J., and Straub, J.A., 1994, Postischemic adminstration of AK275, a calpain inhibitor, provides substantial protection against focal ischemic brain damage, *J. Cereb. Blood Flow Metab.* 14, 537-544.
13. Buroker Kilgore, M., and Wang, K.K.W., 1993, A Coomassie brilliant blue G250-based colorimetric assay for measuring activity of calpain and other proteases, *Anal. Biochem.* 208: 387-392.
14. Wang, K.K.W., Villalobo, A. and Roufogalis, B.D., 1988, Further characterization of the calpain-mediated proteolysis of the human erythrocyte plasma membrane Ca^{2+}-ATPase, *Arch. Biochem. Biophys.* 260: 696-704.
15. Emori, Y., Ohno, S., Tobita, M., and Suzuki, K., 1986, Gene structure of calcium-dependent protease retains the ancestral organization of the calcium-binding protein gene, *FEBS lett.* 194: 249-252.
16. Heussen, C., and Dowdle, E.B., 1980, Electrophoretic analysis of plasminogen activators in polyacrylamide gels containing sodium dodecyl sulfate and copolymerized substrates. *Anal. Biochem.* 102: 196-202.
17. Saido, T.C., Shibata, M., Takenawa, T., Murofushi, H., and Suzuki, K., 1992, Positive regulation of mu-calpain action by polyphosphoinositides, *J. Biol. Chem.* 267: 24585-24590.
18. Weber, M.L., Probert, A.W., Dominick, M.A., and Marcoux, F.W., 1993, Early ultrastructural injury in neuronal cell culture after hypoxia or combined oxygen and glucose deprivation: neuroprotection with 4-(3-phosphonopropyl)-2-piperazinecarboxylic acid (CPP), *Neurodegeneration* 2: 63-72.
19. Marcoux, F.W., Probert, A.W., and Weber, M.L., 1990, Hypoxic neuronal injury in tissue culture is associated with delayed calcium accumulation, *Stroke* 21 (Suppl. III): 71-74.
20. Koh, J.Y., and Choi, D.D., 1987, Quantitative determination of glutamate mediated cortical neuronal injury in cell culture by lactate dehydrogenase efflux assay, *J. Neurosci. Methods* 20: 83-90.

12

STRUCTURAL ASPECTS OF AUTOPHAGY

Per O. Seglen, Trond Olav Berg, Henrietta Blankson, Monica Fengsrud,
Ingunn Holen, and Per Eivind Strømhaug

Department of Tissue Culture
Institute for Cancer Research
The Norwegian Radium Hospital
Montebello
N-0310 Oslo, Norway

SUMMARY

As a first step towards isolation of autophagic sequestering membranes (phagopho-res), we have purified autophagosomes from rat hepatocytes. Lysosomes were selectively destroyed by osmotic rupture, achieved by incubation of hepatocyte homogenates with the cathepsin C substrate glycyl-phenylalanyl-naphthylamide (GPN). Mitochondria and perox-isomes were removed by Nycodenz gradient centrifugation, and cytosol, microsomes and other organelles by rate sedimentation through metrizamide cushions. The purified auto-phagosomes were bordered by dual or multiple concentric membranes, suggesting that autophagic sequestration might be performed either by single autophagic cisternae or by cisternal stacks.

Okadaic acid, a protein phosphatase inhibitor, disrupted the hepatocytic cytok-eratin network and inhibited autophagy completely in intact hepatocytes, perhaps suggesting that autophagy might be dependent on intact intermediate filaments. Vin-blastine and cytochalasin D, which specifically disrupted microtubules and microfila-ments, respectively, had relatively little (25-30%) inhibitory effect on autophagic sequestration.

In a cryo-ultrastructural study, the various autophagic-lysosomal vacuoles were immunogold-labelled, using the cytosolic enzyme superoxide dismutase as an auto-phagic marker, Lgp120 as a lysosomal membrane marker, and bovine serum albumin as an endocytic marker. Vinblastine (50 µM) was found to inhibit both autophagic and endocytic flux into the lysosomes, with a consequent reduction in lysosomal size. Asparagine (20 mM) caused swelling of the lysosomes, probably as a result of the ammonia formation that could be observed at this high asparagine concentration. Autophagosomes and amphisomes (autophagic-endocytic, prelysosomal vacuoles) ac-cumulated in asparagine-treated cells, reflecting an inhibition of autophagic flux that might be a consequence of lysosomal dysfunction.

Intracellular Protein Catabolism, Edited by Koichi Suzuki and Judith Bond
Plenum Press, New York, 1996

INTRODUCTION

Autophagy is a major mechanism for the degradation of protein as well as of other intracellular macromolecules and organelles, yet its structural basis is poorly understood. In rat hepatocytes, autophagy is initiated when the levels of certain amino acids become low (Pösö *et al.*, 1982; Seglen *et al.*, 1980), possibly resulting in the detachment of these amino acids from regulatory cell surface receptors (Kadowaki *et al.*, 1992). The resulting signal activates, by an unknown mechanism, sequestering organelles called phagophores, which have the morphological appearance of electron-dense membrane sheets containing single or multiple (stacked) collapsed membrane cisternae (Seglen, 1987), probably derived from the endoplasmic reticulum (Dunn, 1994). The phagophores envelop whole regions of cytoplasm in a nonselective manner (Kopitz *et al.*, 1990), except in the case of autophagy-dependent regression following peroxisome proliferation, where some organelle specificity can be demonstrated (Luiken *et al.*, 1992). The sequestration of cytoplasm by the phagophore results in the formation of a sealed vacuole, an autophagosome, where the phagophore now forms the wall around a section of completely normal-appearing cytoplasm. It is not known how the extension, curving and vacuole sealing performed by the phagophore membranes is directed, but an involvement of the cytoskeleton has been indicated (Holen *et al.*, 1992; Aplin *et al.*, 1992). Eventually, the autophagosome delivers its contents to a lysosome, probably by fusion, and the contents are degraded by the lysosomal enzymes. There are biochemical data suggesting that at least some autophagosomes may fuse with acidic endosomes before they reach the lysosomes, forming "amphisomes" (Gordon and Seglen, 1988; Strømhaug and Seglen, 1993; Dunn, 1994).

The first step in the autophagic-lysosomal pathaway, i.e., the sequestration of cytoplasm by the phagophore, can be measured as the transfer, in intact cells, of a cytosolic marker from the cytosol to sedimentable autophagic vacuoles (autophagosomes, amphisomes and lysosomes). The marker can be an electroinjected, inert small molecule, e.g., [^{3}H]raffinose (Seglen *et al.*, 1986), or an endogenous, cytosolic enzyme, e.g., lacatate dehydrogenase (LDH), which accumulates if its lysosomal degradation is prevented by leupeptin (Kopitz *et al.*, 1990).

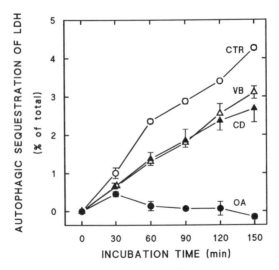

Figure 1. Effects of okadaic acid, vinblastine and cytochalasin D on autophagic sequestration of LDH.

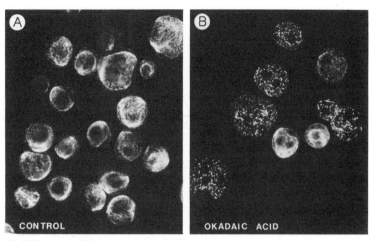

Figure 2. Effect of okadaic acid on the organization of cytokeratin (CK8) intermediate filaments.

Role of the Cytoskeleton in Autophagic Sequestration

By incubating hepatocytes with inhibitors of the various cytoskeletal elements, it should be possible to see whether autophagic sequestration requires the structural integrity of any of these elements.

Hepatocytes were incubated at 37°C for the length of time indicated, in the presence of 0.3mM leupeptin, with no further additions (o); with 30 nM okadaic acid (•); 10 μM vinblastine (VB), or 10μm cytochalasin D (▲). The net amount of cytosolic LDH sequestered at each time point was measured and expressed as per cent of the total cell-associated LDH. Each point is the mean ± S.E. of three independent experiments.

As shown in Fig. 1, neither vinblastine, at a concentration demonstrated by immunostaining to specifically fragment hepatocytic microtubules, nor cyctochalasin D, at a concentration found to specifically fragment actin microfilaments, had much effect on autophagic sequestration. Neither microtubules nor microfilaments would, therefore, seem to be required for autophagic sequestration in hepatocytes. On the other hand, okadaic acid, which disrupted the cytokeratin network (Fig. 2) without affecting microtubules or microfilaments, inhibited autophagic sequestration nearly completely (Fig. 1).

Hepatocytes were incubated for 1h at 37°C with (A), no addition (control), or (B), okadaic acid, 30 nM. After incubation, the cells were sedimented onto glass slides (Cytospin), fixed in 100% methanol and processed for indirect immunofluorescence staining of cytokeratin intermediate filaments, using a monoclonal antibody against CK8.

Several protein kinase inhibitors, including the flavonoid naringin (Gordon *et al.*, 1995) and KN-62, thought to specifically inhibit the Ca^{2+}/calmodulin-dependent protein kinase II (Holen *et al.*, 1992), offered parallel protection of autophagy and cytokeratin against okadaic acid, suggesting that the same protein kinase may control both processes. Although a causal relationship remains to be proven, it would seem reasonable to assume that if there is a cytoskeletal involvement in hepatocytic autophagy, the most likely candidate would be the cytokeratin intermediate filaments. Other cell types may have other cytoskeletal requirements: in a fibroblastic kidney cell line (NRK), autophagosome formation was suppressed by cytochalasins, suggesting a need for intact microfilaments (Aplin *et al.*, 1992). It should be noted that the overall autophagic-lysosomal degradation is strongly suppressed by microtubule inhibitors, but the microtubular requirement apparently relates to the delivery

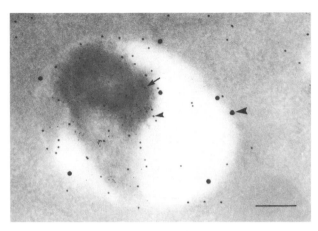

Figure 3. Immunogold labelling of a lysosome.

of autophagocytosed material to lysosomes rather than to the initial autophagic sequestration (Høyvik *et al.*, 1986).

Effects of Autophagic Flux Inhibitors (Vinblastine, Asparagine) on the Distribution of Autophagic Vacuoles, Characterized by Cryoelectron Microscopy and Gold-Immunostaining

Microtubule inhibitors like vinblastine inhibit both autophagic and endocytic flux in hepatocytes, resulting in the accumulation of autophagosomes as well as of endosomes. Asparagine, on the other hand, has been found to interrupt autophagic flux in an apparently selective manner, causing an accumulation of undegraded, autophagocytosed material in autophagosomes (Seglen, 1987) as well as in the prelysosomal autophagic vacuoles called amphisomes (because they are also able to receive endocytosed material) (Gordon and Seglen, 1988; Høyvik *et al.*, 1991). Since the autophagic vacuoles accumulating during different types of inhibitor treatment have been poorly characterized, an ultrastructural study was undertaken, using gold-immunostaining of hepatocyte cryosections to identify the various vacuole categories.

The cytosolic enzyme superoxide dismutase (SOD) is relatively degradation-resistant; autophagocytosed SOD can, therefore, be detected in lysosomes as well as in earlier autophagic vacuoles, making it a perfect marker for autophagic vacuoles in general (Rabouille *et al.*, 1993). As a lysosomal marker, the lysosomal membrane glycoprotein Lgp120 was chosen (Dunn, 1990), whereas endocytosed bovine serum albumin (BSA)-gold was used as an endocytic marker (Fig. 3).

Hepatocytes were incubated for 2h at 37°C, cryosectioned and double-immunolabeled for lgp120 (20-nm gold; large arrowhead) and SOD (10-nm gold; small arrowhead). 5-nm BSA-gold (arrow) was taken up by endocytosis. Bar 200 nm.

Although Lgp120 has been reported to be present in prelysosomal endocytic vacuoles in certain cell types (Green *et al.*, 1987), we found no Lgp120 labelling in hepatocytic endosomes (BSA+/SOD-). Virtually all Lgp120-positive vacuoles (96%) were SOD+, i.e., all hepatocytic lysosomes would appear to be "auto"lysosomes, engaging in autophagy. Most of them (76%) had also received BSA within 2h, indicating that they were "amphi"lysosomes, participating in endocytosis as well.

Treatment with vinblastine strongly suppressed the labelling of autophagic vacuoles with BSA. Small, BSA negative lysosomes (LGP+/BSA-) and autophagosomes (LGP-/BSA) accumulated, consistent with inhibition of both endocytic and autophagic flux (Table 1).

Asparagine, added at a high concentration (20 mM), induced swelling of BSA positive, but not of BSA negative autophagic vacuoles, perhaps suggesting that swelling requires endocytic fluid delivery. Autophagosomes (LGP-/BSA-), amphisomes (LGP-/BSA+) and BSA negative lysosomes (LGP+/BSA-) accumulated (Table 1), indicating that swelling of the BSA positive vacuoles might impair their ability to fuse with vacuoles of other categories. Altogether, the results suggest that the marker proteins used are well suited for the study of autophagic vacuole kinetics.

ISOLATION OF AUTOPHAGOSOMES

Since the inhibition of autophagic processing by vinblastine caused a considerable accumulation of autophagosomes (Table 1), vinblastine-treated hepatocytes were used in an attempt to isolate these organelles in a purified form. As a biochemical marker for auto-phagosomes, autophagically sequestered LDH was used. In the absence of leupeptin, any enzyme reaching the lysosomes will be degraded, causing LDH to accumulate in prelysoso-mal autophagic vacuoles only (i.e., in the presence of vinblastine, in autophagosomes).

Hepatocytes incubated with vinblastine (50μM) for 2h were electropermeabilized and homogenized, and large-granule fractions were prepared and fractionated further on (A) sucrose or (B) Nycodenz density gradients. Gradient fractions were analyzed for LDH (o), acid phosphatase (•) and cytochrome C oxidase (Δ) as markers of autophagosomes, lysosomes and mitochondria, respectively. Enzyme activities per fraction are given as per cent of the respective total cellular acitivities.

In a density gradient study, autophagosomes were found to be well separated from mitochondria in Nycodenz gradients, whereas they coincided with lysosomes in gradients of either sucrose or Nycodenz (Fig. 4). A number of methods were tried in order to achieve separation of autophagosomes from lysosomes; eventually, a procedure for the selective

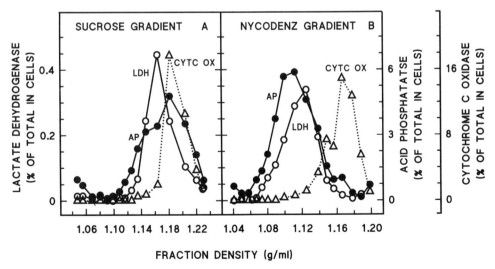

Figure 4. Density distribution of autophagosomes, lysosomes and mitochondria in sucrose and Nycodenz gradients.

Table 1. Effects of vinblastine and asparagine on autophagic vacuole number and size (volume) in rat hepatocytes

Asparagine Autophagic vacuoles (21) profile, mm³/cm³	(SOD⁺)	Vacuole number (no. of vacuole profiles/cell profile)			Vacuole size (vol. fraction/vac.)		
		Control (20)	Vinblastine (10)	Asparagine (21)	Control (20)	Vinblastine (10)	Asparagine (21)
Autophagosomes	(LGP⁻/BSA⁻)	3.3 ± 0.5	16.7 ± 1.1[c]	5.4 ± 0.8[a]	1.8 ± 0.6	2.0 ± 0.4	2.2 ± 0.6
Amphisomes	(LGP⁻/BSA⁺)	3.2 ± 0.5	1.0 ± 0.4[c]	5.4 ± 0.5[a]	2.5 ± 0.9	2.2 ± 1.3	4.3 ± 0.8
"Auto"lysosomes	(LGP⁺/BSA⁻)	4.0 ± 0.5	16.5 ± 1.2[c]	8.3 ± 0.8[c]	2.5 ± 0.7	0.9 ± 0.2[a]	1.7 ± 0.6
"Amphi"lysosomes	(LGP⁺/BSA⁺)	10.1 ± 1.2	1.0 ± 0.3[c]	10.7 ± 1.2	3.1 ± 0.7	1.4 ± 0.9	5.1 ± 0.9[a]

Isolated rat hepatocytes were incubated for 2h at 37°C with gold-conjugated bovine serum albumin (BSA) and vinblastine (50 µM) or asparagine (20 mM) as indicated. The cells were then cryosectioned and immunostained with specific antibody and protein A-gold. Autophagic vacuoles were identified in electron micrographs by positive gold-immunostaining for superoxide dismutase (SOD⁺), lysosomes by positive immunogold staining for lysosomal membrane glycoprotein lgp120 (LGP⁺), and endocytic input by the presence of 2h endocytosed albumin-gold (BSA⁺). Each value is the mean ± S.E. of the number of cell profiles given above each column. $^{a}p<0.05$; $^{b}p<0.01$; $^{c}p<0.001$ for significance vs. control according to the *t-test*.

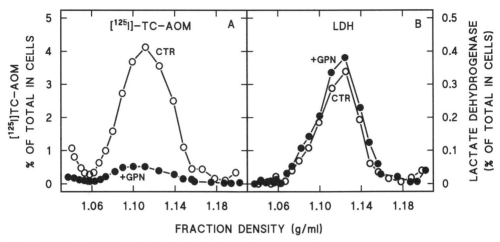

Figure 5. Effect of GPN on the density distributions of lysosomes and autophagosomes in isotonic Nycodenz density gradients.

destruction of lysosomes was developed (Berg *et al.*, 1994). By incubating electrodisrupted hepatocytes with the cathepsin C substrate glycyl-L-phenylalanine-2-naphthylamide (GPN), a selective osmotic rupture of the lysosomes could be achieved (Fig. 5).

Hepatocytes, prepared from a rat that had been injected with 25µg [125I]-TC-AOM 18h prior to cell isolation, were incubated for 2h at 37°C with vinblastine (50µM), then electrodisrupted. The disruptates were incubated either for 5 min at 37°C without additions (o) or for 5 min at 37°C with 0.5 mM GPN (•). Cytosol-free cell corpses were then isolated and homogenized, and large-granule fractions from the corpses were fractionized further on isotonic Nycodenz gradients. [125I]-TC-AOM radioactivity (A) and LDH (B) were measured in the fractions, and expressed as per cent (per fraction) of the corresponding total amounts in the disruptate.

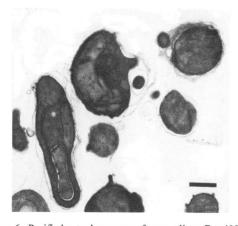

Figure 6. Purified autophagosomes from rat liver. Bar 400 nm.

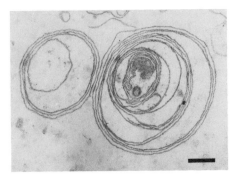

Figure 7. Emptied autophagosomes. The autophagosome walls are seen to consist of many concentric membranes. Bar 200 nm.

In combination with Nycodenz cushions (to remove mitochondria) and several rate sedimentation steps (to remove cytosol, nuclei and microsomes), relatively pure preparations of autophagosomes could be obtained (Fig. 6).

Preliminary attempts have been made to empty the autophagosomes and recover their limiting membranes (the phagophores). The results indicate that the walls of auto-phagosomes are built up of from two to numerous concentric unit membranes (Fig. 7), suggesting that either a single cisterna or a stack of flattened cisternae may fold up to form a multilayered bag. Hopefully, analysis of the molecular composition of the phagophore membranes may eventually provide some clues as to how this "bagging" process is executed.

ACKNOWLEDGEMENTS

This work has been generously supported by The Research Council of Norway and by the Norwegian Cancer Society.

REFERENCES

Aplin, A., Jasionowski, T., Tuttle, D.L., Lenk, S.E., and Dunn, W.A., 1992, Cytoskeletal elements are required for the formation and maturation of autophagic vacuoles, *J. Cell. Physiol.* 152:458-466.

Berg, T.O., Strømhaug, P.E., Berg, T., and Seglen, P.O., 1994, Separation of lysosomes and autophagosomes by means of glycyl-phenylalanine-naphtylamide, a lysosome-disrupting substrate for cathepsin C, *Eur. J. Biochem.* 221:595-602.

Dunn, W.A., 1990, Studies on the mechanisms of autophagy: Maturation of the autophagic vacuole, *J. Cell Biol.* 110:1935-1945.

Dunn, W.A., 1994, Autophagy and related mechanisms of lysosome-mediated protein degradation, *Trends Cell Biol.* 4:139-143.

Gordon, P.B., Holen, I., and Seglen, P.O., 1995, Protection, by naringin and some other flavonoids, of hepatocytic autophagy and endocytosis against inhibition by okadaic acid, *J. Biol. Chem.*, 270:5830-5838.

Gordon, P.B., and Seglen. P.O., 1988, Prelysosomal convergence of autophagic and endocytic pathways, *Biochem. Biophys. Res. Commun.* 151:40-47.

Green, S.A., Zimmer, K.-P., Griffiths, G., and Mellman, I., 1987, Kinetics of intracellular transport and sorting of lysosomal membrane and plasma membrane proteins, *J. Cell Biol.* 105:1227-1240.

Holen, I., Gordon, P.B., and Seglen, P.O., 1992, Protein kinase-dependent effects of okadaic acid on hepatocytic autophagy and cytoskeletal integrity, *Biochem. J.* 284:633-636.

Høyvik, H., Gordon, P.B., and Seglen, P.O., 1986, Use of a hydrolysable probe, [^{14}C]lactose, to distinguish between pre-lysosomal and lysosomal steps in the autophagic pathway, *Exp. Cell Res.* 166:1-14.

Høyvik, H., Gordon, P.B., Berg, T.O., Strømhaug, P.E., and Seglen, P.O., 1991, Inhibition of autophagic-lysosomal delivery and autophagic lactolysis by asparagine, *J. Cell Biol.* 113:1305-1312.

Kadowaki, M., Pösö, A.R., and Mortimore, G.E., 1992, Parallel control of hepatic proteolysis by phenylalanine and phenylpyruvate through independent inhibitory sites at the plasma membrane, *J. Biol. Chem.* 267:22060-22065.

Kopitz, J., Kisen, G.Ø., Gordon, P.B., Bohley, P., and Seglen, P.O., 1990, Non-selective autophagy of cytosolic enzymes in isolated rat hepatocytes, *J. Cell Biol.* 111:941-953.

Luiken, J.J.F.P., Van den Berg, M., Heikoop, J.C., and Meijer, A.J., 1992, Autophagic degradation of peroxisomes in isolated rat hepatocytes, *FEBS Lett.* 304:93-97.

Pösö, A.R., Wert, J.J., and Mortimore, G.E., 1982, Multifunctional control by amino acids of deprivation-induced proteolysis in liver. Role of leucine, *J. Biol. Chem.* 257:12114-12120.

Rabouille, C., Strous, G.J., Crapo, J.D., Geuze, H.J., and Slot, J.W., 1993, The differential degradation of two cytosolic proteins as a tool to monitor autophagy in hepatocytes by immunocytochemistry, *J. Cell Biol.* 120:897-908.

Seglen, P.O., Gordon, P.B., and Poli, A., 1980, Amino acid inhibition of the autophagic/lysosomal pathway of protein degradation in isolated rat hepatocytes, *Biochim. Biophys. Acta* 630:103-118.

Seglen, P.O., Gordon, P.B., Tolleshaug, H., and Høyvik, H., 1986, Use of [^{3}H]raffinose as a specific probe of autophagic sequestration, *Exp. Cell Res.* 162:273-277.

Seglen, P.O., 1987, Regulation of autophagic protein degradation in isolated liver cells, In Lysosomes: Their Role in Protein Breakdown, Glaumann, H., and Ballard, F.J., eds., Academic Press, London, pp. 369-414.

Strømhaug, P.E., and Seglen, P.O., 1993, Evidence for acidity of prelysosomal autophagic/endocytic vacuoles (amphisomes), *Biochem. J.* 291:115-121.

13

MECHANISM OF AUTOPHAGY IN PERMEABILIZED HEPATOCYTES

Evidence for Regulation by GTP Binding Proteins

Motoni Kadowaki,[1] Rina Venerando,[2] Giovanni Miotto,[2]
and Glenn E. Mortimore[3]

[1] Department of Applied Biochemistry
Niigata University
Niigata 950-21, Japan
[2] Dipartimento di Chimica Biologica
Università Degli Studi di Padova
Padova 35121, Italy
[3] Department of Cellular and Molecular Physiology
The Pennsylvania State University
Hershey, Pennsyvania 17033

INTRODUCTION

Autophagic vacuoles in hepatocytes are formed from membranes of rough and smooth endoplasmic reticulum (ER) by processes that are under immediate physiologic control by amino acids, insulin, and glucagon (reviewed in 1). Little, though, is known of the molecular steps involved. Microinjection (2) and electropermeabilization (3) have been used to introduce markers into cells and newly formed vacuoles. But because the pores are transient, observations are restricted to events that occur after membrane resealing. In order to gain access to autophagy under steady state conditions, we permeabilized hepatocytes with α-toxin from *Staphylococcus aureus*, an agent which forms stable ≈1.5 nm channels that limit exchange to molecules of approximately 1000 Da (4). Such pores will admit nucleotides and labeled residualizing probes without loss of cell proteins, a desirable, possibly necessary, condition for evaluating autophagically-mediated proteolysis.

The aims of this communication are: (i) to describe the functioning of *de novo* autophagy in semipermeabilized liver cells, (ii) to discuss the putative role of GTP binding proteins in autophagy, and (iii) to present a preliminary evaluation of autophagy in cells permeabilized to proteins.

Intracellular Protein Catabolism, Edited by Koichi Suzuki and Judith Bond
Plenum Press, New York, 1996

Table 1. Effects of GTPγS and 3-methyladenine (3-MA) on the volume
density of autophagosomes (AVi) in isolated rat hepatocytes
permeabilized by α-toxin (32.5 units/ml).

	Volume density of AVi		
Condition	*% cytoplasm*	*ml/10^8 cells*	*% inhibition*
No additions	0.249	1.611	0
1 mM GTPgS	0.015	0.097	94.0
10 mM 3-MA	0.008	0.052	96.8

Effects of GTPgS and 3-methyladenine (3-MA) on the volume density of
autophagosomes (AVi) in isolated rat hepatocytes permeabilized by α-toxin (32.5
units/ml). The cells were incubated for 45 min with an ATP regenerating system and
the additions listed below; they were then prepared for EM stereology as previously
described (5). The method of sampling was detailed earlier (5) and error was within
3%. Absolute AVi volumes were calculated from the volume of cytoplasm which was
computed to be 647 ml/10^8 cells (5). Control rates of long-lived proteolysis were
representative of rates in Table II. From Kadowaki et al. (5).

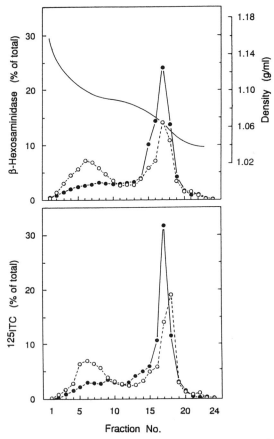

Figure 1. Effect of sample layering and dispersion on the distribution of β-hexosaminidase and [125]ITC in M+L
particles from permeabilized rat liver cells separated in colloidal silica gradients. Isolated cells, treated with
32.5 units/ml of α-toxin, were incubated for 60 min with [125]ITC and an ATP regenerating system. M+L samples
were applied to the gradient before centrifugation by layering (●) or dispersion (○). From Kadowaki et al.(5).

CHARACTERISTICS OF α-TOXIN PERMEABILIZED CELLS

Brief exposure of the isolated rat hepatocyte to highly purified α-toxin from *S. aureus* (a generous gift of Dr. Sydney Harshman, Vanderbilt University School of Medicine) effectively permeabilized the cell to glucose 6-phosphate in a dose-dependent manner (5,6). Since α-toxin pores are large enough to permit nucleotide exchange (7), an ATP regenerating system was routinely employed. At a standard dose of α-toxin (32.5 rabbit hemolytic units/ml) plus ATP, control rates of proteolysis were ≈40% of those in the intact cell (5) but 2-fold higher than rates in the absence of added ATP (Table III). Electron microscopy (EM) revealed no cytotoxicity at 32.5-65 units/ml as judged by the retention of microvilli and lack of plasma membrane blebbing. There was, however, moderate fragmentation and dilatation of the ER, changes that could interfere rather strongly with autophagic vacuole formation. But in fact new vacuoles (AVi) were clearly evident (Table I) although volume densities were decreased to ≈40% of maximal values (5). AVi formation was nearly abolished by the autophagic inhibitor 3-methyladenine (8) and by GTPγS (Table I); the inhibitory effectiveness of amino acids, though, was lost (5).

DISTRIBUTION OF ^{125}I-TYRAMINE -CELLOBIITOL ([125]ITC) AND SUBCELLULAR PARTICLES IN COLLOIDAL SILICA GRADIENTS

The *de novo* formation of autophagic vacuoles in hepatocytes permeabilized with α-toxin was monitored by the vacuolar uptake of [125]ITC, a stable probe that is retained within membrane-bound spaces (5). In these experiments vacuoles were separated on the basis of previously characterized features of colloidal silica-povidone density gradients (9-11). When mitochondrial-lysosomal (M+L) samples from rat liver cells are layered on the gradients, autophagic vacuoles are retained by a heavy midgradient band consisting largely of mitochondria and microsomes (Fig. 1, fractions 14-20). But when the samples are dispersed in the gradient medium, the vacuoles selectively move downward to the dense region (Fig. 1, fractions 1-11) while Golgi vesicles and some endosomes either fail to move or shift upward to lower densities; small lysosomes remain in the dense region. Thus, differences between layering and dispersion in the amount of lysosomal marker of fractions 1-11 will directly reflect autophagy.

The foregoing "sieving" or retention effect has been validated by EM (10) and is believed to result from interactions of an undetermined nature between autophagic vacuoles and particles of the midgradient band (10). Hence, it can be eliminated simply by dispersing the samples in the gradient medium before centrifugation,thereby allowing the vacuoles to equilibrate at their true density before the midgradient band is fully formed (10).

VACUOLAR UPTAKE OF [125]ITC IN PERMEABILIZED HEPATOCYTES AND ITS RELATION TO AUTOPHAGIC PROTEOLYSIS

As shown in Fig. 2, the uptake of [125]ITC by α-toxin permeabilized hepatocytes was remarkably linear over 60 min of incubation, showing that the probe is rapidly taken up by the cell and sequestered within vacuoles that shift into dense fractions after sample dispersion (see Fig. 1). Uptake was also found to be directly proportional to [125]ITC concentration and strongly dependent on external ATP (5). In keeping with the inhibitory effect of GTPγS on AVi formation in Table I, the nonhydrolyzable guanine nucleotides GTPγS and GMP-PNP

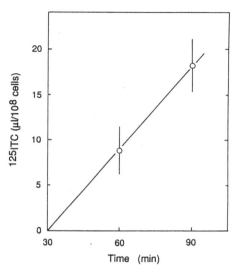

Figure 2. Time course of [125]ITC uptake into autophagic vacuoles in α-toxin-permeabilized rat hepatocytes. The cells were treated as in Fig. 1. After adding [125]ITC at 30 min, samples were taken for the separation of autophagic vacuoles by the layering-dispersion maneuver as shown in Fig. 1. From Kadowaki et al. (5).

(5'-guanylyl-imidodiphosphate) both suppressed [125]ITC uptake by about half (5). As will be discussed below, the failure of GTPγS to inhibit [125]ITC uptake and long-lived proteolysis by more than 50-55%, while strongly suppressing macroautophagy, suggests that the dispersion maneuver shifts both macro- and microautophagic vacuoles.

In addition to its direct uptake by autophagy some [125]ITC could be acquired by vacuoles after fusion with secondary lysosomes or endosomes which in turn had obtained [125]ITC through fluid phase endocytosis (12,13). This possibility was evaluated with the use of [^{14}C] dextran-carboxyl (M_r = 50,000-70,000) in parallel dispersion shift experiments (5). The results showed that only $6.2 \pm 0.09\%$ of [125]ITC was acquired in this way. Interestingly, the proportion remained the same under all conditions of regulation and varying proportions of macro- and microautophagic vacuoles.

Measurement of the inhibitory effect of GTPγS on the volume density of AVi by EM stereology in Table I has made it possible to validate [125]ITC as a marker for autophagic sequestration by an approach independent of ITC uptake. Zhong et al. reported that [125]ITC is mildly acidophilic and lipophilic (14); thus its use could overestimate sequestration. But as shown in Table II the predicted macroautophagic flux computed from the inhibitory effect of GTPγS on autophagic volumes determined stereologically agrees almost exactly with the observed rate of [125]ITC uptake. Since the accuracy of stereologic predictions has been substantiated (1), the agreement very likely means that the foregoing effect (14) is either quite small in magnitude or was rendered negligible in some way by the dispersion shift maneuver.

A coherent, unified mechanism of autophagic sequestration has recently been proposed (reviewed in 1) based in part on the close correspondence that exists between the fractional turnovers of intracellular long-lived proteins and of cytoplasmic volume in Table II. Identical features have been seen in all instances where proteolysis and the volume density of autophagic vacuoles have been measured together (1). Although a complete discussion is beyond the scope of this article, the same pattern is seen in micro- as well as macroautophagy. These findings strongly underscore the idea that the fractions of proteolysis and [125]ITC

Table 2. Comparative effects of GTPγS on cytoplasmic sequestration by macroautophagy, vacuolar uptake of [125]ITC, and long-lived proteolysis in isolated rat hepatocytes permeabilized by α-toxin.

	Control	GTPγS	GTPγS effect	
	(A)	(B)	A - B	% inhibition
Macroautophagic flux				
μl h^{-1}/10^8 cells	7.088	0.427	6.661	94
% total cytoplasm h^{-1}	1.096	0.066	1.030	
[125]ITC uptake				
μl h^{-1}/10^8 cells	12.181	5.577	6.604	54
% total cytoplasm h^{-1}	1.883	0.862	1.021	
Long-lived proteolysis				
μmol Val h^{-1}/10^{8}cells	2.003	0.917	1.086	54
% total protein h^{-1}	1.893	0.867	1.026	

Comparative effects of GTPγS on cytoplasmic sequestration by macroautophagy, vacular uptake of [125]ITC, and long-lived proteolysis in isolated rat hepatocytes premeabilized by α-toxin. Permeable cells were incubated with and without 1mM GTP S as described earlier (5). The macroautophagic flux was computed by dividing the volume density of AVi in Table I by 1.2 and then multiplying the quotient by the autophagic turnover constant 5.28h^{-1} (1,5); 1.2 is the volume ratio of AVi (nascent) to AVd (acidified, degradative) vacuoles (1,5). [125]ITC uptake, corrected for endocytosis, was determined by the gradient dispersion shift maneuver as described earlier (5). Long-lived proteolysis represents total valine release minus the small, nonregulated fraction (largely short-lived) described in the legend to Table III. Total cytoplasmic volume was same as in Table I; the valine content of cell protein was 105.8 mmol/10^8 cells (5).

uptake that remain after macroautophagy is virtually eliminated by GTPγS (Tables I and II) arise from microautophagic vacuoles (1,15).

EFFECTS OF NONHYDROLYZABLE GTP ANALOGS ON PROTEOLYSIS IN PERMEABILIZED HEPATOCYTES: ROLE OF GTP BINDING PROTEINS IN AUTOPHAGY

Table III compares inhibitory effects of GTP and a variety of nonhydrolyzable GTP analogs on proteolysis. External ATP, of course, is required in all conditions. GTPγS is the most effective analog and its dose response is depicted in Fig. 3. Its effects on [125]ITC uptake, proteolysis, and on AVi profiles are fully in accord with a primary inhibition of AVi formation, although additional effects on vacuole fusion, maturation, or acidification cannot be excluded (see below). It is of interest that the maximal effectiveness of GTPγS is similar to that observed with amino acids in the intact liver cell (1).

Since the nature of the GTP binding protein(s) involved in AVi formation is not known, experiments utilizing streptolysin-O (4) were initiated to characterize the behavior of hepatocytes made permeable to proteins. Findings thus far show that cells reconstituted with cytosol in the presence of ATP evoke a sharp increase in protein degradation that can be inhibited by GTPγS. As the increase is suppressed by chloroquine but not by 3-MA, the locus of breakdown is very likely protein sequestered in autophagic vacuoles prior to permeabilization. Because new AVi are not formed, the results suggest that other GTP

Table 3. Effects of ATP, GTP, and nonhydrolyzable GTP analogs
on long-lived proteolysis in α-toxin-permeabilized rat hepatocytes.

Additions	ATP-RS	Long-lived proteolysis	
		mmol Valh^{-1}/10^8 cells	*% inhibition*
None		2.00	0
None	–	0.97	51
GTPgS (1 mM)	+	0.92	54
GMP-PNP (500 mM)	+	1.24	38
GMP-PNP (1 mM)	+	1.04	48
GDPbS (750 mM)	+	1.28	36
GDPbS (2 mM)	+	1.00	50
GTP (2 mM)	+	2.15	0

Effects of ATP, GTP, and nonhydrolyzable GTP analogs on long-lived
proteolysis in a-toxin-permeabilized rat hepatocytes.   Isolated rat
hepatocytes were permeabilized with a-toxin (32.5 units/ml). Rates of
valine release in the presence of cycloheximide were measured as
detailed elsewhere (5) and then corrected for long-lived proteolysis by
subtracting a nonregulated fraction (0.39 mmol Valh^{-1}/10^8 cells) as in
Table II. ATP-RS, ATP regenerating system; GMP-PNP,
5′-guanylyl-imidodiphosphate. Modified from Kadowaki et al. (5).

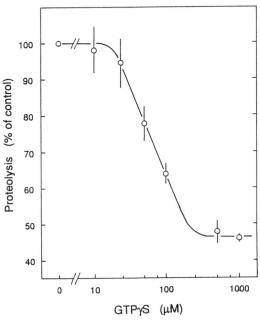

GTPγS (μM)

Figure 3. Proteolytic responsiveness to GTPγS of α-toxin-permeabilized rat hepatocytes. The experimental
procedure was the same as that in Table III and as described previously (5). Half-maximal inhibition was
achieved at 70 μM. From Kadowaki et al. (5).

binding protein(s) regulate the maturation/functioning of vacuoles below sites of AVi assembly (16). But it should be noted that the GTPγS effect could also be explained by lysosomal damage induced by one or more GTPγS-activated proteins in the cytosol, a notion prompted by the recent findings of Sai et al. (17).

REFERENCES

1. Mortimore, G. E., and Kadowaki, M., 1994, Autophagy: Its mechanism and regulation, In: *Cellular Proteolytic Systems*, (Ciechanover, A. J., and Schwartz, A. L., eds) pp.65-87, Wiley-Liss, Inc., New York.
2. Hendil, K. B., 1980, Intracellular degradation of hemoglobin transferred into fibroblasts by fusion with red blood cells, *J. Cell. Physiol.* 105:449-460.
3. Seglen, P. O., 1987, Regulation of autophagic protein degradation in isolated liver cells, In: *Lysosomes: Their Role in Protein Breakdown* (Glaumann, H., and Ballard, F. J., eds) pp.371-414, Academic Press, London.
4. Bhakdi, S., Weller, U., Walev, I., Martin, E., Jonas, D., and Palmer, M., 1993, A guide to the use of pore-forming toxins for controlled permeabilization of cell membranes, *Med. Microbiol. Immunol.* 182:167-175.
5. Kadowaki, M., Venerando, R., Miotto, G., and Mortimore, G. E., 1994, *De novo* autophagic vacuole formation in hepatocytes permeabilized by *Staphylococcus aureus* α-toxin: Inhibition by nonhydrolyzable GTP analogs, *J. Biol. Chem.* 269,3703-3710.
6. McEwen, B. F., and Arion, W. J., 1985, Permeabilization of rat hepatocytes with *Staphylococcus aureus* α-toxin, *J. Cell Biol.* 100:1922-1929.
7. Baldini, G., Hohman, R., Charron, M. J., and Lodish, H. F., 1991, Insulin and nonhydrolyzable GTP analogs induce translocation of GLUT 4 to the plasma membrane in α-toxin-permeabilized rat adipose cells, *J. Biol. Chem.* 266:4037-4040.
8. Seglen, P. O., and Gordon, P. B., 1984, Amino acid control of autophagic sequestration and protein degradation in isolated rat hepatocytes, *J. Cell Biol.* 99:435-444.
9. Surmacz, C. A., Ward, W. F., and Mortimore, G. E., 1982, Distribution of ^{125}I-asialofetuin among liver particles separated on colloidal silica gradients, *Biochem. Biophys. Res. Commun.* 107:1425-1432.
10. Surmacz, C. A., Wert, J. J. Jr., and Mortimore, G. E., 1983, Role of particle interaction on distribution of liver lysosomes in colloidal silica, *Am. J. Physiol.* 245:C52-C60.
11. Surmacz, C. A., Wert, J. J. Jr., and Mortimore, G. E., 1983, Metabolic alterations and distribution of rat liver lysosomes in colloidal silica, *Am. J. Physiol.* 245:C61-C67.
12. Gordon, P. B., and Seglen, P. O., 1988, Prelysosomal convergence of autophagic and endocytic pathways, *Biochem. Biophys. Res. Commun.* 151:40-47.
13. Tooze, J., Hollinshead, M., Ludwig, T., Howell, K., Hoflack, B., and Kern, H., 1990, In exocrine pancreas, the basolateral endocytic pathway converges with the autophagic pathway immediately after the early endosome, *J. Cell Biol.* 111:329-345.
14. Zhong, Z.-D., Jadot, M., Wattiaux-De Coninck, S., and Wattiaux, R., 1992, Uptake of tyramine cellobiose by rat liver, *Biochim. Biophys. Acta* 1106,311-316.
15. Mortimore, G. E., Lardeux, B. R., and Adams, C. A., 1988, Regulation of microautophagy and basal protein turnover in rat liver: Effects of short-term starvation, *J. Biol. Chem.* 263:2506-2512.
16. Sai, Y., and Ohkuma, S., 1992, Small GTP-binding proteins on rat liver lysosomal membranes, *Cell Struc. Func.* 17:363-369.
17. Sai, Y., Arai, K., and Ohkuma, S., 1994, Cytosol treated with GTPγS disintegrates lysosomes *in vitro*, *Biochem. Biophys. Res. Commun.* 198:869-877.

14

LYSOSOMAL PROTEINOSIS BASED ON DECREASED DEGRADATION OF A SPECIFIC PROTEIN, MITOCHONDRIAL ATP SYNTHASE SUBUNIT C

Batten Disease

J. Ezaki, L. S. Wolfe,[2] K. Ishidoh,[1] D. Muno,[1] T. Ueno,[1] and E. Kominami[1]

[1] Department of Biochemistry
Juntendo University of School of Medicine
2-1-1 Hongo, Bunkyo-ku
Tokyo 113, Japan
[2] Montreal Neurological Institute and Hospital
McGill University, 3801 University Street
Montreal
Quebec H3A-2B4, Canada

INTRODUCTION

Lysosomal storage diseases are a group of more than 35 genetically determined heterogenous metabolic disorders. Biochemically, they are characterized by deficient activity of specific lysosomal hydrolases and intralysosomal accumulation of one or more substrates in affected cells. They are often categorized according to the chemical nature of the most prominent storage compounds. They include lipid storage diseases, glycogen storage diseases, mucopolysaccharidoses, mucolipidoses and gangliosidoses. However, there are no reports of protein storage diseases (proteinoses) based on deficiency of specific lysosomal proteases. Experimentally, inhibition of lysosomal cysteine proteinases by administration of leupeptin and E-64 causes intralysosomal storage of cytoplasmic proteins in cells (1).

Neuronal ceroid lipofuscinosis (NCL, Batten disease) is probably the most common group of lysosomal storage diseases. The nature of the deficient enzyme is not known. The term 'ceroid lipofuscinosis' has been given to this group of diseases because the storage substance is stainable with Sudan stains and luxol fast blue, which stain phospholipids, is autofluorescent, and is not extractable with lipid soluvents (2, 3). It is, however, distingush-able from the aging pigment. Recently, the nature of the major storage compound in Batten diseases was demonstrated. Palmer *et al.* disclosed that in the ovine model of NCL the major protein stored is identical to the dicyclohexylcarbodiimide-reactive proteolipid, subunit c, of the mitochondrial ATP synthase complex (F0F1-ATPase) (4). The subunit c of mitochondrial ATP synthase is very hydrophobic, and is classified as a proteolipid because of its lipid-like solubility in chloroform-methanol mixtures. It is composed of 75 amino acids and normally found as an intrinsic inner mitochondrial membrane component of the FoF1 oligomeric ATP synthase complex (4). This complex contains 14 different polypeptides and subunit c is an essential component of the transmembrane proton channel. Amino acid sequence analysis of human storage bodies has also suggested that the normal subunit c is stored in the juvenile and late infantile diseases (5). We demonstrated immunochemically that a particularly high concentration of subunit c is specifically stored in lysosomes in the brains and in fibroblast cell types from the late infantile cases of this disease (6, 7).

In this review we describe and discuss recent advances in the biochemical and molecular studies of the mechanism for lysosomal proteinosis in Batten disease.

BATTEN DISEASE

The neuronal ceroid lipofuscinoses represent a group of recessively inherited neurogenerative disease of infants, children and young adults that lead to blindness, seizures, dementia and premature death. The late infantile and juvenile forms are often referred to as Batten disease. It is clear now that this group of diseases represents different mutations and recent linkage studies have shown localization of the infantile disease to chromosome region 1p32 (8), the juvenile onset disease to chromosome region 16 p12.1-p11.2 (9), and a variant form of late infantile form to chromosome region 13q21.1-q32 (10). The gene locations of other forms of the disease have not been found. Storage bodies isolated from all forms of ceroid-lipofuscinosis studied so far have a high protein content. The major storage materials in late infantile, juvenile and adult forms are subunit c of mitochondrial ATP synthase, whereas in infantile form the sphingolipid activator proteins (saposins) are the most prominent stored protein. Saposins A and C are also hydrophobic proteins without the carbohydrate side chains.

LYSOSOMAL LOCALIZATION OF SUBUNIT C IN BATTEN DISEASE

Our previous immunolocalization studies showed that antiserum against subunit c reacted very strongly with the storage material in fibroblasts from patients with the late infantile type of Batten disease (6). Strong staining for a lysosomal marker, lgp 120, was observed largely in association with the storage material and partially with distinct granular deposits in the cytoplasm. However, only faint staining for subunit c was detected in control fibroblasts. The localization of various mitochondrial proteins containing subunit c and lysosomal markers was also examined biochemically in control and patient fibroblasts (11). After fibroblasts from controls and patients were fractionated in a Percoll density gradient into mitochondrial and lysosomal fractions, the two fractions were analysed for localization of subunit c, β subunit of ATP synthase, and cathepsin B. There was no significant difference in distribution of β-hexosaminidase activity, a lysosomal marker, and succinic dehydrogenase activity, a mitochondrial marker, between control and diseased cell types.The majority of subunit c from control cells was found in mitochondrial fractions. In contrast, about 30-70% of subunit c from patient cells was found in the higher density fractions. The antigens recognized by anti-β subunit distributed exclusively in mitochondrial fractions from both control and patient cells, while cathepsin B antigens were detected only in lysosomal fractions from control and diseased cell types. We also found that subunit IV of cytochrome oxidase, a mitochondrial inner membrane protein and ornithine aminotransferase, a mitochondrial matrix protein from patient cells, were found only in mitochondrial fractions. These results indicate that in fibroblasts from patients with NCL, subunit c as well as other mitochondrial proteins exist in the mitochondrion, and that only subunit c, not other mitochondrial proteins, are found in lysosomes.

LEVELS OF LYSOSOMAL PROTEASES IN BATTEN DISEASE

To test the possibility that a defect in lysosomal proteases causes the accumulation of subunit c in lysosomes, various lysosomal protease activities in fibroblasts from patients with the late infantile form of Batten disease and controls were measured (12). There were no large differences in activities of cathepsins L, B, H, dipeptidyl peptidase II, or lysosomal carboxypeptidase A (cathepsin A) between patient and control cells. Only dipeptidyl peptidase I (cathepsin C) activities were markedly lower in some cell types from patients with the late infantile form Batten disease. Accumulation of subunit c was found in all cells from patients with the late infantile form tested which suggests that the decrease in cathepsin C activity is not related to lysosomal accumulation of subunit c. Western blot analysis using anti-cathepsin antibodies confirmed those results. Although the possibility of a defect in other lysosomal proteases not tested here is not ruled out, it is unlikely.

SYNTHESIS, IMPORT AND ASSEMBLY OF SUBUNIT C IN BATTEN DISEASE

Human and cattle have two expressed genes for subunit c, P1 and P2 that code a precursor protein containing the same mature protein (13-15). Medd et al (16) showed with Northern blotting experiments, no significant differences in the levels of mRNAs for P1 and P2 genes between normal sheep and affected sheep. To confirm these results in human fibroblasts, the expression levels of P1 gene and P2 gene were examined by RNA blot hybridization analyses (11). Both mRNAs for P1 and P2 gene transcripts do not differ in length or amounts between three fibroblast preparations from patients with the late infantile form of NCL and two controls. It was also shown that the P2 gene was expressed predominantly and the P1 gene was expressed in trace amounts in both affected and unaffected cells. These results are consistent with an earlier observation that there was no gross difference in the levels of the mRNAs of subunit c in normal and diseased animals (16).

The synthetic rates of subunit c in fibroblasts from patients with Batten disease was investigated by pulse-chase and immunoprecipitation (11). The mature form of subunit c of apparent 4.4 kDa could be detected at 10 min after pulse-labeling and incorporation of radiolabel into subunit c increased with time both in control and patient fibroblasts. There were no apparent differences in the rates of synthesis of subunit c and total cellular proteins between control and patient cells from late infantile patients and so subunit c is unlikely to accumulate because of excessive transcription of either genes. Furthermore, detection of the mature subunit c at 10 min after pulse-labeling both in control and patient cells suggests that subunit c is normally imported and assembled into mitochondrial ATP synthase. Protein sequencing of the stored subunit c in the late infantile and juvenile forms of Batten disease (16) and in the sheep and cattle disease (3, 17) show that the presequences have been removed at the site of specificity of the mitochondrial import system, supporting the above possibility.

In addition, there is no evidence of a disruption of mitochondrial function. Mean P/O ratios for mitochondria as in those obtained from kidneys of affected and normal sheep did not differ significantly (3). Taken together, these findings suggest a specific failure in the degradation of subunit c after its normal inclusion into mitochondria, and subsequently abnormal accumulation in lysosomes.

SPECIFIC DELAY OF DEGRADATION OF SUBUNIT C IN BATTEN DISEASE

It is possible that a slower degradation rate of subunit c causes its storage in the brain or fibroblasts of patients with the late infantile form of Batten disease. Catabolism of subunit c was investigated. Fibroblasts from patients with the late infantile form of NCL and controls were labeled for 24 hours with [^{35}S] methionine and chased for 1, 3, 5 days. There was a clear decrease in the quantity of prelabeled subunit c in control cells (apparent $t_{1/2}$ 40-50 h), whereas no apparent degradation of subunit c was found in fibroblasts with the disease. The mitochondrial inner membrane protein, subunit IV of cytochrome oxidase and total cellular proteins were degraded at similar rates in both control and patient cells. These results indicate clearly that degradation of subunit c is delayed markedly in cells from patients with the late infantile form of Batten disease (11).

LYSOSOMAL APPEARANCE OF LABELED SUBUNIT C IN BATTEN DISEASE

Investigations were conducted to determine whether intra-mitochondrial degradation of subunit c is delayed and when the radio-labeled subunit c appeares in lysosomal fractions from diseased cells (11). Fibroblasts were labeled with [^{35}S] methionine for 24 hours and chased for 0, 7, and 14 days. The post-nuclear supernatant fractions were fractionated with percoll density gradients into mitochondrial and lysosomal fractions as described before and two pooled fractions were analysed on SDS-PAGE after immunoprecipitation. During a 14 day chase period, the radioactivity of prelabelled subunit c in mitochondrial fractions from control cells decreased with time, whereas a marked delay of degradation of subunit c was found in the mitochondrial fractions from all cell types with the late infantile form of Batten disease. With loss of radiolabel of subunit c in mitochondrial fractions, lysosomal appearance of labelled subunit c was found in two cell types with NCL among three cell types used after 7 and 14 days of chase. These results indicate that subunit c accumulates in mitochondria due to retarded degradation and is then transfered to lysosomes. To confirm these results in various patient and control cell types, fibroblasts from 6 patients with the late infantile form of Batten disease and 3 controls were subjected to pulse and long chase (14 day) experiments. In control cells the radiolabelled subunit c almost disappeared in both mitochondrial and lysosomal fractions after the long chase time, whereas clear signals for subunit c were found in mitochondrial fractions from all cell types with the late infantile form of Batten disease. The amount of radiolabelled subunit c in lysosomes after 2 weeks of chase varied greatly between cells tested. Radiolabeled subunit c was high in three cell types, and low in the other three cell types among fibroblasts with the late infantile form of Batten disease. The difference in lysosomal storage of subunit c among diseased cell types may be due to the difference in cellular ability of autophagocytosis or a difference in lysosomal degradative capacity after autophagy.

POSSIBLE DEGRADATIVE PATHWAY OF SUBUNIT C IN NORMAL AND BATTEN DISEASE

It has been assumed that mitochondrial proteins are degraded by lysosomes because of the presence of mitochondria inside autophagic vacuoles (18). More direct evidence that mitochondrial proteins are detected in autolysosomes isolated from livers of leupeptin-treated rats was obtained by Ueno and Kominami (19). Macroautophagy is most prominent under deprivation conditions such as starvation of animals or during serum withdrawal from cells in culture. This macroautophagy appears to be rather nonselective in that several different proteins including mitochondrial matrix and inner membrane proteins can be taken up at similar rates (1, 19). Thus, selectivity of mitochondrial protein degradation may not be explained by lysosomal autophagy. The present analysis in Batten disease showed that an apparently normal sized subunit c accumulated in mitochondria and is transferred to lysosomes, suggesting that mitochondrial subunit c is taken up normally by macroauto-phagic-lysosomal pathway, but that its lysosomal proteolysis is impaired. Lysosomal storage of proteins is caused by blocking of a class of cysteine proteinases by E-64 or leupeptin, but a variety of cytoplasmic proteins including mitochondrial proteins accumulate in lysosomes (1, 19). Administration of CA-074, a selective inhibitor of cathepsin B or pepstatin-asia-lofetuin, a potent inhibitor of cathepsin D, had much less effect on the degradation of endogenous proteins and on accumulation of autolysosomes (20). However, the possibility that lysosomal storage of a specific or a group of proteins could occur due to deficiency or impairment of a specific lysosomal protease is not ruled out.

Certain proteins, which retain KFERQ and related sequences in the molecule, may be able to directly traverse lysosomal membranes with the help of a 73 kDa heat shock protein as proposed by Dice (21). The KFERQ-related peptide is found in the leader sequence of subunit c, but not in the mature region, suggesting that targetting of mature subunit c to lysosomes by this pathway is unlikely.

Some mitochondrial proteins may be catabolized within the mitochondria. Unassembled subunit proteins and a fraction of mitochondrial translation products from Hela cells and hepatoma cells are rapidly degraded in growing cells (22). An ATP-dependent, vanadate-sensitive protease from rat liver mitochondria, which could degrade newly synthesized proteins as well as incomplete proteins, was isolated (23). This proteinase may function to eliminate abnormal or unneeded proteins. We found no significant difference between control and diseased fibroblasts in the contents of the mitochondrial ATP dependent protease (12). It has been proposed, on the basis of morphometric and immunocytochemical calculations (18, 24), that at least 50% of mitochondrial proteins of rat liver must be degraded by means other than autophagy. Short-lived mitochondrial proteins may be degraded by mitochondrial proteinases while long-lived proteins are degraded by lysosomes. Mitochondrial proteins may also be degraded normally in a piecemeal fashion by intramitochondrial proteinases. Although several mitochondrial proteinases have been identified (25, 26), understanding of their role in the mitochondrial pathway is very poor.

Turnover rates of proteins of the mitochondrial inner membrane of hepatoma cells and hepatocytes were studied and results showed that proteins are selectively degraded with different half lives (27-29). Subunit c has distinct properties in the mitochondrial inner membrane: First, subunit c is an abundant component of mitochondrial membranes and probably has 10-12 copies per ATP synthase complex (16). Second, it is one of the most hydrophobic proteins that has been studied. However, the half-life of subunit c, estimated as 40-50 h from our results, is almost the same as the average half lives of proteins of the mitochondrial inner membrane (48h). Thus, macroautophagy and subsequent intralysosomal degradation is the most plausible mechanism for the degradation of subunit c in normal cells.

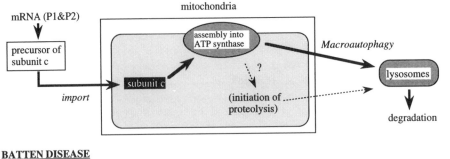

Figure 1. Possible degradative pathway of subunit c in normal cells and cells from patients withBatten disease. Closed circle, posttranslational modification.

CONCLUSIONS AND PERSPECTIVES

A specific delay of the degradation of subunit c in cells from patients with Batten disease was demonstrated directly (11). Measurement of radiolabelled subunit c in mitochondrial and lysosomal fractions supports the idea that subunit c is transported to lysosomes by autophagocytosis. Mitochondrial accumulation of subunit c seen in some cell types may be due to a delay of turnover of subunit c within the mitochondrion or may be due to a decrease in degradation after distribution of subunit c to the lysosomal compartment.

The following possibilities could be considered as lesion (s) for the specific impairment of degradation of subunit: [1] Mutation of subunit c. Recently, the sequences of cDNAs for P1 and P2 genes of subunit c from affected sheep was found to be identical to those in healthy control animals (16). In addition, the gene for juvenile form of NCL maps to chromosome 16 (10), whereas location of the human P1 and P2 genes is reported to be on chromosomes 17 and 12, respectively (16) Therefore, this possibility is unlikely. [2] A lack of a protein that may affect the structure or microenvironment of subunit c within mitochondria. [3] Post-translational modification of subunit c. This possibility has been reported by Katz *et al.* (30). They suggested that the storage body proteins in affected animals is rich in methylated lysines and methionines. [4] Accumulation of other storage materials such as lipids may affect intralysosomal proteolysis of the proteolipid, subunit c. [5] A failure of the proteolytic system initiating the degradation of subunit c in lysosomes or in mitochondria is also possible.

Studies on mechanism for intracellular degradation of mitochondrial proteins including subunit c in normal cells will help the understanding of the mutation(s) underlying the late infantile and juvenile forms of Batten disease.

Table 1. Classification of Neuronal Ceroid Lipofuscinoses

Locus	Form	Major storage material(s)	Chromosomal localization
CLN1	infantile	SAPs (sphingolipid activator proteins)	1p32[8]
CLN2	late infantile	subunit c of ATP synthase	?
CLN3	juvenile	subunit c of ATP synthase	16p12[9]
CLN4	adult	subunit c of ATP synthase	?
CLN5	Finnish variant	subunit c of ATP synthase	13q21.1-q32[10]
mnd	mouse model	subunit c of ATP synthase subunit c of vacuolar H⁺-ATPase	8 or 19[31]
	ovine model	subunit c of ATP synthase	?

ACKNOWLEDGEMENTS

This work was suppoted in part by a Grant-in Aid for Scientific Research (6281114) from the Ministry of Education, Science and Culture of Japan, The Science Research Promotion Fund from Japan Private School Promotion Foundation, and a grant MT-1345 from the Medical Research Council of Canada.

REFERENCES

1. Kominami, E., Hashida, S., Khairrallah, E.A., and Katunuma, N., 1983, Sequestration of cytoplasmic enzymes in an autophagic vacuole-lysosomal system induced by injection of leupeptin, *J. Biol. Chem.* 258 : 6093-6100.
2. Carpenter, S., 1988, Morphological diagnosis and misdiagnosis in Batten-Kufs disease, *Am. J. Med. Genet. Supple.* 5: 85-91.
3. Palmer, D.N., Fearnley, I. M., Walker, J. E., Hall, N. A., Lake, B. D., Wolfe, L. S., Haltia, M., Martinus, R. D., and Jolly, R. D., 1992, Mitochondrial ATP synthase subunit c storage in the ceroid lipofuscinosis (Batten's disease), *Am. J. Med. Genet.* 42: 561-567.
4. Palmer, D. N., Martinus, R. D., Cooper, S. M., Midwinter, G. G., Reid, J. C., and Jolly, R.D., 1989, Ovine ceroid lipofuscinosis: The major lipopigment protein and the lipid binding subunit of mitochondrial ATP synthase have the same NH₂ -terminal sequence, *J. Biol. Chem.* 264: 5736-5740.
5. Fearnley, I. M., Walker, J. E., Martinus, R. D., Jolly, R. D., Kirkland, K. B., Shaw, G. J., and Palmer, D. N., 1990, The major protein stored in ovine ceroid lipofuscinosis is identical to the DCCD-reactive proteolipid of mitochondrial ATP synthase, *Biochem. J.* 268: 751-758.
6. Kominami, E., Ezaki, J., Muno, D., Ishido, K., Ueno, T., and Wolfe., L. S., 1992, Specific storage of subunit c of mitochondrial ATP synthase in lysosomes of neuronal ceroid lipofscinosis (Batten's disease), *J. Biochem.* (Tokyo) 111: 278-282.
7. Hall, N. A., Lake, B. D., Dewji, N. N., and Patrick, A. D., 1991, Lysosomal storage of subunit c of mitochondrial ATP synthase in lysosomes of neuronal ceroid lipofuscinosis (Batten's disease), *Biochem. J.* 275: 269-272.
8. Järvelä, I., Schleutker, J., Haataja, L., Santavuori, P., Puhakka, L., Manninen, T., Palotie, A., Sandkuijl, L.A., Renlund, M., White, R., Aula, P. and Peltonen, L., 1991, Infantile form of neuronal ceroid lipofuscinosis (CLN1) maps to the short arm of chromosome 1, *Genomics* 9: 170-173.
9. Mitchison HM, Thompson AD, Mulley JC, Kozman HM, Richards RI, Callen DF, Stallings, RL, Doggett, NA, Attwood J, McKay TR, Sutherland GR, and Gardiner RM, 1993, Fine genetic mapping of Batten disease locus (CLN3) by haplotype analysis and demonstration of alleic association with chromosome 16p mixrosatellite loci, *Genomics* 16: 455-460.
10. Savukoski M, Kestilä M, Williams R, Järvelä I, sharp J, Harris, J., Santavuori P, Gardiner M, and Peltonen L, 1994, Difined chromosomal assignment of CLN5 demonstrates that at least four genetic loci are involved in the pathogenesis of human ceroid lipofuscinoses, *Am. J. Hum. Genet.* 55: 695-701, 1994.

11. Ezaki, J., Wolfe, L.S., Higuti, T., Ishidoh, K., and Kominami, E., 1995, Specific Delay of Degradation of Mitochondrial ATP Synthase Subunit c in Late Infantile Neuronal Ceroid Lipofuscinosis (Batten Disease), *J Neurochem*,64: 733-41.

12. Ezaki, J., Wolfe, L.S., Ishidoh, K., and Kominami, E., 1995, Abnormal Degradative Pathway of Mitochondrial ATP Synthase Subunit c in Late Infantile Neuronal Ceroid-Lipofuscinosis (Batten Disease), *Am. J. Med. Genet.* , 57: 254-259.

13. Dyer, M.R., Gay, N.J. and Walker, J.E., 1989, DNA sequences of a bovine gene and of two related pseudogenes for the proteolipid subunit of mitochondrial ATP synthase, *Biochem. J.* 260: 249-258.

14. Dyer, M.R., and Walker, J.E., 1993, Sequences of members of the human gene family for the c subunit of mitochondrial ATP synthase, *Biochem. J.* 293: 51-64.

15. Higuti T, Kawamura Y, Kuroiwa K, Miyazaki S, and Tsujita H., 1993, Molecular cloning and sequence of two cDNAs for human subunit c of H^+-ATP synthase in mitochondria, *Biochim Biophys Acta* 1173: 87-90.

16. Medd, S.M., Walker, J.E., and Jolly, R.D., 1993, Characterization of the expressed genes for subunit c of mitochondrial ATP synthase in sheep with ceroid lipofuscinosis, *Biochem. J.* 293: 65-73.

17. Palmer, D.N., Fearnley, I. M., Medd, S. M., Walker, J. E., Martinus, R. D., Bayliss, S. L., Hall, N. A., Lake, B. D., Wolfe, L. S., and Jolly, R. D., 1990, Lysosomal storage of the DCCD reactive proteolipid subunit of mitochondrial ATP synthase in human and ovine ceroid lipofuscinoses, *Adv. Exp. Med. Biol.* 266: 211-224.

18. Pfeifer, U., 1978, Inhibition by insulin of the formation of autophagic vacuoles in rat liver, *J .Cell. Biol.* 78: 152-167.

19. Ueno T, and Kominami E, 1991, Mechanism and regulation of lysosomal sequestration and proteolysis, *Biomed. Biochim. Acta* 50: 365-371.

20. Kominami, E., Ueno, T., Muno, D., and Katunuma, N., 1991, The selective role of cathepsins B and D in the lysosomal degradation of endogenous and exogenous proteins, *FEBS Lett.*, 287: 189-192.

21. Dice, J.F., and Chiang, H.L., 1989, Lysosomal degradation of microinjected proteins, Cell Biology Reviews 20: 13-33.

22. Kalnov, S.L., Novikova, L.A., Zubatov, A.S., and Luzikov, V.N., 1979, *Biochem. J.*, 182: 195-202.

23. Desautels, M., and Goldberg, A.L., 1982, Liver mitochondria contain an ATP-dependent vanadate-sensitive pathway for the degradation of proteins, *Proc Natl Acad Sci USA* 79: 1869-1877.

24. Grisolia, S., Hernandez-Yago, J., and Knecht, E., 1985, Regulation of mitochondrial protein concentration: a plausible model which may permit assessing protein turnover, *Curr. Top. Cell. Regul.* 27: 387-396.

25. Marcillat, O., Zhang, Y., Lin, S.W., and Davies, K.J.A., 1988, Mitochondria contain a proteolytic system which can recognize and degrade oxidatively-denatured proteins, *Biochem. J.*, 254: 677-683.

26. Bharadwaj, D., Roy, M.S., Bose, D., and Hati, R., 1994, A new blood-coagulating protease in mitichondrial membranes of rat submaxillary glands, *J. Biol. Chem.* 269, 16229-16235.

27. Hare, J.F., and Hodges, R., 1982, Turnover of mitochondrial inner membrane proteins in hepatoma monolayer cultures, *J. Biol. Chem.* 257: 3575-3580.

28. Russel, S.M., and Mayer, R.J., 1983, Degradation of transplanted rat liver mitochondrial-outer membrane-proteins in hepatoma cells, *Biochem. J.* 216, 163-175.

29. Perez-Paster, E., Wallance, R., and Grisolia, S., 1982, Turnover of adenosine triphosphatase from rat liver mitochondria. Effect of high protein and low protein diets, *Eur. J. Biochem.* 127: 275-278.

30. Katz, M.L., Christianson, J.S., Norbury, N.E., Gao, C.-L., Siakotos, A.N., Koppang, N. ,1994, Lysine methylation of mitochondrial ATP synthase subunit c stored in tissues of dogs with hereditary ceroid lipofuscinosis, *J Biol Chem* 269: 9906-9911.

31. Bronson, R.T., Lake B.D., Cook, S, Taylor, S., Davisson, M.T., 1993, Motor neuron degeneration of mice is a model of neuronal ceroid lipofuscinosis (Batten's disease), *Annals of Neurology*, 33: 381-385.

15

THE NEURONAL CEROID LIPOFUSCINOSES (BATTEN DISEASE)

A Group Of Lysosomal Proteinoses

D. N. Palmer and J. M. Hay

Centre for Molecular Biology, AVSG
Lincoln University
P.O. Box 84
Canterbury, New Zealand

INTRODUCTION

The neuronal ceroid lipofuscinoses (NCL), also called Batten disease, are a group of neurodegenerative lysosomal storage diseases affecting humans and other animals. They are inherited as autosomal recessive traits. Affected children start life normally and then develop increasing dementia, blindness and seizures, culminating in premature death. There are three major forms of human NCL, the infantile form (INCL or CLN1) which occurs most frequently in Finland, and the late infantile (LINCL or CLN2) and juvenile (JNCL or CLN3) forms which occur worldwide. A number of other forms have been described including a rare adult form, Kufs disease (CLN4) and a Finnish late infantile variant (CLN5) (Dyken, 1988). Each form is distinguished by the age of onset and the clinical course of the disease, and a confusing number of eponyms have been used to describe them (Dyken, 1988; Kohlschütter *et al.*, 1993). Partly because of this, the collective incidence of these diseases is unknown. However they are relatively common and an estimate of about 1 in 12,500 live births world-wide has been made (Rider and Rider, 1988).

The ceroid lipofuscinoses were originally grouped together with the other familial amaurotic idiocies, although there had been some suggestions that they were a separate group of diseases (for example, Speilmeyer, 1906). Their distinctive identity became apparent as the enzyme defects underlying the other amaurotic familial idiocies, the sphingolipidoses and gangliosidoses, were discovered in the 1960s and 1970s. Because of this history as well as the histological properties of the storage bodies it was presumed that the defects underlying them would be in lipid metabolism. The formulation of the lipid peroxidation hypothesis to explain the histologically similar lipopigments, lipofuscin and ceroid, reinforced this expectation. A number of defects in lipid metabolism and lack of control of fatty acid oxidation were postulated. None of these ideas proved worthwhile and the enzyme defects causing the ceroid lipofuscinoses remain unknown.

Intracellular Protein Catabolism, Edited by Koichi Suzuki and Judith Bond
Plenum Press, New York, 1996

A more fruitful line of enquiry has been the analysis of storage body composition. This has lead to the discovery that the NCLs are proteinoses in which specific hydrophobic proteins are stored in lysosome derived organelles. They are not lipidoses. This article reviews the work leading to this conclusion and discusses the implications.

STORAGE BODY COMPOSITION

The initial discovery was made in a thoroughly described form of NCL in a flock of sheep deliberately bred and maintained as a model of the human diseases. Analyses of storage bodies isolated from fresh tissues revealed that more than two-thirds of their mass is protein (Palmer et al., 1986a, b, 1988), the rest being lipids typical of lysosome-derived organelles.

The major storage body protein is not lysosomal. It was difficult to identify because it is insoluble in all but a few solvents, the most useful being lithium dodecyl sulphate, and it stains poorly with Coomassie blue. A specially developed silver stain (described in Fearnley et al., 1990) revealed a major storage body specific component with an apparent molecular mass of 3.5 kDa on polyacrylamide gels. Another specific component at 14.8 kDa and a minor band at 24 kDa are frequently detected. Attempts at purifying each of these components proved unsuccessful but the identity of the major band was discovered by direct N-terminal sequencing of total storage body proteins which gave a clear sequence, identical to the first 40 amino acids of subunit c (the DCCD reactive proteolipid) of mitochondrial ATP synthase (Palmer et al., 1989).

At the time it was hoped that the stored protein would be a partially degraded product of complete subunit c and identification of the C-terminus would point the way to the missing lysosomal protease. This strategy had worked well for discovering the deficient hydrolases underlying many of the gangliosidoses and sphingolipidoses. This was not to be. Further sequencing established that it was the complete and normal 75 amino acid protein that is stored and that the 14.8 and 26 kDa bands are oligomers of it (Fearnley et al., 1990). A comparison of the yield from the sequencer and total storage body protein established that this protein makes up at least 50% of the storage body mass. These results established that ovine ceroid lipofuscinosis is a proteinosis in which subunit c of ATP synthase, an inner mitochondrial membrane protein, is abnormally stored in lysosome derived organelles.

Subsequent studies, using the same strategy, showed that complete subunit c is also stored in the late infantile and juvenile human diseases (Palmer et al., 1990, 1992). N-terminal sequencing also established subunit c storage in affected English Setter, Border Collie and Tibetan Terrier dogs (Jolly et al., 1992), Devon cattle (Martinus et al., 1991) and "mnd" mice (Faust et al., 1994). However the relative amounts of subunit c in storage bodies varies in these different diseases and there is evidence of co-storage of other proteins in some forms. The relative amount of subunit c is highest in the late infantile form whereas storage bodies from the "mnd" mice contain up to equal mass proportions of another hydrophobic protein, subunit c of vacuolar ATPase, depending on the tissues of origin (Faust et al., 1994). Traces of this molecule had previously been observed in sheep pancreas storage bodies (Palmer et al., 1989). Storage bodies in different subunit c storage diseases also react differently to subunit c antibodies in situ. Some are more susceptible to prolonged fixation than others and some more susceptible to different antibodies than others suggesting that different epitopes are exposed (Westlake et al., 1995). Either minor components mask the subunit c differently or the subunit c is packed in different configurations.

Subunit c has unusual physical properties for a protein. It is extremely insoluble in most solvents but can be extracted with lipids with chloroform/ methanol and is classified as a proteolipid, even though it contains no covalently attached lipid. This explains the anomalous staining pattern of the storage bodies in situ. They stain poorly with protein stains but well with lipid ones so it is easy to see how these protein storage diseases were confused with lipidoses on histological grounds.

This protein is not stored in the infantile human disease, but other hydrophobic proteins are. N-terminal sequencing of storage bodies, combined with a careful analysis of aggregate bands on blots, indicated that these are the sphingolipid activator proteins (SAPs or saposins) A and D (Tyynelä et al., 1993). Subunit c is not stored in significant amounts in an adult onset form in Miniature Schnauzer dogs either. Preliminary results indicate SAP storage in this disease. Some minor SAP storage has been reported in the subunit c storage diseases also. Western blotting of storage body proteins and immunocytochemical studies indicate that SAP D and perhaps SAP A are components of storage bodies in the late infantile, juvenile and adult human diseases (Tyynelä et al., 1995). Experiments aimed at determining the contributions of vacuolar ATPase subunit c and SAPs to the storage body mass in all the different ceroid lipofuscinoses are underway.

Any defects in protein turnover underlying the ceroid lipofuscinoses have little influence on protein trafficking in cells. Only a few fine differences were found in a rigorous two-dimensional polyacrylamide gel electrophoresis survey of proteins in liver subcellular fractions from normal and affected sheep. There appeared to be an accumulation of ferritin light chains in microsomes from affected sheep and rather less of senescence marker protein in the cytosolic fraction (Moroni-Rawson et al., 1995). Whether or not these changes are caused by the disease or are consequences of it remains an open question.

GENE LINKAGE STUDIES

Normally different forms of the same basic storage disease arise from different mutations in the same gene. This is not the case for the ceroid lipofuscinoses where mapping studies have placed the genes at different loci (Järvelä et al., 1992). The juvenile form (CLN3) has been mapped to chromosome 16p12.1-11.2, interval *D16S288-D16S383* (Mitchison et al., 1994). CLN2, the late infantile form, has been excluded from this region and the Finnish late infantile variant, CLN5, has been mapped to chromosome 13q21.1-q32 (Williams et al., 1994). CLN1 is the best mapped of the ceroid lipofuscinoses and has been localised to 1p32, but despite the sequencing of a candidate gene in this region the lesion remains unknown (Vesa et al., 1994). The genetic lesion in *"mnd"* mice is not syntenic with any of the human loci mapped so far (Messer et al., 1992). It is also unlikely that the wide range of disease forms that have been observed in dogs (Jolly et al., 1994) result from mutations in a single gene. The fact that mutations in so many different genes are involved in the ceroid lipofuscinoses helps explain the relatively high collective incidence of these diseases.

THE NORMAL STRUCTURE, FUNCTION AND LOCATION THE STORED PROTEINS

In all the mammals studied, mitochondrial ATP synthase subunit c is coded for on at least two nuclear genes, P1 and P2. Sequences of P1 and P2 from humans (Dyer and Walker, 1993), sheep (Medd et al., 1993), cattle (Gay and Walker, 1985; Dyer et al., 1989), mice (Piko et al., 1994) and rat (Higuti et al., 1993) have been documented. Human P1 and P2 are on chromosomes 17 and 12, respectively. Recently a third gene, P3, possibly located on chromosome 2, has been reported to encode subunit c in humans (Yan et al., 1995).

Each of these genes encodes a precursor protein containing the same mature protein of 75 amino acids but with different amino-terminal presequences of 61, 68 and 67 amino acids, respectively. The mature proteins from different species are identical and the lead sequences are related. They are thought to direct the importation of the mature subunit c into the ATP synthase complex in the inner mitochondrial membrane, being cleaved off the mature protein in two steps during this process (Hartl *et al.*, 1989; Hendrick *et al.*, 1989). This complex, also called the F_0F_1 ATPase complex contains at least 14 different polypeptides and there are multiple copies of some of these, including 6-12 of subunit c (Walker *et al.*, 1991).

ATP synthase is the ATP generating enzyme at the end of the mitochondrial respiratory chain. It has two major parts, F_1 and F_0. Subunit c is part of the intramembrane F_0 portion and accounts for 2-4% of the total membrane protein. It forms the channels for proton translocation across the membrane into the mitochondrial matrix which powers ATP generation. The bound F_1 portion intrudes into the mitochondrial matrix and contains the catalytic portion of the enzyme.

The sphingolipid activator proteins, SAPs or saposins, A and D are coded for on the *proSAP* gene which is on human chromosome 10 q21-22 (Kao *et al.*, 1988). ProSAP is cleaved into four products, SAPs A,B,C and D. Ammonium chloride inhibition studies indicate that this probably occurs in the lysosome or late endosome (Fujibayashi and Wenger, 1986). Their site of action is in the lysosome where they activate glycosphingolipid degradation by lifting the substrates from the plane of the membrane so they are exposed to the water soluble exohydrolases responsible for their digestion (Sandhoff and Klein, 1994). Vacuolar subunit c is part of the vacuolar ATPase complex and forms a similar proton translocating channel to mitochondrial subunit c. This complex works in the opposite direction to the mitochondrial ATP synthase, pumping protons into vesicles to acidify them and consuming ATP in the process. It is normally found in endosomes, late endosomes, lysosomes, and other vesicles that are acidified (Mandel *et al.*, 1988, Mellman *et al.*, 1986).

WHERE ARE THE DEFECTS SITED?

The ceroid lipofuscinoses can be considered as lysosomal proteinoses because proteins are abnormally stored in lysosomes. What is surprising is the specificity of that storage. No other ATP synthase or other mitochondrial components are major components of storage bodies in the subunit c storage diseases. These results lead to the concept that underlying these diseases are lesions that specifically affect the turnover of subunit c in such a way that it accumulates in lysosome-derived organelles (Palmer *et al.*, 1992).

One possibility was that subunit c is over-produced but this is unlikely. The quantities of mRNA transcribed from P1 and P2 are similar in both NCL affected and normal sheep eliminating excessive transcription and subsequent translation of either gene (Medd *et al.*, 1993). P1 and P2 cDNA precursor sequences from affected and normal animals are identical (Medd *et al.*, 1993) and so the disease does not arise from a mutation in the mitochondrial import sequence and mistargeting of subunit c.

Isolated mitochondria from affected sheep have normal respiration indicating that sufficient subunit c is incorporated into the ATPase complex to facilitate ATP generation (Palmer *et al.*, 1990). Biochemical and immunocytochemical studies in affected sheep did not reveal more subunit c in affected mitochondria than in controls (Fearnley *et al.*, 1990; Westlake *et al.*, 1995) and two dimensional gel analyses of control and affected mitochondrial proteins revealed no differences in their composition (Moroni-Rawson *et al.*, 1995). On the other hand, subcellular fractionation studies suggest that subunit c turnover through mitochondria of late infantile affected fibroblasts might be slowed compared to controls (Ezaki et al., 1995).

It has been suggested that the stored subunit c is covalently modified by a disease specific trimethylation of one of the lysines (lys 43) (Katz and Rodrigues, 1991; Katz and Gerhardt, 1992; Katz et al., 1994) or S-methylation of methionine (Katz and Gerhardt, 1990), but this is not so. Similar amounts of trimethyllysine were found in proteolipid extracted from normal mitochondria and storage bodies from affected sheep and late infantile affected humans (Palmer et al., 1993). Subunit c from normal mitochondria and storage bodies have the same immunoreactivity (Palmer et al., 1995) and mass spectral studies have shown that they have the same mass (I.M. Fearnley and J.E. Walker, pers. comm.).

Lysosomes possess many proteases with a wide range of specificities and ingest whole mitochondria by autophagocytosis (Pfeifer, 1987). However it is unlikely that subunit c storage is the remnant of defective lysosomal proteolysis, at least in all forms of the disease. Too many genes are implicated given the relative lack of specificity of known lysosomal proteases and the relatively small number of them. It has been known for some time that different mitochondrial proteins have different turnover rates suggesting an alternative degradation route to autophagy. Mitochondrial proteases may contribute to this (Hare, 1990). The importation of nuclear encoded proteins into the inner mitochondrial membrane is a complicated process, involving a number of factors (Glover and Lindsay, 1992; Hannavy et al., 1993). The pathway for subunit c transport to its normal site of degradation is also likely to be complicated, especially given the highly hydrophobic nature of subunit c and its tendency to aggregate. Lack of a factor to prevent aggregation could result in a non-degradable structure. It is not known if the lysosome is the normal site of subunit c degradation and it is possible that storage there is a "default value". Lesions underlying the subunit c storing ceroid lipofuscinoses could occur in a cytosolic transport or degradation complex or in the lysosome itself. Since different genes are implicated in different forms of disease they could affect different subcellular sites. The primary biochemical lesion may be sited in mitochondria or the cytosol or lysosomes in different forms of the disease.

One system that could be involved in subunit c degradation is the ubiquitin-dependent proteolytic pathway, as this is one of the major routes by which intracellular proteins are destroyed. Initially proteins are marked for degradation by conjugation to ubiquitin and subsequently recognised by a high molecular mass multicatalytic protease (26S or 1300 kDa; MCP) located in the cytosol. The proteolytic core of the MCP is a 19S protease which is composed of a number of subunits and contains the enzymic activities necessary for intracellular protein turnover (for a review, see Hershko and Ciechanover, 1992). This system is inviting as a candidate for the basis of NCL because it is an integrated system composed of a number of highly regulated activities. The system may sustain mutations in genes encoding any one of these activities or constitutive subunits from ubiquitin tagging to specific MCP encoded proteolysis, causing the different forms of NCL. Furthermore, elements of it have been detected along the route from mitochondria to lysosomes (Laszlo et al., 1990; Magnani et al., 1991; Schwartz et al., 1992).

The storage of other proteins and disturbances of protein turnover may provide useful clues for delineating the subunit c turnover pathway and the sites of lesions. Vacuolar ATPase and mitochondrial ATP synthase subunit c are highly related proteolipids with similar functions and physical properties, including an outstanding hydrophobicity, but arise from different subcellular sites. It is possible that the apparent storage of the vacuolar ATPase subunit c in "mnd" mice arises from impurities in the preparations but this is unlikely. Preliminary results suggest different levels of storage of both vacuolar and mitochondrial subunit c in different forms of the disease. It is likely that the defect in mice affects the catabolism of subunit c from both sources at some point after a junction in the specific pathways for the two. The cleavage of proSAP to SAPs A, B, C and D is thought to be an intralysosomal process but only SAPs A and D are stored in the infantile disease. Therefore the proteolytic lesion here is likely to be intralysosomal. SAPs A and D are the most hydrophobic of the SAPs so this could be at the lysosomal end of a pathway for hydrophobic molecules. It is not clear if the minor extent of SAP storage observed in the subunit c storage diseases is specific. Incidental SAP storage has been observed in the sphingolipidoses where hydrophobic substrates accumulate (Kishimoto *et al.*, 1992) and they are normal lysosome components.

Unravelling the lesions underlying the ceroid lipofuscinoses should lead to a better understanding of these diseases and point the way to treatments. It will also lead to a better understanding of the normal proteolysis of hydrophobic proteins in cells.

ACKNOWLEDGEMENT

This work was supported by the United States National Institute of Neurological Disorders and Stroke, grant NS32348.

REFERENCES

Dyer, M. R., Gay, N. J. and Walker, J. E., 1989, DNA sequences of a bovine gene and of two related pseudogenes for the proteolipid subunit of mitochondrial ATP synthase, *Biochem. J.* 260:249-258.

Dyer, M. R. and Walker, J. E., 1993, Sequences of members of the human gene family for the c subunit of mitochondrial ATP synthase, *Biochem. J.* 293:51-64.

Dyken, P. R., 1988, Reconsideration of the classification of the neuronal ceroid-lipofuscinoses, *Am. J. Med. Genet., Suppl.* 5:69-84.

Ezaki, J., Wolfe, L. S., Higuti, T., Ishidoh, K. and Kominami, E., 1995, Specific delay of degradation of mitochondrial ATP synthase subunit c in late infantile neuronal ceroid-lipofuscinosis (Batten disease), *J. Neurochem.* In press.

Faust, J. R., Rodman, J. S., Daniel, P. F., Dice, J. F. and Bronson, R. T., 1994, Two related proteolipids and dolichol-linked oligosaccharides accumulate in motor neuron degeneration mice (*mnd/mnd*), a model for neuronal ceroid lipofuscinosis, *J. Biol. Chem.* 269:10150-10155.

Fearnley, I. M., Walker, J. E., Martinus, R. D., Jolly, R. D., Kirkland, K. B., Shaw, G. J. and Palmer, D. N., 1990, The sequence of the major protein stored in ovine ceroid lipofuscinosis is identical to that of the dicyclohexylcarbodiimide-reactive proteolipid of mitochondrial ATP synthase, *Biochem. J.* 268:751-758.

Fujibayashi, S. and Wenger, D. A., 1986, Synthesis and processing of sphingolipid activator protein-2 (SAP-2) in cultured human fibroblasts, *J. Biol. Chem.* 261:15339-15343.

Gay, N. J. and Walker, J. E., 1985, Two genes encoding the bovine mitochondrial ATP synthase proteolipid specify precursors with different import sequences and are expressed in a tissue-specific manner, *EMBO J.* 4:3519-3524.

Glover L. A. and Lindsay J. G., 1992, Targeting proteins to mitochondria: a current overview, *Biochem. J.* 284:609-620.

Hannavy K., Rospert S. and Schatz G., 1993, Protein import into mitochondria: a paradigm for the translocation of polypeptides across membranes, *Curr. Opin. Cell Biol.* 5:694-700.

Hare, J. F., 1990, Mechanisms of membrane protein turnover, *Biochim. Biophys. Acta* 1031:71-90.

Hartl, F-U., Pfanner, N., Nicholson, D. W. and Neupert, W., 1989, Mitochondrial protein import, *Biochim. Biophys. Acta* 988:1-45.

Hendrick, J. P., Hodges, P. E. and Rosenberg, L. E., 1989, Survey of amino-terminal proteolytic cleavage sites in mitochondrial precursor proteins: Leader peptides cleaved by two matrix proteases share a three-amino acid motif, *Proc. Natl. Acad. Sci. (USA)* 86:4056-4060.

Hershko, A. and Cienchanover, A., 1992, The ubiquitin system for protein degradation, *Ann. Rev. Biochem.* 61:761-807.

Higuti, T., Kuroiwa, K., Kawamura, Y., Morimoto, K. and Tsujita, H., 1993, Molecular cloning and sequence of cDNAs for the import precursors of oligomycin sensitivity conferring protein, ATPase inhibitor protein, and the subunit c of H⁺-ATP synthase in rat mitochondria, *Biochim. Biophys. Acta* 1172:311-314.

Järvelä, I., Vesa, J., Santavuori, P., Hellsten, E. and Peltonen, L., 1992, Molecular genetics of the neuronal ceroid lipofuscinoses, *Pediat. Res.* 32:645-648.

Jolly, R. D., Martinus, R. D. and Palmer, D. N., 1992, Sheep and other animals with ceroid-lipofuscinoses: Their relevance to Batten disease, *Am. J. Med. Genet.* 42:609-614.

Jolly, R. D., Palmer, D. N., Studdert, V. P., Sutton, R. H., Kelly, W. R., Koppang, N., Dahme, G., Hartley, W. J., Patterson, J. S. and Riis, R. C., 1994, Canine ceroid-lipofuscinoses: A review and classification, *J. Small Anim. Pract.* 35:299-306.

Kao, F. T., Law, M. L., Hartz, J., Jones, C., Zhang, X-L., Dewji, N. N., O'Brien, J. S. and Wenger, D. A., 1988, Regional localization of the gene coding for sphingolipid activator protein (SAP-1) on human chromosome 10, *Somat. Cell Molec. Genet.* 13:685-688.

Katz, M. L., Christianson, J. S., Norbury, N. E., Gao, C.-L., Siakotos, A. N. and Koppang, N., 1994, Lysine methylation of mitochondrial ATP synthase subunit c stored in tissues of dogs with hereditary ceroid lipofuscinosis, *J. Biol. Chem.* 269:9906-9911.

Katz, M. L. and Gerhardt, K. O., 1990, Storage protein in hereditary ceroid lipofuscinosis contains S-methylated methionine, *Mech. Ageing Dev.* 53:277-290.

Katz, M. L. and Gerhardt, K. O., 1992, Methylated lysine in storage body protein of sheep with hereditary ceroid-lipofuscinosis, *Biochim. Biophys. Acta* 1138:97-108.

Katz, M. L. and Rodrigues, M., 1991, Juvenile ceroid-lipofuscinoses. Evidence for methylated lysine in neural storage body protein, *Am. J. Pathol.* 138:323-332.

Kishimoto, Y., Hiraiwa, M. and O'Brien, J. S., 1992, Saposins: structure, function, distribution, and molecular genetics, *J. Lipid Res.* 33:1255-1267.

Kohlschütter, A., Gardiner, R. M. and Goebel, H. H., 1993, Human forms of neuronal ceroid lipofuscinosis (Batten disease): consensus on diagnostic criteria, Hamburg 1992, *J. Inher. Metab. Dis.* 16:241-244.

Laszlo, L., Doherty, F. J., Osborn, N. U. and Mayer, R. J., 1990, Ubiquinated protein conjugates are specifically enriched in the lysosomal system of fibroblasts, *FEBS Lett.* 261:365-368.

Magnani, M., Serafini, G., Antonelli, A., Malatesta, M. and Gazzanelli, G., 1991, Evidence for a particulate location of ubiquitin conjugates and ubiquitin-conjugating enzymes in rabbit brain, *J. Biol. Chem.* 266:21018-21024.

Mandel, M., Moriyama, Y., Hulmes, J. D., Pan, Y. C. E., Nelson, H. and Nelson, N., 1988, cDNA sequence encoding the 16 kDa proteolipid of chromaffin granules implies gene duplication in the evolution of H⁺-ATPases, *Proc. Natl. Acad. Sci. (USA)* 85:5521-5524.

Martinus, R. D., Harper, P. A. W., Jolly, R. D., Bayliss, S. L., Midwinter, G. G., Shaw, G. J. and Palmer, D. N., 1991, Bovine ceroid-lipofuscinosis (Batten's disease): The major component stored is the DCCD-reactive proteolipid, subunit c, of mitochondrial ATP synthase, *Vet. Res. Commun.* 15:85-94.

Medd, S. M., Walker, J. E. and Jolly, R. D., 1993, Characterization of the expressed genes for subunit c of mitochondrial ATP synthase in sheep with ceroid lipofuscinosis, *Biochem. J.* 293:65-73.

Mellman, I., Fuchs, R. and Helenius, A., 1986, Acidification of the endocytic and exocytic pathways, *Ann. Rev. Biochem.* 55:663-700.

Messer, A., Plummer, J., Maskin, P., Coffin, J. M. and Frankel, W. N., 1992, Mapping of the motor neuron degeneration (*Mnd*) gene, a mouse model of amyotrophic lateral sclerosis (ALS), *Genom.* 18:797-802.

Mitchison, H. M., Taschner, P. E. M., O'Rawe, A. M., De Vos, N., Phillips, H. A., Thompson, A. D., Kozman, H. M., Haines, J. L., Schlumpf, K., D'Arigo, K., Boustany, R.-M. N., Callen, D. F., Breuning, M. H., Gardiner, R. M., Mole, S. E. and Lerner, T. J., 1994, Genetic mapping of the Batten disease locus (CLN3) to the interval *D16S288-D16S383* by analysis of haplotypes and allelic association, *Genom.* 22:465-468.

Moroni-Rawson, P., Palmer, D. N., Jolly, R. D. and Jordan, T. W., 1995, Variant proteins in ovine ceroid-lipofuscinosis, *Am. J. Med. Genet.* In press.

Palmer, D. N., Barns, G., Husbands, D. R. and Jolly, R. D., 1986b, Ceroid lipofuscinosis in sheep. II. The major component of the lipopigment in liver, kidney, pancreas and brain is low molecular weight protein, *J. Biol. Chem.* 261:1773-1777.

Palmer, D. N., Bayliss, S. L., Clifton, P. A. and Grant, V. J., 1993, Storage bodies in the ceroid lipofuscinoses (Batten disease): Low-molecular-weight components, unusual amino acids and reconstitution of fluorescent bodies from non-fluorescent components, *J. Inher. Metab. Dis.* 16:292-295.

Palmer, D. N., Fearnley, I. M., Medd, S. M., Walker, J. E., Martinus, R. D., Bayliss, S. L., Hall, N. A., Lake, B. D., Wolfe, L. S. and Jolly, R. D., 1990, Lysosomal storage of the DCCD reactive proteolipid subunit of mitochondrial ATP synthase in human and ovine ceroid lipofuscinoses. In *Lipofuscin and Ceroid Pigments,* Ed. Porta, E. A., Plenum Press, New York, pp. 211-223.

Palmer, D. N., Fearnley, I. M., Walker, J. E., Hall, N. A., Lake, B. D., Wolfe, L. S., Haltia, M., Martinus, R. D. and Jolly, R. D., 1992, Mitochondrial ATP synthase subunit c storage in the ceroid-lipofuscinoses (Batten disease), *Am. J. Med. Genet.* 42:561-567.

Palmer, D. N., Husbands, D. R., Winter, P. J., Blunt, J. W. and Jolly, R. D., 1986a, Ceroid lipofuscinosis in sheep I. Bis(monoacylglycero)phosphate, dolichol, ubiquinone, phospholipids, fatty acids and fluorescence in liver lipopigment lipids, *J. Biol. Chem.* 261:1766-1772.

Palmer, D. N., Martinus, R. D., Barns, G., Reeves, R. D. and Jolly, R. D., 1988, Ovine ceroid-lipofuscinosis I. Lipopigment composition is indicative of a lysosomal proteinosis, *Am. J. Med. Genet., Suppl.* 5:141-158.

Palmer, D. N., Martinus, R. D., Cooper, S. M., Midwinter, G. G., Reid, J. C. and Jolly, R. D., 1989, Ovine ceroid lipofuscinosis. The major lipopigment protein and the lipid-binding subunit of mitochondrial ATP synthase have the same NH_2-terminal sequence, *J. Biol. Chem.* 264:5736-5740.

Palmer, D. N., Bayliss, S. L. and Westlake, V. J., 1995, Batten disease and the ATP synthase subunit c turnover pathway: Raising antibodies to subunit c, *Am. J. Med. Genet.* In press.

Pfeifer, U., 1987, Functional morphology of the lysosomal apparatus. In *Lysosomes: Their Role in Protein Breakdown*, Eds. Glaumann H., Ballard, F. J., Academic Press, London, pp. 3-59.

Piko, L., Nofziger, D. E., Western, L. M. and Taylor, K. D., 1994, Sequence of a mouse embryo cDNA clone encoding proteolipid subunit 9 (P1) of the mitochondrial H^+-ATP synthase, *Biochim. Biophys. Acta* 1184:139-141.

Rider, J. A. and Rider, D. L., 1988, Batten disease: Past, present and future, *Am. J. Med. Genet., Suppl.* 5:21-26.

Sandhoff, K. and Klein, A., 1994, Intracellular trafficking of glycosphingolipids: role of sphingolipid activator proteins in the topology of endocytosis and lysosomal digestion, *FEBS Lett.* 346:103-107.

Schwartz, A. L., Trausch, J. S., Ciechanover, A., Slot, J. W. and Geuze, H., 1992, Immunoelectron microscopic localization of the ubiquitin-activating enzyme E1 in HepG2 cells, *Proc. Natl. Acad. Sci. (USA)* 89:5542-5546.

Speilmeyer, W., 1906, Uber eine besondere Form von familiarer amaurotischer Idiotie, *Neurol. Zbl.* 25:51-55.

Tyynelä, J., Palmer, D. N., Baumann, M. and Haltia, M., 1993, Storage of saposins A and D in infantile neuronal ceroid-lipofuscinosis, *FEBS Lett.* 330:8-12.

Tyynelä, J., Baumann M., Henseler, M., Sandhoff, K., and Haltia, M ., 1995, Sphingolipid activator proteins in the neuronal ceroid-lipofuscinoses: An immunological study, *Acta Neuropathol. (Berl.)* In press.

Vesa, J., Hellsten, E., Barnoski, B. L., Emanuel, B. S., Billheimer, J. T., Mead, S., Cowell, J. K., Strauss III, J. F. and Peltonen, L., 1994, Assignment of sterol carrier protein X/sterol carrier protein 2 to 1p32 and its exclusion as the causative gene for infantile neuronal ceroid lipofuscinosis, *Hum. Molec. Genet.* 3:341-346.

Walker J. E., Lutter R., Dupuis A. and Runswick M.J., 1991, Identification of subunits of F_1F_0-ATPase from bovine heart mitochondria, *Biochemistry* 30:5369-5378.

Westlake, V. J., Jolly, R. D., Bayliss, S. L. and Palmer, D. N., 1995, Immunocytochemical studies in the ceroid-lipofuscinoses (Batten disease) using antibodies to subunit c of mitochondrial ATP synthase, *Am. J. Med. Genet.* 57: 177-181.

Williams R., Santavuori P., Peltonen L., Gardiner R. M. and Järvelä I., 1994, A variant form of late infantile neuronal ceroid lipofuscinosis (CLN5) is not an allelic form of Batten (Speilmeyer-Vogt-Sjögren, CLN3) disease: Exclusion of linkage to the CLN3 region of chromosome 16, *Genom.* 20:289-290.

Yan, W. L., Lerner, T. J., Haines, J. L. and Gusella, J. F., 1995, Sequence analysis and mapping of a novel human mitochondrial ATP synthase subunit 9 cDNA, *Genom.* 24: 375-377.

16

ENERGY-DEPENDENT DEGRADATION OF A MUTANT SERINE:PYRUVATE/ALANIN: GLYOXYLATE AMINOTRANSFERASE IN A PRIMARY HYPEROXALURIA TYPE 1 CASE

Toshiaki Suzuki, Kozo Nishiyama, Tsuneyoshi Funai, Keiji Tanaka,[2] Akira Ichihara,[2] and Arata Ichiyama[1]

[1] Department of Biochemistry
Hamamatsu University School of Medicine
Hamamatsu 431-31, Japan
[2] Institute for Enzyme Research
The University of Tokushima
Tokushima 770 , Japan

PRODUCTION AND METABOLISM OF GLYOXYLATE

The subject of this study is serine:pyruvate/alanine:glyoxylate aminotransferase (SPT/AGT). This enzyme has been called serine:pyruvate aminotransferase (SPT, EC 2.6.1.44) or alanine:glyoxylate aminotransferase (AGT, EC 2.6.1.51), and both the SPT and AGT activities are believed to be of physiological importance. Therefore, this enzyme is called SPT/AGT in this paper. SPT/AGT in the liver is an enzyme of dual organelle localization and presumably of dual function. In the rat, SPT/AGT is located in both mitochondria and peroxisomes, and only the mitochondrial enzyme is markedly induced by glucagon (1-6). In carnivores, localization of this enzyme is mostly mitochondrial (7, 8) and the enzyme in mitochondria is probably involved in serine metabolism. In herbivores and man, this enzyme locates exclusively in peroxisomes (7-9), where glyoxylate, an immediate precursor of oxalate, is mainly produced as shown in Fig. 1. It is well known that the calcium salt of oxalic acid is hardly soluble in aqueous solutions, and frequently forms calculi in urinary and other tissues. Since plants are rich in oxalate, precursors of oxalate may also be included in this staple food of herbivores. Therefore, SPT/AGT in peroxisomes is thought to be essential, especially for herbivores, to remove glyoxylate by conversion to glycine and to keep animals from harmful overproduction of oxalate (Fig. 1).

In fact, primary hyperoxaluria type 1 (PH1), a congenital metabolic disease characterized by increased production of oxalate and precipitation of calcium oxalate crystals in many tissues, is caused by a defect in SPT/AGT (10).

Intracellular Protein Catabolism, Edited by Koichi Suzuki and Judith Bond
Plenum Press, New York, 1996

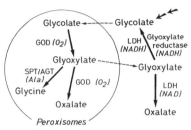

Figure 1. Production and metabolism of glyoxylate. Abbreviations used in this figure: GOD, glycolate oxidase; LDH, lactate dehydrogenase.

PROFILE OF A PRIMARY HYPEROXALURIA TYPE 1 CASE

In 1982, we encountered a case of this disease. This patient (male) had had recurrent urinary calculi since the age of 6, and reached the end-stage renal failure at age 38, which necessitated hemodialysis. The oxalate concentration in the plasma of this patient was considerably higher than that in controls who were undergoing hemodialysis due to chronic glomerulonephritis, not only before but also after hemodialysis (11). His parents are consanguineous to each other, and three out of his five children also suffered from juvenile-onset urolithiasis. The patient suddenly died at age 45 in 1989. On autopsy, massive deposition of calcium oxalate crystals was observed in the kidney and bone, and to a lesser extent, in all tissues examined, other than the brain, liver, parathyroid and adrenal. We assume that calcium oxalate calculi in the heart may have blocked the conduction system. We studied the SPT/AGT deficiency in this PH1 patient, using a piece of liver obtained on autopsy. The SPT activity in the cytoplasmic extract of the liver was specifically decreased to about one- hundredth of that in control livers, confirming that the patient is indeed PH1.

SPT/AGT DEFICIENCY IN THE PRIMARY HYPEROXALURIA TYPE 1 CASE

We cloned SPT/AGT-cDNA from the patient's liver and demonstrated a point mutation of T to C in exon 6, encoding a Ser to Pro substitution at residue 205. The T to C

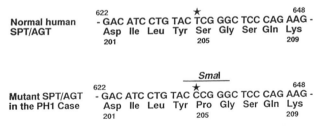

Figure 2. Nucleotide and deduced amino acid sequences around mutation site in the PH1 case. Nucleotide numbers in cDNAs for normal human SPT/AGT and mutant SPT/AGT in the PH1 case are relative to respective 5-ends. Deduced amino acid residues are numbered from the initiation Met. This figure is taken from reference (11).

conversion created a new *SmaI* site (Fig. 2), which enabled us to confirm that the mutation exists in the patient's gene and that the patient is homozygous with respect to this mutation (11).

A remarkable feature of the SPT/AGT deficiency in this PH1 case was that mutant SPT/AGT appeared to be actively synthesized in the patient's liver, but the enzyme was very low with respect to not only the activity but also the protein detectable on Western blot analysis (12). On RNA blot analysis, a single 1.7 kb band of SPT/AGT-mRNA was observed in both the control and the patient's liver, and the level of the mRNA in the patient's liver was even higher than normal. The mRNA from the patient's liver was translated as effectively as that from control liver in an *in vitro* system with a reticulocyte lysate and in transfected COS cells and transformed *E. coli*. On Western blot analysis, on the other hand, the 43 kDa band representing SPT/AGT was clearly observed only for the control liver, when 10 mg protein of liver extract was subjected to electrophoresis. Mutant SPT/AGT was detectable when as much as 500 mg protein of the extract was applied. Transfected COS cells and transformed *E. coli* also failed to accumulate mutant SPT/AGT. Immunocyto-chemically detectable SPT/AGT labeling was also low in the patient's liver, although it was detected predominantly in peroxisomes (12).

ENERGY-DEPENDENT DEGRADATION OF MUTANT SPT/AGT

The discrepancy between the synthesis and the intracellular steady state level of mutant SPT/AGT was explained, at least in part, by the finding that the mutant enzyme was decomposed much faster than normal in transfected COS cells, transformed *E. coli* and *in vitro* (12). In addition, the *in vitro* degradation of mutant SPT/AGT with a reticulocyte lysate system was clearly dependent on the presence of ATP and Mg^{2+} (Fig. 3). Normal SPT/AGT was stable irrespective of the presence or absence of ATP and Mg^{2+}, under the conditions

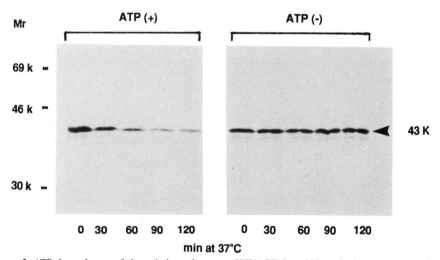

Figure 3. ATP-dependency of degradation of mutant SPT/AGT in rabbit reticulocyte extract. Detailed experimental conditions are given in reference (12). ATP (+): The reaction for degradation of ^{35}S-labeled patients SPT/AGT was carried out in the presence of 5 mM ATP and 5 mM $MgCl_2$. ATP (-): 20 mM 2-deoxyglucose and 25 U/ml of hexokinase were added instead of ATP to deplete ATP from the reaction mixture. This figure is taken from reference (12).

used. On the other hand, mutant SPT/AGT was stable when either ATP or Mg^{2+} was depleted, but rapidly decomposed in the presence of both ATP and Mg^{2+}. Deoxy-ATP, GTP, CTP, UTP and ADP were also effective in place of ATP, but AMP, a,b-methylene ATP and b,g-methylene ATP were without effect, indicating that the degradation is energy-dependent.

It appears that a single amino acid substitution in SPT/AGT found in the PH1 case caused a mis-folded structure, which is recognized in cells as that of abnormal protein to be eliminated by degradation. With respect to the mechanism of the mutant SPT/AGT degradation, it is well documented that 26S proteasome plays a major role in non-lysosomal, energy-dependent proteolysis in mammalian cells. However, mutant SPT was broken down in a reticulocyte lysate from which proteasomes had been removed by immunoprecipitation. Under the same conditions, degradation of ^{125}I-labeled lysozyme and succinyl-LLVY-MCA, substrates of the proteasome, was almost abolished by the immunoremoval of proteasomes. Further experiments are necessary before any conclusion, but these results suggest a possibility that some proteolytic system(s) other than 26S proteasome is responsible for the energy-dependent degradation of the mutant SPT/AGT.

Although the biological and pathophysiological importance of intracellular degradation systems to eliminate misfolded proteins has long been suspected, this is still one of the few cases in which rapid degradation of a mutant protein is demonstrated in a congenital metabolic disease.

REFERENCES

1. Noguchi, T., Minatogawa, Y., Takada, Y., Okuno, E., and Kido, R. 1978 *Biochem. J.* 170: 173-175.
2. Oda, T., Yanagisawa, M., and Ichiyama, A. 1982 *J. Biochem.* 91: 219-232.
3. Yokota, S. and Oda, T. 1984 *Histochemistry* 80: 591-595.
4. Oda, T., Funai, T., and Ichiyama, A. 1990 *J. Biol. Chem.* 265: 7513-7519.
5. Yokota, S., Funai, T., and Ichiyama, A. 1991 *Biomed. Res.* 12: 53-59.
6. Uchida, C., Funai, T., Oda, T., Ohbayashi, K., and Ichiyama, A. 1994 *J. Biol. Chem.* 269: 8849-8856.
7. Takada, Y. and Noguchi, T. 1982 *Comp. Biochem. Physiol.* 72B: 597-604.
8. Danpure, C.J., Guttridge, K.M., Fryer, P., Jennings, P.R., Allsop, J., and Purdue, P.E. 1990 *J. Cell Sci.* 97: 669-678.
9. Noguchi, T. and Takada, Y. 1978 *J. Biol. Chem.* 253: 7598- 7600.
10. Danpure, C.J., and Jennings, P.R. 1986 *FEBS Lett.* 201: 20-24.
11. Nishiyama, K., Funai, T., Katafuchi, R., Hattori, F., Onoyama, K., and Ichiyama, A. 1991 *Biochem. Biophys. Res. Commun.* 176: 1093-1099.
12. Nishiyama, K., Funai, T., Yokota, S., and Ichiyama, A. 1993 *J. Cell Biol.* 123: 1237-1248.

17

ENDOPEPTIDASE-24.11 (NEPRILYSIN) AND RELATIVES

Twenty Years On

A. J. Turner, L. J. Murphy, M. S. Medeiros, and K. Barnes

Department of Biochemistry and Molecular Biology
University of Leeds
Leeds LS2 9JT, United Kingdom

INTRODUCTION

The renal and intestinal brush border membranes are an abundant source of peptidase and proteinase activities, most of which are zinc metalloenzymes (see e.g. Kenny *et al.*, 1987). These membranes have proved valuable as starting material for purification of these enzymes as well as for investigating their likely physiological functions. Such studies have been much aided by a range of selective and potent peptidase inhibitors, mainly of microbial origin. In the human renal brush border there is only one identified endopeptidase, an activity now referred to as endopeptidase-24.11 (E-24.11) or neprilysin (EC 3.4.24.11). E-24.11 was first purified to homogeneity and characterised some twenty years ago (Kerr & Kenny, 1974) as an insulin B chain degrading activity, although the physiological role of the enzyme was unknown. In the intervening time, several key substrates for the enzyme have been described, the localization of the enzyme explored in considerable detail and other members of the mammalian E-24.11 gene family have recently been reported.

LOCALIZATION AND FUNCTIONS OF ENDOPEPTIDASE-24.11

E-24.11 is a mammalian cell-surface zinc peptidase of Mr approx. 90 000 which constitutes some 4% of the renal brush border membrane protein. The enzyme was shown to have a specificity similar to the bacterial enzyme thermolysin and, most significantly, was inhibited by the fungal metabolite phosphoramidon (K_i=2nM) (see e.g. Turner, 1987 for review). The primary requirement for peptide hydrolysis was shown to be a large, hydrophobic residue in the P_1' position of the substrate, the enzyme therefore being rather non-specific in its actions. It was one year after the purification of E-24.11 that Hughes and colleagues identified the first two opioid peptides in the brain, [Met]- and [Leu]-enkephalin (Hughes *et al.*, 1975). It was not until 1978 that these two fields of research converged with the report that the opioid peptides were inactivated principally through hydrolysis of the Gly^3-Phe^4 bond by a membrane bound peptidase ('enkephalinase') in brain (Malfroy et al., 1978). The subsequent observation that phosphoramidon inhibited the hydrolysis of this bond led us to infer that brain 'enkephalinase' might be identical with renal endopeptidase (Fulcher et al., 1982). Immunological comparisons of the two activities subsequently confirmed identity (Matsas et al., 1983). The rapid identification of many other peptide neurotransmitters in the brain widened the physiological role for E-24.11, substance P, cholecystokinin and neurotensin being some of the putative substrates for the enzyme in vivo. Immunohistochemical studies in the pig nervous system have shown that E-24.11 is concentrated in the striatonigral pathway, co-localising with substance P and with enkephalin immunoreactivity in striosomes (Barnes et al., 1988a). Immunoperoxidase electron microscopy of the globus pallidus (Barnes et al., 1988b) has revealed E-24.11 to be localised to synaptic terminals as well as to axonal membranes but the enzyme is not present on glial cells, although it is expressed in certain glioma lines (Medeiros et al., 1991). Percoll gradient fractionation of homogenates of pig substantia nigra have suggested that E-24.11 is present both on pre- and post-synaptic membranes where it can efficiently hydrolyse and inactivate peptides released into the synaptic cleft (Barnes et al., 1992). Further ultrastructural studies have strongly suggested that E-24.11 plays a role in the hydrolysis of substance P. Double-labelling studies, using immunogold and immunoperoxidase methods at the electron microscopic level, have established the presence of substance P within some terminals in the substantia nigra that reveal positive staining for E-24.11 at the plasma membrane. This provides direct evidence for co-localisation of a target neuropeptide with its degrading enzyme (Barnes et al., 1993). More recently, E-24.11 has also been found to be present in the hippocampus, localised to the stratum oriens and stratum radiatum where it is postulated to play a role in terminating the actions of the neuropeptide somatostatin (Barnes et al., 1995). There are therefore now numerous examples of E-24.11 functioning at peptidergic synapses in a manner analogous to that of acetylcholinesterase at cholinergic terminals. Parallel research has also revealed important roles for E-24.11 in the periphery especially in the maintenance of cardiovascular homeostasis. Thus, E-24.11 is the primary enzyme involved in initiating the hydrolysis of the potent vasodilator atrial natriuretic peptide (Kenny & Stephenson, 1988). E-24.11 also plays a major role in the metabolism of gastrointestinal peptides, peptide YY, a member of the 'pancreatic polypeptide' family, being a notable example (Medeiros and Turner, 1994). This peptide is subject to post-secretory processing by the serine peptidase dipeptidylpeptidase IV to generate a receptor-selective agonist lacking the N-terminal dipeptide. E-24.11, on the other hand, cleaves the parent peptide into an inactive form. Thus, these two peptidases, both located as ectoenzymes at cell-surfaces, play pivotal roles in the processing and metabolism of peptide YY (Medeiros and Turner, 1994). A similar mechanism appears to operate in the case of the related peptide, neuropeptide Y. E-24.11 does not function exclusively in the inactivation of circulating peptides. For example, it can cleave calcitonin gene-related peptide to generate a tetrapeptide of potent chemotactic activity (Davies et al., 1992). In summary, therefore, twenty years of studies on E-24.11 have revealed it as a broadly distributed and promiscuous enzyme which plays distinct roles at different tissue and cellular locations. Molecular cloning has established E-24.11 as a type II integral membrane protein which exists as a plasma membrane ectoenzyme (Devault et al., 1987). It is identical with the common acute lymphoblastic leukemia antigen (CALLA; CD10) and, in the immune system, is presumed to play a role in the differentiation or maturation of hemopoietic cells by hydrolysis of as yet unidentified regulatory peptides (Turner, 1993). E-24.11 possesses the typical zinc motif, HEXXH, and the critical catalytic residues have been identified by site-directed mutagenesis. Until recently, E-24.11 was regarded as the only mammalian phosphoramidon-sensitive zinc

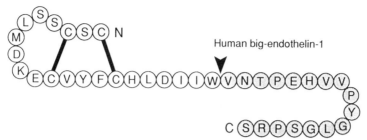

Figure 1. Cleavage of big endothelin-1 to generate mature endothelin-1. The site of hydrolysis catalysed by ECE is indicated by the arrow. The N-terminal portion (open circles) represents ET-1.

endopeptidase. The discovery of the endothelins and the novel processing event involved in their biosynthesis has changed that viewpoint dramatically.

BIOSYNTHESIS OF THE ENDOTHELINS

The endothelins comprise a new family of potent vasoconstrictor peptides which may regulate vascular tone and blood pressure (Masaki & Yanagisawa, 1992). They have been implicated in the pathophysiology of cardiovascular as well as renal and respiratory diseases. Endothelin was originally identified in 1988 as a vasoconstrictor compound in the culture supernatant of porcine aortic endothelial cells and was shown to be a 21-amino acid peptide (Yanagisawa et al., 1988). Subsequently, three distinct endothelin genes have been identified encoding three closely related peptides, endothelins -1, -2 and -3 (ET-1, ET-2, ET-3) (Inoue et al., 1989). Endothelins are synthesized constitutively by vascular endothelial or smooth muscle cells and appear to act as locally produced peptide hormones in a paracrine fashion. Like many biologically active peptides, the endothelins are initially synthesized as a much larger precursor, prepro-ET, of approx 200 amino acid residues (Yanagisawa et al., 1988). After removal of the signal peptide early in biosynthesis the propeptide is cleaved at pairs of basic amino acids to generate the intermediate big ET whose vasoconstrictor activity is approx two orders of magnitude less than that of ET itself. This processing step is probably carried out by furin, a prohormone convertase of the constitutive secretory pathway. The conversion of big ET-1 to ET-1 and its C-terminal fragment occurs via a unique processing event which involves cleavage at the Trp^{21}-Val^{22} bond, catalysed by an activity referred to as endothelin converting enzyme (ECE) (Yanagisawa et al., 1988) (Fig. 1).

The key to identification of a metalloprotease ECE as the physiologically relevant activity was the use of the compound phosphoramidon. Numerous studies have now shown that phosphoramidon at micromolar concentrations is able to inhibit exogenous big ET conversion in vivo, and also in vitro in cultured cells (see e.g. Sawamura et al., 1991; McMahon et al., 1991). The majority of studies have failed to show an effect of other E-24.11 inhibitors (particularly thiorphan) on big ET conversion in vivo or in vitro suggesting that ECE is a phosphoramidon-sensitive enzyme, structurally and catalytically similar to E-24.11, but relatively insensitive to thiorphan.

Is Endopeptidase-24.11 an Endothelin Converting Enzyme?

The specificity of E-24.11, hydrolysing on the amino side of a hydrophobic residue, indicates that it might have the capacity to serve as an endothelin converting activity.

However, E-24.11 degrades ET-1 efficiently (e.g. Sokolovsky *et al.*, 1990) and recombinant E-24.11 was reported to be unable to generate ET-1 from big ET-1 (Abassi *et al.*, 1993). However, we have shown unequivocally that immunoaffinity purified porcine E-24.11 (as well as thermolysin) can catalyse the conversion process (Murphy *et al.*, 1994), although further degradation of the ET-1 product does follow. Thus, care must be taken in the assay of ECE activity in crude preparations since a portion of the detectable activity may be attributable to E-24.11. In lung membrane preparations this can amount to as much as 40% of the total phosphoramidon-sensitive endothelin converting activity unless E-24.11 is inhibited by 10 μM thiorphan (L. J. Murphy and A. J. Turner, unpublished observations).

Purification and Molecular Cloning of Ece

ECE exhibits a number of properties consistent with it being an amphipathic integral transmembrane protein like E-24.11. In particular, the enzyme from porcine lung partitions predominantly into the detergent-phase after phase-separation in Triton X-114 (Murphy LJ and Turner AJ, unpublished observations). Initial purification of ECE to homogeneity was reported from rat lung, from porcine aortic endothelium and from bovine adrenal cortex (Takahashi *et al.*, 1993; Ohnaka *et al.*, 1993; Xu *et al.*, 1994). All three studies report that the purified protein was apparently homogeneous on reducing SDS-PAGE with an M_r in the range 120 000 to 130 000. On sucrose density centrifugation or gel filtration the enzyme appeared to be monomeric. In a further study, Schmidt et al (1994) have purified ECE from a membrane preparation of a permanent bovine endothelial cell line, FBHE. In contrast to the earlier studies, these authors established by SDS-PAGE under reducing and non-reducing conditions that bovine endothelial ECE may exist as a disulphide-linked homodimer, each subunit being of M_r 120 000. The reasons for these discrepancies are, at present, unclear. Several studies have now reported the molecular cloning and sequencing of ECE based on partial amino acid sequence data obtained from the purified enzyme. All predicted protein sequences are very similar apart from some differences in the N-terminal region. Like E-24.11, ECE is a type II integral membrane protein with a short N-terminal cytoplasmic tail, a transmembrane hydrophobic region which represents the uncleaved signal peptide, and a large cytoplasmic domain containing the catalytic site and the motif HEXXH typical

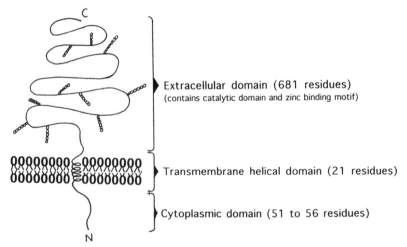

Figure 2. Proposed structure of ECE. The native enzyme may exist as a disulphide-linked dimer.

Table 1. Comparison of the properties of ECE and E-24.11.

Characteristic	E-24.11	ECE
Localisation (major sites)	Kidney, intestine, lymph nodes, (brain)	Endothelial and smooth muscle cells; lung; adrenal cortex, pancreas
Cell compartment	Plasma membrane	Plasma membrane, Golgi (?), secretory vesicles (?)
Specificity	↓ -O-●- (●, hydrophobic)	↓ -Trp- Val-
Physiological substrates	Enkephalins, tachykinins, atrial natriuretic peptides, somatostatin	big ET-1 > big ET-2
pH optimum	7.4	6.8
Subunit Mr	90 000	120 - 130 000
N-linked glycosylation sites	6	10
Inhibition by: EDTA o-phenanthroline phosphoramidon thiorphan	 + + $K_i = 2$ nM $K_i = 2.3$ nM	 + + $K_i = 3.5$ μM $I_{50} > 100$ μM
Sequence around zinc motif (rat enzymes)	-VIGHEITHGF-	-VVGHELTHAF-

of many zinc peptidases (Jongeneel *et al.*, 1989). ECE is a highly glycosylated protein (approx 33% carbohydrate) with 10 putative N-linked glycosylation sites and shows significant homologies, particularly in the C-terminal region, both to E-24.11 and the erythrocyte blood group antigen Kell (Lee *et al.*, 1991). All three proteins also contain a cluster of four conserved cysteine residues in a 32 amino acid sequence immediately following the predicted transmembrane domain, a structural feature also seen in the brush border hydrolases, sucrase-isomaltase and γ-glutamyl transpeptidase. The predicted topology of ECE is shown in Fig. 2.

In addition to the conserved HEXXH motif, some additional essential catalytic residues are conserved between E-24.11 and ECE. Analysis of the tissue distribution of ECE-1 mRNA by Northern blot analysis (Shimada *et al.*, 1994; Xu *et al.*, 1994; Ikura *et al.*, 1994; Schmidt *et al.*, 1994) reveals abundant expression of the protein in lung, pancreas, placenta, adrenal gland, ovary and testis. *In situ* hybridisation analysis has confirmed the presence of ECE mRNA in vascular endothelial cells in various tissues including heart, lung, liver, brain, pancreas, kidney and adrenal (Xu *et al.*, 1994). Studies on ECE from various cell lines and tissues reveal some distinctive features of its specificity. In all cases, the conversion of big endothelin substrate is limited to the production of endothelin and no subsequent degradation occurs. No other biologically active peptides or their precursors have yet been identified as substrates for ECE. There are marked differences reported in the ability of purified, or cloned and expressed, ECE to process the three different big ET isoforms. Most of the cloned and expressed ECE proteins showed a similar specificity, converting big ET-2 poorly and failing to convert big ET-3 (Shimada *et al.*, 1994; Schmidt *et al.*, 1994). Some of the properties of ECE are compared with E-24.11 in Table 1.

CONCLUSIONS AND CONTROVERSIES

The last few years has seen new additions to the E-24.11 family as a result of molecular cloning and sequence comparisons. The Kell erythrocyte blood group antigen may well be a zinc peptidase or proteinase in view of the presence of an HEXXH motif and some other conserved residues, although no functional activity has yet been reported for the protein. On the other hand, ECE has a highly specific biological activity, catalysing exclusively the formation of ET-1 from its precursor big ET-1. Like E-24.11, ECE is a phosphoramidon-sensitive enzyme although it is insensitive to a number of other potent E-24.11 inhibitors. The structural differences between E-24.11 and ECE that produce these major differences in substrate specificity and inhibitor sensitivity remain to be resolved. Some other aspects of ECE structure and function also remain controversial at the present time. The precise subcellular location of ECE remains unclear. Although some of the activity appears to be present on the plasma membrane as an ectoenzyme (Waxman *et al.*, 1994), the presence of an intracellular pool of mature ET-1 suggests that some conversion takes place within the cell, perhaps in the Golgi or in secretory vesicles (Xu *et al.*, 1994). The oligomeric structure of the enzyme also needs to be clarified. Development of specific and potent ECE inhibitors will allow further exploration of the physiological roles of ECE and such compounds may have therapeutic application in cardiovascular disease. Finally, are there yet further members of the mammalian E-24.11/ECE/Kell family to be discovered?

ACKNOWLEDGEMENTS

This work was supported by the British Heart Foundation, The Medical Research Council and The Wellcome Trust.

REFERENCES

Abassi, Z. A., Golomb, E., Bridenbaugh, R.. and Keiser, H. R., 1993, Metabolism of endothelin-1 and big endothelin-1 by recombinant neutral endopeptidase EC 3.4.24.11., *Brit J Pharmacol.*109:1024-1028.

Barnes, K., Matsas, R., Hooper, N. M., Turner, A. J., and Kenny, A. J., 1988a, Endopeptidase-24.11 is striosomally ordered in pig brain and, in contrast to aminopeptidase N and peptidyl dipeptidase A, is a marker for a set of striatal efferent fibres, *Neuroscience* 27:799-817.

Barnes, K., Turner, A. J., and Kenny, A. J., 1988b, Electron microscopic immunocytochemistry of pig brain shows that endopeptidase-24.11 is localised in neuronal membranes, *Neurosci. Lett.,* 94:64-69.

Barnes, K., Turner, A. J., and Kenny, A. J., 1992, Membrane localisation of endopeptidase-24.11 and peptidyl dipeptidase A in the pig brain: a study using subcellular fractionation and electron microscopic immunocytochemistry, *J. Neurochem.* 58:2088-2096.

Barnes, K., Turner, A. J., and Kenny, A. J., 1993, An immunoelectron microscopic study of pig substantia nigra shows co-localisation of endopeptidase-24.11 with substance P, *Neuroscience,* 53:1073-1082.

Barnes, K., Doherty, S., and Turner, A.J., 1995, Endopeptidase-24.11 is the integral membrane peptidase initiating degradation of somatostatin in the hippocampus. *J. Neurochem.,* 64, in press.

Davies, D., Medeiros, M. S., Keen, J. N., Turner, A. J. and Haynes, L. W., 1992, Endopeptidase-24.11 cleaves a chemotactic factor from α-calcitonin gene-related peptide. *Biochem. Pharmacol.* 43:1753-1762.

Devault, A., Lazure, C., Nault, C., Le Moual, H., Seidah, N. G., Chretien, M., Kahn, P., Powell, J., Mallet, J., Beaumont, A., Roques, B. P., Crine, P., and Boileau, G., 1987, Amino acid sequence of rabbit kidney neutral endopeptidase-24.11 (enkephalinase) deduced from a complementary DNA, *EMBO J.,* 6:1317-1322.

Fulcher, I. S., Matsas, R., Turner, A.J., and Kenny, A.J., 1982, Kidney neutral endopeptidase and the hydrolysis of enkephalin by synaptic membranes show similar sensitivity to inhibitors, *Biochem. J.,* 203:519-522.

Hughes, J., Smith, T. W., Kosterlitz, H.W., Fothergill, L.H., Morgan, B. A., and Morris, H. R., 1975, Identification of two related pentapeptidase from the brain with potent opiate agonist activity, *Nature* 258:577-579.

Ikura, T., Sawamura, T., Shiraki, T., Hosokawa, H., Kido, T., Hoshikawa, H., Shimada, K., Tanzawa, K., Kobayashi, S., Miwa, S., and Masaki, T., 1994, cDNA cloning and expression of bovine endothelin converting enzyme, *Biochem Biophys Res Commun.* 203:1417-1422.

Inoue, A., Yanagisawa, M., Kimura, S., Kasuya, Y., Miyauchi, T., Goto, K., and Masaki, T., 1989, The human endothelin family: three structurally and pharmacologically distinct isopeptides predicted by three separate genes, *Proc Natl Acad Sci USA* 86:2863-2867.

Jongeneel, C. V., Bouvier, J., and Bairoch, A. A., 1989, Unique signature identifies a family of zinc-dependent metallopeptidases, *FEBS Lett* 242: 211-214.

Kenny, A. J., and Stephenson, S. L., 1988, Role of endopeptidase-24.11 in the inactivation of atrial natriuretic peptide. *FEBS Lett.*, 232:1-8.

Kenny, A. J., Stephenson, S. L., and Turner, A. J., 1987, Cell surface peptidases. In: *Mammalian ectoenzymes* (Eds. Kenny AJ and Turner AJ), pp. 169-210. Elsevier, Amsterdam.

Kerr, M. A., and Kenny, A. J., 1974, The purification and specificities of a neutral endopeptidase from rabbit kidney brush border. *Biochem. J.,* 137:477-488.

Lee, S., Zambas, E. D., Marsh, W. L., and Redman, C. M., 1991, Molecular cloning and primary structure of Kell blood group protein, *Proc Natl Acad Sci USA* 88: 6353-6357.

Malfroy, B., Swerts, J. P., Guyon, A., Roques, B. P., and Schwartz, J-C., 1978, High-affinity enkephalin degrading peptidase is increased after morphine, *Nature* 276:523-526.

Masaki, T., and Yanagisawa, M., 1992, Endothelins, *Essays Biochem* 27:79-89.

Matsas, R., Fulcher, I. S., Kenny, A. J., and Turner, A. J., 1983, Substance P and[Leu]enkephalin are hydrolyzed by an enzyme in pig caudate synaptic membranes that is identical with the endopeptidase of kidney microvilli, *Proc. Natl. Acad. Sci. U.S.A.* 80:3111-3115.

McMahon, E. G., Palomo, M. A., Moore, W. M., McDonald, J. F., and Stern, M. K., 1991, Phosphoramidon blocks the pressor activity of porcine big endothelin-1-(1-39) in vivo and conversion of big endothelin-1-(1-39) to endothelin-1-(1-21) in vitro, *Proc Natl Acad Sci USA* 88:703-707.

Medeiros, M. S., Balmforth, A. J., Vaughan, P. F. T., and Turner, A. J., 1991, Hydrolysis of atrial and brain natriuretic peptides by the human astrocytoma clone D384 and the neuroblastoma line SHSY-5Y. *Neuroendocrinol.,* 54:295-302.

Medeiros, M. S., and Turner, A. J., 1994, Processing and metabolism of peptide YY: pivotal roles for dipeptidyl peptidase IV, aminopeptidase P and endopeptidase-24.11, *Endocrinology* 134: 2088-2094.

Murphy, L. J., Corder, R., Mallet, A. I., and Turner, A. J., 1994, Generation by the phosphoramidon-sensitive peptidases, endopeptidase-24.11 and thermolysin, of endothelin-1 and C-terminal fragment from big endothelin-1. *Brit J Pharmacol* 113: 137-142.

Ohnaka, K., Takayanagi, R., Nishikawa, M., Haji, M., and Nawata, H., 1993, Purification and characterization of a phosphoramidon-sensitive endothelin-converting enzyme in porcine aortic endothelium, *J Biol Chem* 268:26759-26766.

Sawamura, T., Kasuya, Y., Matsushita, Y., Suzuki, N., Shinmi, O., Kishi, N., Sugita, Y., Yanagisawa, M., Goto, K., Masaki, T., and Kimura, S., 1991, Phosphoramidon inhibits the intracellular conversion of big endothelin-1 to endothelin-1 in cultured endothelial cells, *Biochem Biophys Res Commun* 174:779-784.

Schmidt, M., Kröger, B., Jacob, E., Seulberger, H., Subkowski, T., Otter, R., Meyer, T., Schmalzing. G., and Hillen, H., 1994, Molecular characterization of human and bovine endothelin converting enzyme (ECE-1), *FEBS Lett.* 356:238-243.

Shimada, K., Takahashi, M., and Tanzawa, K., 1994, Cloning and functional expression of endothelin-converting enzyme from rat endothelial cells, *J Biol Chem* 269:18275-18278.

Sokolovsky, M., Galron, R., Kloog, Y., Bdolah, A., Indig, F. E., Blumberg, S., and Fleminger, G., 1990, Endothelins are more sensitive than sarafotoxins to neutral endopeptidase: possible physiological significance, *Proc Natl Acad Sci USA* 87: 4702-4706.

Takahashi, M., Matsushita, Y., Iijima, Y., and Tanzawa, K., 1993, Purification and characterization of endothelin-converting enzyme from rat lung, *J Biol Chem* 268: 21394-21398.

Turner, A. J., 1987, Endopeptidase-24.11 and neuropeptide metabolism. In: *Neuropeptides and their peptidases* (Ed. Turner AJ), pp. 183-201. Ellis Horwood, Chichester.

Turner, A. J., 1993, Membrane peptidases of the nervous and immune systems, *Adv Neuroimmunol* 3:163-170.

Waxman, L., Doshi, K. P., Gaul, S. L., Wang, S., Bednar, R. A. and Stern, A. M., 1994, Identification and characterization of endothelin converting activity from EAHY 926 cells: Evidence for the physiologically relevant human enzyme, *Arch Biochem Biophys* 308:240-253.

Xu, D., Emoto, N., Giaid, A., Slaughter, C., Kaw, S., deWit, D., and Yanagisawa, M., 1994, ECE-1: A membrane-bound metalloprotease that catalyzes the proteolytic activation of big endothelin-1. *Cell* 78:473-485.

Yanagisawa, M., Kurihara, H., Kimura, S., Tomobe, Y., Kobayishi, Y., Mitsui, Y., Yazaki, Y., Goto, K. and Masaki, T., 1988, A novel potent, vasoconstrictor peptide produced by vascular endothelial cells, *Nature* 332:411-415.

18

FUNCTION OF CALPAINS

Possible Involvement in Myoblast Fusion

M. Hayashi,[1] M. Inomata,[1] and S. Kawashima[2]

[1] Tokyo Metropolitan Institute of Gerotology
35-2 Sakae-cho, Itabashi-ku
Tokyo 173, Japan
[2] Tokyo Metropolitan Institute of Medical Science
3-18-22 Honkomagome, Bunkyo-ku
Tokyo 113, Japan

INTRODUCTION

Calpains are major intracellular proteases that are activated by Ca^{2+}, an important modulator of cell function. Although their structures and enzymatic properties are well characterized (1, 2) and new tissue-specific calpains are being found successively (3), their physiological functions are not elucidated. From many studies, it seems likely now that the major locus of calpain activation is just beneath the cytoplasmic membrane, and that cytoskeletal proteins such as membrane lining proteins are physiological substrates of calpains (4). Since alterations of membrane lining proteins increase the membrane fluidity and disrupt the asymmetry of membrane phospholipids (5, 6) that is prerequisite for membrane fusion, we examined the possibility that calpains trigger membrane fusion through degradation of cytoskeletal proteins. So far, we have shown that A2C/Ca^{2+}-induced erythrocyte fusion was accompanied by activation of intracellular m-calpain and degradation of spectrin, and that inhibition of intracellular m-calpain by a cell-permeable calpain inhibitor, benzyloxycarbonyl-leucyl-leucyl-leucinal (Z-Leu-Leu-Leu-al) which had been developed in our laboratory (7), resulted in blockage of erythrocyte fusion (8). In addition to this, calpain localizes to coated vesicles, which are products of Ca^{2+}-dependent intracellular membrane fusion, i.e. endocytosis, on stimulation of proerythroblastic K562 cells by phorbol ester (9). In this paper, using several calpain inhibitors, we show the possible involvement of calpain in myoblast fusion to multinucleated myotubes, a physiological cell fusion occuring in muscle differentiation.

Intracellular Protein Catabolism, Edited by Koichi Suzuki and Judith Bond
Plenum Press, New York, 1996

MATERIALS AND METHODS

Materials

The C2C12 mouse skeletal muscle cells, a subclone of C2 line which was isolated by Yaffe and Saxel (10) and established by Blau *et al.* (11), was purchased from ATCC Rockville, MD. Leupeptin and Calpain Inhibitor-1 (acetyl-leucyl-leucyl-norleucinal) were obtained from Peptide Institute, Inc., and Boehringer-Mannheim, respectively. E-64c and E-64d were kind gifts from Taisho Pharmaceutical Co. Benzyloxycarbonyl-leucyl-leucyl-leucinal (Z-Leu-Leu-Leu-al) and calpeptin (benzyloxycarbonyl-leucyl-norleucinal) were synthesized as described before (12). Dulbecco's Modified Eagle Medium (D-MEM), fetal bovine serum (FBS) and horse serum were products of Gibco Laboratories. Insulin was purchased from Collaborative Research Inc.

Cell culture and differentiation

C2C12 cells were maintained in D-MEM containing 10% FBS at 37°C in humidified 5% CO_2-95% air mixture. Confluent cells were induced to differentiate by replacing the media with D-MEM containing 1% horse serum and 5 mg/ml insulin (differentiation medium). To examine the effects of protease inhibition, C2C12 cells were cultured on LAB-TEK chamber slides (Nunc) and induced to differentiate in the differentiation medium containing various concentrations of protease inhibitors. Cells were fixed after 48 hr with ethanol, stained with hematoxylin-eosin Y (Muto Chemicals), and photographed under a light microscope.

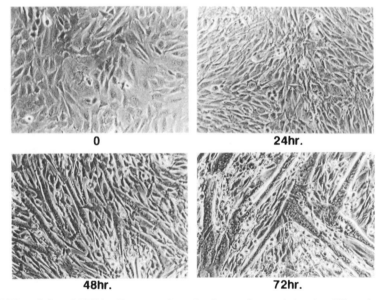

Figure 1. Differentiation of C2C12 cells to myotubes. Confluent cells were induced to differentiate at 0 hr by replacing the media with the differentiation medium.

RESULTS

On induction of C2C12 cells to myotubes, fusion of cells began after 24 hr, continued at 48 hr, and reached the maximum at 72 hr but some cellls started to detach from the substrate at this time (Fig. 1). Therefore, 48 hr was chosen to examine the effects of protease inhibitors on myoblast fusion.

As seen in Fig. 2(A), E-64d, a cell-permeable derivative of E-64, and calpeptin were slightly effective in inhibiting myoblast fusion at a high concentration of 100 mM. Calpain Inhibitor-1, acetyl-leucyl-leucyl-norleucinal, was effective at 1 mM concentration but toxic at higher concentrations. The most effective and the least cytotoxic one was Z-Leu-Leu-Leu-al.; the inhibitor blocked myoblast fusion completely at a concentration as low as 0.05 mM (50 nM) and was not toxic at higher concentrations.

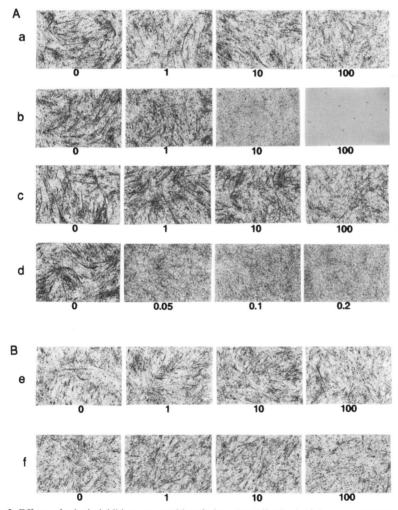

Figure 2. Effects of calpain inhibitors on myoblast fusion. (A) Effective inhibitors. (a), E-64d; (b), Calpain Inhibitor-1; (c), calpeptin; (d), Z-Leu-Leu-Leu-al. (B) Non-effective inhibitors. (e), E-64c; (f), leupeptin. Numbers below are concentrations used (mM).

On the contrary, E-64c and leupeptin, both of which are potent inhibitors of calpains *in vitro*, were not effective at all in blocking myoblast fusion and the latter was cytotoxic at high concentrations. This suggests that these inhibitors are poor in cell-permeability and cannot inhibit intracellular calpains. Leupeptin probably exerts its effects on lysosomal cathepsins rather than on calpains.

DISCUSSION

The results described above strongly suggest that calpain plays a crucial role in myoblast fusion to myotubes. Using this sytem, it was also possible to evaluate the effectiveness of so-called calpain inhibitors on intracellular calpains. The compounds examined here are all potent inhibitors of calpains *in vitro*. Using a caseinolytic assay, the ID50 values (concentrations required for 50% inhibition) were 1 mM for leupeptin (2), 3 mM for E-64c (2), 0.05 mM for Calpain Inhibitor-1 (13), 0.04 mM for calpeptin (14), and 0.1 mM for Z-Leu-Leu-Leu-al (Kawashima, unpublished result). The most potent inhibitor in blocking myoblast fusion was Z-Leu-Leu-Leu-al, although the inhibition of calpains *in vitro* is moderate among the inhibitors examined. The inhibitor was also shown to be the

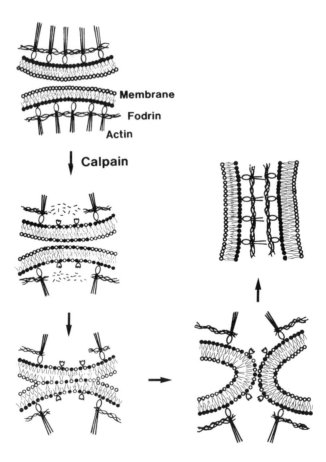

Figure 3. Facilitation scheme of membrane fusion through degradation of membrane lining proteins by calpain.

most effective among a series of similar inhibitors in inhibiting the autolytic activation of intracellular calpain of erythrocytes (7) and to block the Sendai-virus-induced cell fusion very efficiently (Taira *et al.*, unpublished result). Thus, as a calpain inhibitor to be used exogenously to inhibit intracellular calpains, high cell permeability and low toxicity are required in addition to inhibitory potency. In this respect, Z-Leu-Leu-Leu-al is the best compound as far as we know.

Immunohistochemical and biochemical localization of calpains in cells by us and others also suggest its involvement in membrane fusion: localization of calpains in coated vesicles, a vehicle for membrane trafficking (Sato *et al.*, unpublished result), on the cytoplasmic membrane of fusible myoblasts (15), and in the sperm acrosome (16), a high content of m-calpain in megakaryocytes which produce platelets by pinching off cytoplasm (17), and disappearance of m-calpain during enucleation of mammalian erythrocytes (Kawashima et al, unpublished result). It was reported that an anti-calpain antibody inhibits the exocytosis of granule content from neutrophils (18). Since calpain digests the N-terminal portion of lipocortin I in a limited manner and makes this Ca^{2+}-dependent phospholipid-binding protein more Ca^{2+}-sensitive (19), calpain could possibly function in exocytosis through limited proteolysis of annexins.

Thus, the accumulated data suggest that calpains are involved in intercellular and intracellular membrane fusions. Illustrated in Fig. 3 is a scheme in which calpain degrades membrane lining proteins, after which the membranes lose their phospholipid assymetry and become more fusible with each other. The myoblast is a cell in which phospholipid assymetry is not preserved well (20) suggesting facilitation of the coming fusion to myotubes. However, it is not known if loss of membrane assymetry in just one of the fusing cells is sufficient, or if it is required in both counteparts.

REFERENCES

1. Suzuki,K., 1990, The structure of calpains and calpain gene. In "Intracellular Calcium-Dependent Proteolysis" (Mellgren,R.L., and Murachi, T., eds.), pp25-35, CRC Press, Boca Raton.

2. Inomata,M., Nomoto,M., Hayashi,M., Nakamura,M., Imahori, K., and Kawashima,S., 1984, Comparison of low and high calcium requiring forms of the calcium-activated neutral protease (CANP) from rabbit skeletal muscle. *J. Biochem.* 95:1661-1670.

3. Sorimachi,H., Saido,T.C., and Suzuki,K., 1994, New era of calpain research- Discovery of tissue-specific calpains. *FEBS Lett.* 343:1-5.

4. Inomata,M., Hayashi,M., Nakamura,M., Saito,Y., and Kawashima,S., 1989, Properties of erythrocyte membrane binding and autolytic activation of calcium-activated neutral protease. *J. Biol. Chem.* 264:18838-18843.

5. Haest,C.W., Plasa,G., Kamp,D., and Deuticke,B., 1978, Spectrin as a stabilizer of the phospholipid asymmetry in the human erythrocyte membrane. *Biochim. Biophys. Acta* 509:21-32.

6. Williamson,P., Bateman,J., Kozarsky,K., Mattocks,K., Hermanowicz,N., Choe,H.-R., and Schlegel,R.A., 1982, Involvement of spectrin in the maintenance of phase-state asymmetry in the erythrocyte membrane. *Cell* 30:725-733.

7. Hayashi,M., Inomata,M., Saito,Y., Ito,H., and Kawashima,S., 1991, Activation of intracellular calcium-activated neutral proteinase in erythrocytes and its inhibition by exogenously added inhibitors. *Biochim. Biophys. Acta*, 1094:249-256.

8. Hayashi,M., Saito,Y., and Kawashima,S., 1992, Calpain activation is essential for membrane fusion of erythrocytes in the presence of exogenous Ca^{2+}. *Biochem. Biophys. Res. Commun.* 182: 939-946.

9. Nakamura,M., Mori,M., Morishita,Y., Mori,S., and Kawashima, S., 1992, Specific increase in calcium-activated neutral protease with low calcium sensitivity (m-calpain) in proerythroblastic K562 cell line cells induced to differentiation by phorbol 12-myristate 13-acetate. *Exp. Cell Res.* 200:513-522.

10. Yaffe,D., and Saxel,O., 1977, Serial passaging and differentiation of myogenic cells isolated from dystrophic mouse muscle. *Nature*, 270:725-727.

11. Blau,H.M., Chiu,C.-H., and Webster,C., 1983, Cytoplasmic activation of human nuclear genes in stable heterocaryons. *Cell* 32:1171-1180.

12. Ito,A., Takahashi,R., Miura,C., and Baba,Y., 1975, Synthesis study of peptide aldehyde. *Chem. Pharm. Bull.* 23:3106-3113.

13. Kajikawa,Y., Tsujinaka,T., Sakon,M., Kambayashi,J., Ohshiro, T., Murachi,T., and Mori,T., 1987, Elucidation of calpain dependent phosphorylation of myosin light chain in human platelets. *Biochem. Int.* 15:935-944.

14. Tsujinaka,T., Kajikawa,Y., Kambayashi,J., Sakon,M., Higuchi, N., Tanaka,T., and Mori,T., 1988, Synthesis of a new cell penetrating calpain inhibitor (calpeptin). *Biochem. Biophys. Res. Commun.* 153:1201-1208.

15. Schollmeyer,J.E., 1986, Possible role of calpain I and calpain II in differentiating muscle. *Exp. Cell Res.* 163:413-422.

16. Schollmeyer,J.E., 1986, Identification of calpain II in porcine sperm. *Biol. Reprod.* 34:721-731.

17. Nakamura,M., Mori,M., Nakazawa,S., Tange,T., Hayashi,M., Saito,Y., and Kawashima,S., 1992, Replacement of m-calpain by m-calpain during maturation of megakaryocytes and possible involvement in platelet formation. *Thrombos. Res.* 66:757-764.

18. Pontremoli,S., Melloni,E., Daminani,G., Salamino,F., Sparatore, B., Michetti,M., and Horecker,B.L., 1988, Effects of a monoclonal anti-calpain antibody on responses of stimulated human neutrophils. *J. Biol. Chem.* 263:1915-1919.

19. Ando,Y., Imamura,S., Hong,Y.-M., Owada,M.K., Kakunaga,T., and Kannagi,R., 1989, Enhancement of calcium sensitivity of lipocortin I in phospholipid binding induced by limited proteolysis and phosphorylation at the amino terminus as analyzed by phospholipid affinity column chromatography. *J. Biol. Chem.* 264:6948-6955.

20. Sessions,A., and Horwitz,A.F., 1981, Myoblast amino-phospholipid asymmetry differs from that of fibroblasts. *FEBS Lett.* 134:75-78.

19

CLEAVAGE OF THE HUMAN C5A RECEPTOR BY PROTEINASES DERIVED FROM *PORPHYROMONAS GINGIVALIS*

Cleavage of Leukocyte C5a Receptor

Mark A. Jagels,[1] Julia A. Ember,[1] James Travis,[2] Jan Potempa,[3] Robert Pike,[2] and Tony E. Hugli[1]

[1] Department of Immunology
The Scripps Research Institute
La Jolla, California 92037
[2] Department of Biochemistry
The University of Georgia
Athens, Georgia 30602
[3] Department of Microbiology and Immunology
Institute of Molecular Biology, Jagiellonian University
Krakow, Poland

ABSTRACT/SUMMARY

The anaerobic bacteria *P. gingivalis* has been implicated as a primary causative agent in adult periodontitis. Several proteinases are produced by this bacteria and it is suggested that they contribute to virulence and to local tissue injury resulting from infection by *P. gingivalis*. Collagenases and cysteine proteinases (i.e., **the gingipains**)[*] have been characterized as the predominant vesicular enzymes produced by this bacterium. It has been shown that an arginine-specific cysteine proteinase from *P. gingivalis*, called gingipain-1 or Arg-gingipain, can selectively cleave complement components C3 and C5. In the case of C5, cleavage by Arg-gingipain results in the generation of **C5a**, a potent chemotactic factor for PMNs. Since these bacterial proteinases are capable of generating pro-inflammatory factors at sites of infection, we examined the possibility that gingipains or other proteinases from

---

[*] Abbreviations: *P. gingivalis*, *Porphyromonas gingivalis;* Arg-gingipain (gingipain-1), a cysteinyl proteinase from *P. gingivalis* that is specific for Arg-X bonds; Lys-gingipain (gingipain-2), a cysteinyl proteinase from *P. gingivalis* that is specific for Lys-X bonds; PMA, Phorbol-myristate-acetate; PMN, polymorphonuclear leukocytes; C3 and C5, blood complement components; C3a and C5a, bioactive fragments derived from C3 and C5, respectively, and MOPS, 3-(N-Morpholino)propanesulfonic acid; AP, adult periodontitis.

this bacterium might attack or destroy cell surface proteins, such as receptor molecules. Using an affinity-purified rabbit antibody raised against residues 9-29 of the C5a **receptor** (i.e., C5aR; CD88), the **signal transmitting element** for the pro-inflammatory mediator C5a, we demonstrated that the mixture of proteinases in *P. gingivalis* vesicles cleaves the C5a **receptor** on human neutrophils. This vesicular proteinase activity did not require cysteine activation which indicates that proteinases other than the gingipains may be responsible for cleavage of the C5aR molecule. In addition, the purified Lys-gingipain, but not Arg-gingipain, also cleaved C5aR on the human neutrophils. The N-terminal region of C5aR (residues 9-29, PDYGHYDDKDTLDLNTPVDKT) was readily cleaved by chymotrypsin, but not by trypsin, despite the presence of potential trypsin (i.e., lysyl-X) cleavage sites. The specific sites of C5aR 9-29 peptide cleavage were determined by mass spectroscopy for both chymotrypsin and Lys-gingipain. These studies suggest that the proteolytic activity in the bacterial vesicles that is responsible for cleaving C5aR is primarily a non-tryptic proteinase, distinct from either Arg- or Lys-gingipain. Consequently, there appear to be additional proteinase(s) in the vesicles that attacks the cell surface molecule C5aR which are not the same (i.e., Arg- and Lys-gingipain) as were shown to generate pro-inflammatory activity from complement components C3 and C5. Evidence that the proteinases which attack the inflammatory precursor molecules (i.e., C3 and C5) exhibit different specificities than those that attack receptors to these bioactive complement products makes a particularly interesting story of how this bacteria avoids major host defense mechanisms.

It is well known that generation of pro-inflammatory factors such as C3a and C5a at extra-vascular sites can promote edema, leukocyte recruitment and cellular activation responses that could lead to the release of toxic oxygen products and to phagocytosis of the bacteria. Destruction of receptors to these cellular activating factors generated by bacterial proteinases may eliminate the ability of these (i.e., complement-derived) and other mediators to carry out their anti-bacterial actions and thereby limit the host's defense mechanisms in responses to the infecting bacteria. The concept of anti-bacterial responses (i.e., oxygen radical generation and phagocytosis) being effectively eliminated at the injury site, by bacterial proteinases acting at the cellular receptor level, has not been studied in detail. In this case, the situation is particularly unusual because, once the bacterial gingipains generate potent plasma-derived inflammatory factors that can enhance edema and deliver essential nutrients to the bacteria, other bacterial proteinases may destroy their cellular receptors. These receptors transmit the signal activation mechanisms in the infiltrating cells that elicit bacterial killing. It is this series of events which might explain the ability of these anaerobes to persist and flourish in gingival tissue.

INTRODUCTION

Severe adult periodontitis is characterized by an acute inflammatory response with excessive granulocyte infiltration into the gingival tissue which become a chronic disease typified by extensive tissue degradation and eventual bone erosion and loss of teeth (White and Mayrand, 1981). Adult periodontitis (AP) affects 36% of Americans over the age of 19 and severity of the disease increases significantly with age (Mayrand and Holt, 1988). About half of all Americans have lost their teeth to AP by the age of 65 and approximately 10-20% of AP patients are refractory to present day treatment including antibiotics. In third world countries, where poor nutrition is a major contributing factor, AP is even more pervasive and tooth loss is the norm for adults over 30 years of age. It is agreed that gram negative anaerobes are mainly responsible for AP and that the bacterium ***Porphyromonas gingivalis*** is a major pathogen in this form of oral disease. Gingival crevicular fluid accumulates in periodontitis as gingival tissue degradation progresses at the foci of infection. Initially, the bacterial proteinases promote collagen and connective tissue degradation and induce granulocytic infiltration (White and Mayrand, 1981). Granular enzymes from invading PMNs also contribute to local gingival injury as they are recruited to the site of inflammation by bacterial chemotactic factors and by newly generated humoral factors and host cell chemokines. As the tissue injury advances, edema fluid increases and plasma proteins including complement components are delivered to the injury site for processing by the bacterial enzymes. Hence, it is bacterial proteinases that appear to play a central role in initiating and sustaining the progressively advancing tissue injury so characteristic of AP.

A number of granular enzymes from *P. gingivalis* have been examined and several of the cysteine proteinases have been fully characterized. Two granular proteinases exhibiting trypsin-like specificities have been identified and isolated. These proteinases are cysteine proteinases, functionally but not structurally similar to the calpains, and have consequently been called gingipains. The arginine-specific proteinase or Arg-gingipain (gingipain-1) was characterized by Chen *et al.* (1992) and the lysine-specific gingipain or Lys-gingipain (gingipain-2) was isolated and characterized by Pike *et al.* (1994). Previous studies have shown that Arg-gingipain will cleave both C3 and C5 to generate C3a- and C5a-like fragments (Wingrove *et al.*, 1992). In the case of C3, Arg-gingipain selectively cleaves the alpha chain, at or near the C3a/C3b bond junction (i.e., residues 77-78, Arg-Thr), reducing the size of the digested alpha chain by approximately 10,000 Da. Further degradation of the alpha' chain occurs versus time with little evidence of beta chain degradation when 1% (w/w) levels of the enzyme are used. There was no evidence of C3a activity in the C3 digests indicating that cleavage may have either occurred N-terminal to the C3a/C3b scission site at Arg 77 and/or any C3a generated was rapidly degraded by the enzyme. When C3a was incubated with 1% Arg-gingipain (w/w), extensive degradation was evident versus time, especially when glycyl-glycine was added to enhance enzyme activity. Since degradation of C3a was relatively slow at 1% enzyme levels, we speculate that no C3a activity was found in the C3 digest because these enzymes prefer to cleave the Arg-Ala bond at residue positions 69-70 in the C3 alpha chain, rather than at Arg 77. Cleavage at Arg 69 would release a C3a-like fragment having no biologic activity.

The story was quite different for C5 cleavage by Arg-gingipain. When C5 was incubated with 1-4% (w/w) of enzyme the alpha chain of C5 was not cleaved initially at the C5a/C5b bond junction, but rather at an alternative internal site. The first alpha chain scission site was Arg-Gly at positions 715-716, a known plasmin cleavage site in C5, generating an 86 and a 30 kDa fragment of the alpha chain. There was no evidence of degradation of the more stable beta chain. As degradation continued, we obtained functional evidence that C5a activity was generated which indicated that the C5a/C5b scission site between residues 74-75 (Arg-Gly bond) in the C5 alpha chain was cleaved. Direct incubation of the C5a with Arg-gingipain indicated that this fragment of C5 was relatively resistant to further degradation by the enzyme. These results provided evidence that at least one of the bacterial granular enzymes could generate chemotactic activity (i.e., C5a) from plasma proteins in the crevicular edema fluid. The implication of these results tends to show that infection by *P. gingivalis* can lead to proteolytic events that generate fragments from plasma components capable of recruiting granulocytes and monocytes to the site of tissue injury and thus promote or amplify the inflammatory response. Generation of C5a by *P. gingivalis* proteinases appears to confirm the hypothesis that these bacterial enzymes actually participate directly in promoting the cellular inflammatory response so characteristic of AP.

More recent studies using Lys-gingipain have shown degradation patterns with C3 and C5 similar to those observed with Arg-gingipain. As with Arg-gingipain, there was no C3a activity generated by the lysine-specific enzyme. This result was predicted according to the known specificity of the lysyl proteinase which should not cleave C-terminal to the essential Arg-77 in C5a. However, C5 cleavage did result in low levels of C5a-like activity. This was explained by the fact that C5 alpha chain fragments that were cleaved C-terminal to Arg 74 of C5a have been shown to generate C5a-like activity (Wetzel and Kolb, 1982). When C5 molecules were first oxidized by H_2O_2, much as they could be in the presence of activated PMNs, the C5a-like activity generated by Lys-gingipain was markedly enhanced (DiScipio et al.,1995). These results indicated that plasma proteins modified by oxidation, such as in the presence of activated PMNs, actually released enhanced quantities of inflammatory mediator when digested by the bacterial enzymes.

In the study presented here, we asked whether the *P. gingivalis* enzymes could also attack receptor molecules (i.e., signal transduction molecules) expressed on the surface of infiltrating cells. If the bacterial proteinases can degrade these molecules on the cell's surface, then they have the capacity to alter or eliminate responses to activating factors that are potentially cellular functions involved in bacterial killing.

RESULTS AND DISCUSSION

The key reagent required in performing the following experiments was access to an antibody that can detect a cellular receptor and can discern when the receptor is degraded. Advantage was taken of an anti- C5a receptor antibody that was raised to a synthetic fragment of C5aR (i.e. residues 9-29) based on the N-terminal sequence of the neutrophil C5a receptor (C5aR). This particular fragment was chosen because it represents a known extracellular region of the receptor (see Figure 1). There are seven transmembrane regions in the C5a receptor with four extra-cellular regions directly exposed to the cell surface. The antibody that was raised to C5aR 9-29, a synthetic peptide, not only binds to the receptor on the intact PMN but also blocks C5a binding and inhibits C5a-induced cellular responses (Morgan et al., 1993). These results indicated that the N-terminal region of the receptor was directly involved in the ligand/receptor interaction. These important characteristics of the antibody can be exploited using the reagent as a tool for monitoring proteolytic attack on this region of the receptor. Consequently, we were able to evaluate proteolytic effects of the *P. gingivalis* proteinases on cell surface molecules. An initial series of experiments were designed to establish that this region of the C5a receptor was susceptible to proteolysis.

Since the fragment C5aR 9-29, to which the antibody binds, contained both potential trypsin and chymotrypsin cleavage sites (see Figure 2), it was expected that either trypsin or chymotrypsin could degrade the receptor and destroy the antibody binding site. In fact, we found that the C5aR on PMNs was relatively resistant to trypsin, but was readily cleaved in the N-terminal region by chymotrypsin (see Figure 3), based on our analysis by flow cytometry. This result might be explained because the potential trypsin cleavage site in the (N-terminal) TDDK-↓-DTLD region is surrounded by anionic aspartyl side chains and the site located at positions 28-29 (K-↓-T) may already be protected by proximity to, or insertion in, the membrane. In any case, it was the chymotrypsin cleavage sites that were readily attacked (in the receptor) and cleavage at positions 14-15 (Y-↓-D) and 20-21 (L-↓-D) were preferred over the other potential sites at 12-13 (Y—G) and 22-23 (L—N), based on our mass spectral analysis (see Figure 4). When a similar experiment was carried out using mixed proteinases from the *P. gingivalis,* we observed that there were proteases in the bacterial vesicles that attacked the C5aR on PMNs. Exposure of PMNs to the vesicles eliminated binding of the anti-receptor antibody to the N-terminal region of C5aR based on our flow cytometry data (Figure 5). Digestion of the C5aR by the vesicle enzymes did not require cysteine (i.e., reduction) for activation, suggesting that the primary proteinases responsible for this degradation are different than the Arg- or Lys-gingipains. Like trypsin, the isolated

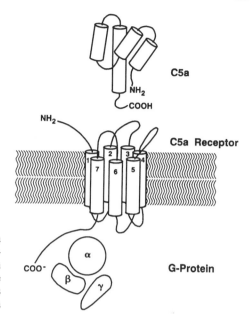

Figure 1. A schematic representation of the C5a receptor with the seven transmembrane regions illustrated as helices. The extracellular N-terminal region (1-29) and the extracellular loops connecting the transmembrane regions offer the most likely regions of the receptor molecule to interacts with the C5a ligand.

arginine-specific proteinase (gingipain-1) failed to destroy the receptor, but the lysine-specific proteinase (gingipain-2) did degrade the N-terminal region of the receptor on PMNs (see Figure 5). We concluded that selectivity of Lys-gingipain for the Lys-X bonds in C5aR 9-29 must be distinctly different from that of trypsin which failed to cleave the N-terminal region of the receptor. When the C5aR 9-29 peptide is digested by Lys-gingipain, we found evidence of cleavage at both positions 17-18 (K-↓-D) and 28-29 (K-↓-T) based on mass spectral analysis (Figure 6).

These results indicate that this bacteria elaborates proteinases capable of inactivating cell surface molecules whose role it is to transmit activation signals to the cell. The implications of such a process suggests that signal mechanisms, elicited from invading cells by pro-inflammatory mediators, can be destroyed by the bacterial proteinases. The net effect may be a decrease cell response in activation of phagocytic

SEQUENCE OF THE N-TERMINAL PORTION OF THE C5a RECEPTOR
PEPTIDE (RESIDUES 9-29) USED AS THE ANTIGEN[a]

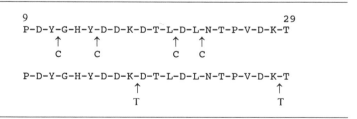

[a] The potential chymotrypsin (C) and trypsin (T) cleavage sites have been identified in the synthetic C5aR peptide.

Figure 2. The sequence of C5aR 9-29 is shown with the potential cleavage sites for chymotrypsin-like and trypsin-like proteinases indicated (C- chymotrypsin-like and T - trypsin-like).

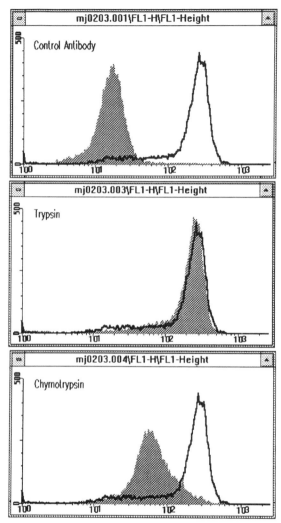

Figure 3. Flow cytometric analysis of PMNs indicates that binding of the anti-C5aR 9-29 antibody to the cell surface can be disrupted by proteolytic attack on the receptor. Purified human PMNs were treated with trypsin or chymotrypsin (0.5 mg/ml) in MOPS-buffered Earl's medium for 30 min at pH 7.4 and 37° C. Cells were then incubated either with non-immune rabbit IgG (shaded area in top panel) or with 5 μg/ml anti-C5aR (9-29) (receptor-positive cells are indicated by right-shifted histogram in top panel) for 45 min at 4° C. All cells were washed and incubated with FITC-labeled goat anti-rabbit IgG prior to analysis. The shaded histograms in the middle and bottom panels indicate fluorescence of proteinase-treated PMN in comparison with the untreated cells (non-shaded histograms). Exposure of these cells to trypsin (middle panel) had little, if any, effect on the receptor (i.e., antibody binding). Cells exposed to chymotrypsin (bottom panel) loose their ability to bind anti-receptor antibody indicating that the N-terminal region of the receptor has undergone extensive degradation.

functions and lower local levels of cell-derived toxic products (i.e., such as oxygen radicals) that would help contain bacterial growth. We examined the effects of proteolysis on C5a-induced PMN activation. C5a-induced enzyme release (i.e., myeloperoxidase) was decreased or eliminated by gingipain-2 and by proteinases from the isolated vesicles (Figure 7). No spontaneous release was observed by proteinase treat-

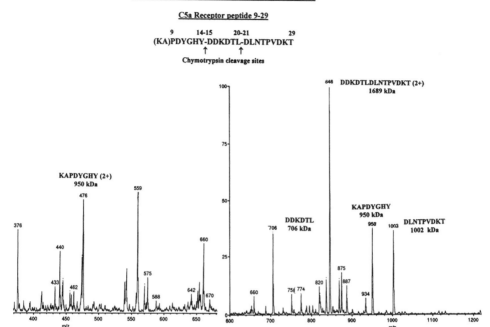

Figure 4. Mass spectral analysis of the chymotrypsin digest of C5aR 9-29. The C5aR 9-29 peptide was digested with chymotrypsin (1:50 w/w) for 3° min at 37° C. The digest was then analyzed by ion-spray mass spectrometry. The fragments KAPDYGHY, DDKDTL and DLNTPVKT were identified indicating that the enzyme had cleaved only the Y-D (14-↓-15) and L-D (20-↓-21) bonds in the receptor peptide.

ment alone. Since proteolysis of cell surface receptors can affect other receptors or pathways of activation, such as PMN stimulation via cleavage of the thrombin receptor (Vu *et al.*, 1991 a and b), it was important to show that proteolysis alone does not cause the cells to be activated. We now know that there are *P. gingivalis* proteinases that have an ability to destroy surface receptors on granulocytes and presumably numerous signal proteins on the surface of other cells in the environment, but we do not yet know whether this process leads to increased or decreased cell-mediated tissue injury or serves as a means of self-protection for the bacterial to avoid phagocytic attack.

Another aspect of this study was to explore the effects of these bacterial proteinases on the surrounding infected tissue. We examined cultured oral epithelial cells for C5a receptors and found evidence of a surface molecule on these cells that binds the anti-C5aR 9-29 antibody (data not shown). Further evidence of a reactive surface molecule on epithelial cells was obtained by immunostaining gingival tissue thin slices with the antibody. The epithelial layer appears to be differentially stained by the immunoreagent. When the cultured cells are treated with the *P. gingivalis* proteinases the anti-C5aR binding protein is destroyed by the vesicle proteinases and by trypsin. We remain cautious in calling this epithelial cell molecule a C5a receptor. However, there is a protein on the epithelial cell surface that is detected by a specific anti-C5aR reagent and that is susceptible to bacterial proteinase degradation. If this molecule, like the C5a receptor on PMNs, represents a signal receptor for transmitting cellular activation events, then the sensory mechanism of this resident tissue

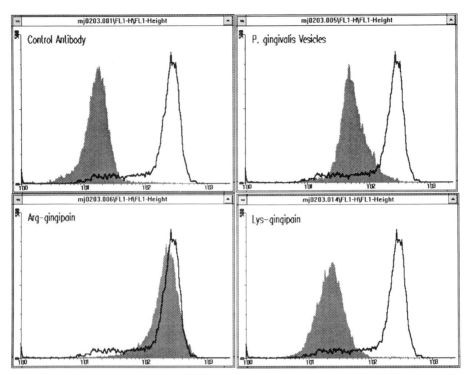

Figure 5. Flow cytometry analysis of human PMNs after exposure to gingivalis vesicle enzymes, Arg-gingipain or Lys-gingipain. Cells were incubated in the presence of *P. gingivalis* vesicles (strain W50; 100 μg protein/ml), or either cysteine-activated Arg-gingipain (100 μg/ml) or Lys-gingipain (50 μg/ml), for 60 min. at 37° C. The C5a receptor on these cells was monitored using the anti-C5aR 9-29 antibody as described in Fig. 3. Only the Arg-gingipain failed to degrade the C5a receptor on PMNs.

cell may also be eliminated by the bacteria at the tissue injury site. Receptor degradation renders these cells inoperative for responding to their environment to either control or promote the ongoing inflammatory process. It is clear that we need to know more about what functions are lost or altered by proteolysis of surface molecules, but these results tell us that selective proteolytic attack of surface receptors may significantly modify host defense responses from both the infiltrating and resident cells.

Application of investigative tools such as anti-receptor antibodies that can detect the integrity of molecules on cells, and whose role it is to sense signals from the environment, can provide valuable information concerning the responsiveness of cells or tissues to these hormone-like activators. The role of bacterial proteinases in promoting tissue inflammatory responses, by perhaps altering normal defense mechanisms in the host, needs to be more carefully examined at the effector/receptor level. Proteolysis of exposed regions of cellular receptor molecules can lead to an inability of tissue to respond to or resist injury, and could consequently limit the normal repair processes. The results reported here indicate that an entire class of receptors, known as the Rhodopsin family of G protein-associated receptors (Savarese and Fraser, 1992), may be particularly susceptible to limited digestion by bacterial proteinases. These bacterial proteinases may render signal activation molecules on host cell surfaces impotent and thereby disarm important anti-bacterial mechanisms and other protective functions of these cells designed purposefully to prevent further damage to the tissue.

MASS SPECTRAL ANALYSIS OF C5aR PEPTIDES

C5a Receptor Peptide 9-29

17-18   28-29
(KA)PDYGHYDDK-DTLDLNTPVDK-T 2621 kD
↑   ↑
Lys-Gingipain cleavage sites

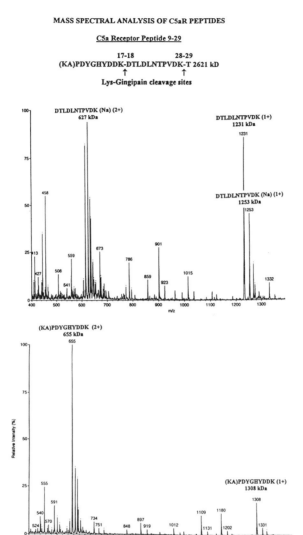

Figure 6. Mass spectral analysis of the Lys-gingipain digest of C5aR 9-29. The C5aR 9-29 peptide was treated with Lys-gingipain (1:50 w/w) for 60 min at 37° C. The digest mixture was then separated by reverse phase HPLC on a C18 column and fractions representing the two most prominent peaks detected by 220 nm absorbance were collected and analyzed separately by ion-spray mass spectrometry. Fragments KAP-DYGHTDDK and DTLDLNTPVDK were identified in the HPLC-eluted fractions indicating that both the Lys-Asp (17-↓-18) and Lys-Thr (28-↓-29) bonds were cleaved by the *P. gingivalis* proteinase.

ACKNOWLEDGMENTS

This work was supported in part by the National Institutes of Health Grants AI 17354 (TEH), HL26148 and HL37090 (JT), NSF grant MCB-9306501 (JAE), and an Arthritis Foundation Fellowship (MAJ). Use of Mass Spectrometry facilities was supported in part by the Lucille P. Markey Charitable Trust and by the National Institute of Health Shared Instrumentation Grant S10 RPR07273-01. We wish to thank Drs. R.G. DiScipio and P.

Effect of P. gingivalis enzymes on PMN MPO release

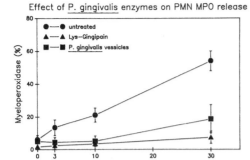

Figure 7. Isolated PMN were incubated with either *P. gingivalis* vesicles, or purified Lys-gingipain (50 µg/ml), for 60 min at 37°C. The cells were then washed and treated for 15 min at 37° with cytochalasin B (5 µg/ml) and stimulated for 30 min with C5a. Myeloperoxidase was then measured in supernatants by reduction of H_2O_2 in the presence of o-dianisidine and the absorbance at 460 nm was recorded. Exposure to either Lys-gingipain or the mixed proteinases of the vesicles reduced the ability of these cell to respond to C5a stimulation.

Daffern for providing some of the results that are summarized in this report. The technical assistance Marleen Kawahara is acknowledged and we thank Ms. Mary Glass for her aid in preparing the manuscript.

REFERENCES

Chen, Z., Potempa, J., Polanowski, A., Wikstrom, M., and Travis, J., 1992, Purification and characterization of a 50-kDa cysteine proteinase (Gingipain) from *Porphyromonas gingivalis. J. Biol. Chem.* 267: 18896-18901.

DiScipio, R.G., Daffern, P.J., Kawahara, M., Pike, R., Potempa, J., Travis, J. and Hugli, T.E., 1995, Cleavage of human complement component C5 by cysteine Proteinases from *Porphyromonas (Bacteroides) gingivalis.* Prior oxidation of C5 augments neutrophil activating capacity in Arg-gingipain and Lys-gingipain digests of C5. *Immunology* (In press).

Mayrand, D., and Holt, S.C., 1988, Biology of asaccharolytic black-pigmented bacteroides species. *Microbiol. Rev.* 52: 134-152.

Morgan, Edward L., Ember, Julia A., Sanderson, Sam D., Scholz, Wolfgang, Buchner, Robert, Ye, Richard D., and Hugli, T.E., 1993, Anti-C5a Receptor Antibodies. Characterization of Neutralizing Antibodies Specific for a Peptide, C5aR-(9-29), Derived from the Predicted Amino-Terminal Sequence of the Human C5a Receptor. *J. Immunol.*151:377-388.

Pike, R., McGraw, W., Potempa, J. and Travis, J., 1994, Lysine- and arginine specific proteinases from *Porphyromonas gingivalis.* Isolation, characterization, and evidence for existence of complexes with hemagglutinins. *J. Biol. Chem.* 269: 406-411.

Savarese, T. M and Fraser, C. M., 1992, *In Vitro* mutagenesis and the search for structure-function relationships among G protein-coupled receptors. *Biochem.* J. 283:1-19.

Vu, T.-K., H. Wheaton, V. I., D. T. Hung, I. Charo, and Coughlin, S. R., 1991, Molecular cloning of a functional thrombin receptor reveals a novel proteolytic mechanism of receptor activation. *Cell* 64:1057-1068.

Vu, T.-K., H. Wheaton, V. I. Hung, I. Charo, Coughlin, S. R., 1991, Domains specifying thrombin-receptor interaction. *Nature* 353:674-677.

Wetzel, R. A. and Kolb, W.P., 1982, Complement-independent activation of the fifth component (C5) of human complement: Limited trypsin digestion resulting in the expression of biological activity. *J. Immunol.* 128: 2209-2219.

White, D. and Mayrand, D., 1981, Association of oral *Bacteriodes* with gingivitis and adult periodontitis, *J. Periodontal Res.* 16: 259-265.

Wingrove, J.A., DiScipio, R.G., Chen, Z., Potempa, J., Travis, J., and Hugli, T.E. 1992, Activation of complement components C3 and C5 by a cysteine proteinase (Gingipain-1) from *Prophyromonas (Bacteroides) gingivalis. J. Biol. Chem.* 267: 18902-1890.

20

REGULATION OF PROGRAMMED CELL DEATH BY INTERLEUKIN-1β-COVERTING ENZYME FAMILY OF PROTEASES

Masayuki Miura[1] and Junying Yuan[2]

[1] Cardiovasular Research Center
Massachusetts General Hospital-East
Charlestown, Maryland 02129
[2] Department of Medicine
Harvard Medical School
Boston, Maryland 02115

INTRODUCTION

Programmed cell death plays a significant role in morphogenesis and histogenesis during animal development (Hinchliffe, 1981). Well-known examples include cell death in chick limb development and during metamorphosis of the tadpole tail. In these examples, programmed cell death is directly involved in morphogenesis. Programmed cell death is also involved in the generation of specific tissues and organs including kidney (Coles *et al.*, 1993; Koseki *et al.*, 1992), lens epithelial cells (Ishizaki *et al.*,1993) and cartilage cells (Ishizaki *et al.*, 1994). In addition, programmed cell death is also important for establishment of neural and immune system. During neural development, as much as 50% of originally generated neurons die (Cowan1984; Hamburger and Oppenheim, 1982; Oppenheim, 1991). In the immune system, cell death occurs constantly to eliminate cells that may react against self-antigens (Duvall and Wyllie 1986; Cohen *et al.*, 1992). Thus, programmed cell death during animal development may be as important as cell proliferation, growth and differentiation. Abnormally controlled programmed cell death may be underlaying causes of many diseases including neurodegenerative diseases.

Although the phenomenon of cell death during development was established in the 1950's (Glücksman 1951), little was known about the molecular mechanisms of programmed cell death. The important finding that programmed cell death was strictly determined in cell lineage during development of the nematode, *Caenorharbditis elegans* (*C.elegans*), brought up the idea that programmed cell death is genetically controlled as a specific cell differentiation fate (Sulston and Horvitz, 1977). Recently, intensive molecular genetic studies of programmed cell death in *C. elegans* have led to the identification of the central players of programmed cell death (Ellis *et al.*, 1991; Hengartner and Horvitz, 1994; Yuan and Horvitz, 1992; Yuan *et al.*, 1993). Progress in studies of the molecular mechanisms of cell death

Intracellular Protein Catabolism, Edited by Koichi Suzuki and Judith Bond
Plenum Press, New York, 1996

strongly suggest that the genes controlling programmed cell death ARE evolutionally conserved from worm to mammals (Hengartner and Horvitz 1994; Miura *et al.*, 1993; Vaux, *et al.*, 1992; Yuan *et al.*, 1993; Gagliardini *et al.*, 1994; Kumar *et al.*, 1994; Wang *et al.*, 1994).

PROGRAMMED CELL DEATH IN *C. ELEGANS*

Among the 1090 somatic cells generated during *C. elegans* hermaphrodite development, 131 die in every animal at specific developmental stages and positions (Sulston and Horvitz 1977; Sulston, *et al.*, 1983). A wide variety of cell types including intestinal cells, epithelial cells, muscle cells, neurons and gonadal cells are programmed to die (Horvitz *et al.*,1982). In *C. elegans*, detailed morphological changes of cells undergoing death have been described by ultrastructual studies (Robertson and Thomson 1982). Characteristic features include the nuclear condensation, nuclear membrane breakdown, cytoplasm contraction, membranous whole formation and the appearance of autophagic vacuoles . These features are similar to the charactisitics of apoptosis observed in mammals (Wyllie,1981). To date, 14 genes that specifically affect programmed cell death have been identified. Among them, 11 genes have been identified which affect all programmed cell death (Ellis *et al.*, 1991). This suggests that the common genetic pathway is controlling all the programmed cell death in *C. elegans*. Genetic analyses have placed these genes into a genetic pathway of programmed cell death. These genes can be placed into 4 distinct steps: determination of cells undergoing programmed cell death, execution of programmed cell death, engulfment of dead cells by neighboring cells, and degradation of the engulfed cells. Genes involved in execution of programmed cell death are the central players for cell death. Two genes, *ced-3* and *ced-4* (cell death abnormal), are identified for the execution of all programmed cell deaths in *C. elegans* (Ellis and Horvitz 1986). If one of these genes is mutated, all programmed cell deaths are inhibited. Genetic mosaic analysis demonstrated that the products of both *ced-3* and *ced-4* most likely function within dying cells (Yuan and Horvitz 1990). This supports the idea that programmed cell death in *C. elegans* is suicide, not murder.

Since the products of *ced-3* and *ced-4* genes appear either to be cytotoxic themselves or control proteins that are cytotoxic, the activities of *ced-3* and *ced-4* genes must be tightly regulated during development. The *ced-9* gene is involved in this process by repressing *ced-3* and *ced-4* activities (Hengartner *et al.*, 1992). The *ced-9* mRNA encodes a 280 amino acid protein which is 23% identical to the human Bcl-2 proto-oncogene product. Overexpression of *bcl-2* has been shown to protect cells from apoptosis in a number of systems including deaths of certain hematopoietic cell lines induced by cytokine deprivation and neuronal cell death induced by neurotrophic factor withdrawal (Reed 1994). Overexpression of human *bcl-2* can also partially prevent the ectopic cell death of *ced-9 (lf)* mutant as well as the normal programmed cell death (Hengartner and Horvitz 1994; Vaux *et al.*, 1992).

CED-3 AND INTERLEUKIN-1ß-CONVERTING ENZYME (ICE)

The *ced-3* gene encodes a 2.8 kb mRNA which is most abundantly transcribed during embryogenesis when most programmed cell deaths occur. The level of *ced-3* mRNA is very high (comparable to that of actin 1). This suggests that *ced-3* may not be transcribed only in dying cells since there are usually no more than two or three cells dying at any given time during embryonic development. Analysis of the *ced-3* genomic and cDNA sequence revealed that Ced-3 protein is 503 amino acids in length. The C-terminal region (amino acid 206-503) is well conserved among the three nematode species (84% identical). All eight EMS-induced

ced-3 missense mutations altered amino acid residues that are conserved among the three different nematode species and six of eight mutations alter residues within the C-terminal 100 amino acid region. These results suggest that the carboxyl half of the Ced-3 protein is important for *ced-3* function.

The amino acid sequence of Ced-3 protein has significant homology with human and murine interleukin-1ß-converting enzyme (ICE) (Cerretti *et al.*, 1992; Thornberry *et al.*, 1992). ICE has been identified as a substrate specific cysteine protease that cleaves the 31 kDa pro-IL-1ß between Asp-116 and Ala-117 to produce mature 17.5 kDa IL-1ß (Black *et al.*, 1989; Kostura *et al.*, 1989). The overall amino acid identity between *C. elegans* Ced-3 protein and human ICE protein is 29%. The carboxy terminal region of the Ced-3 protein is most similar to ICE. A stretch of 115 residues of Ced-3 (amino acids 246-360 of Ced-3) is 43% identical to amino acids 166-287 of human ICE. This region contains a conserved pentapeptide QACRG which surrounds a cysteine known to be essential for ICE function. Specific modifications of this cysteine in human ICE results in complete loss of activity (Thornbery *et al.*, 1992). The *ced-3* mutation *n2433* altered the conserved glycine in this pentapeptide to a serine and eliminated *ced-3* function, suggesting that this glycine is important for *ced-3* activity and might be an integral part of the active site of ICE. In 5 out of 8 *ced-3* mutations identified, single amino acids were altered that are conserved between ICE and Ced-3. Proteolytic cleavage of ICE from a precursor 45 kDa into P20 and P10 subunit is required for enzymatic activation of ICE. Two of these cleavage sites (Asp/X dipeptide) are conserved in Ced-3. The structure of human ICE has been determined by X-ray diffraction (Wilson *et al.*, 1994; Walker *et al.*, 1994). The active site spans both P20 and P10 subunits of ICE. The holoenzyme is a homodimer of catalytic domains formed by P20/P10 heterodimer. The catalytic residues His-237 and Cys-285, and the residues that form the P1 carboxylate-binding pocket are all conserved with Ced-3. These observations strongly support the hypothesis that ICE and Ced-3 may function as cysteine proteases in controlling programmed cell death by proteolytically cleaving proteins that are crucial for initiation of cell death or cell viability.

Overexpression of the murine *Ice* (*mIce*) gene or the *C. elegans ced-3* gene in Rat-1 fibroblast cells caused programmed cell death (Miura *et al.*, 1993). Point mutations in a region homologous between mICE and *ced-3* gene eliminated the cell death activities of these genes. The cowpox virus serpin gene *crmA* can specifically inhibit ICE activity (Ray *et al.*, 1992; Komiyama *et al.*, 1994). The expression of CrmA protein can inhibit mICE-induced programmed cell death, indicating that the protease activity of mICE is essential to its ability to kill cells (Miura, et al, 1993). Since *ced-9* suppresses the activity of *ced-3* and *ced-4* to protect cells from programmed cell death, *Ice* induced cell death by analogy should be suppressed by *bcl-2*. As expected, overexpression of *bcl-2* in Rat-1 cells prevents cell death induced by *mIce*. These results indicate that cell death induced by overexpression of *mIce* is likely to be caused by the activation of normal programmed cell death mechanisms in mammals (Miura *et al.*, 1993).

In *C. elegans* hermaphrodites, most programmed cell death is observed in neuronal lineages (105 of 131 programmed cell deaths) (Horvitz *et al.*, 1982). Effects of *mIce* expression was tested in vertebrate neuronal cells. Overexpression of *mIce* in NG108-15 neuronal cell line (Wang *et al.*, 1994; M. Miura, unpublished results) or chicken dorsal root ganglia (DRG) neurons can induce cell death (Gagliardini *et al.*, 1994). These results suggest that mICE may cause programmed cell death in the nervous system. The survival of cultured DRG neurons depends upon neurotrophic factors, especially on NGF. Deprivation of NGF from the culture of DRG neurons mimics the normal situation in the developing nervous system where up to 50% of neurons may die because of failure in competition for trophic factors. When NGF is removed from the culture media, DRG neurons die in 3 days or less. The cell death induced by deprivation of NGF can be prevented by overexpression of *bcl-2* or *crmA* gene (Gagliardini *et al.*, 1994). Thus, the genes in *Ice/ced-3* family may play a key role in neuronal cell death during normal development. ICE activity is required for the production of mature IL-1ß. IL-1ß is produced not only by peripheral blood monocytes but also by astrocytes (Fontana *et al.*, 1982), microglia (Giulian *et al.*, 1986), sympathetic neurons (Freiden *et al.*, 1992). Interestingly, the level of IL-1 has been found to be elevated in the brain of patients with Alzheimer's disease (Griffin *et al.*, 1989) and the high level of IL-1 may promote the expression of the amyloid precursor protein (APP) (Forloni *et al.*, 1992). These findings suggest that *Ice/ced-3* family may contribute to the pathogenesis of Alzheimer's disease.

After murine peritoneal macrophages were stimulated with lipopolysaccharide (LPS) and induced to undergo programmed cell death by exposure to extracelluar ATP, mature active IL-1ß was released into culture supernatant; in contrast, when cells were injured by scraping, IL-1ß was released exclusively as the unprocessed inactive form (Hogoquist, *et al.*, 1991). These results suggest that ICE might be activated upon induction of programmed cell death. *Ice* mRNA has also been detected in a variety of tissues that do not produce IL-1ß including resting and activated peripheral blood T lymphocytes, placenta and the B lymphoblastid line CB23 (Cerretti *et al.*, 1992). The distribution of ICE suggests that ICE may mediate programmed cell death by cleaving substrates other than pro-IL-1ß. Specific cleavage of the nuclear enzyme poly (ADP-ribose) polymerase or the 70 kDa protein component of the U1 small nuclear ribonucleoprotein is observed in apoptotic cells. The inhibitory characteristics of the protease suggests that ICE-like protease activity is involved in this process (Lazebnik *et al.*, 1994; Casciola-Rosen *et al.*, 1994). Since ICE cannot cleave those substrates, other *Ice/ced-3* family genes may be responsible for this process.

Two other members of the *Ice/ced-3* family, *Ich-1/Nedd2* and *CPP32*, have been identified (Kumar *et al.*, 1992; Kumar *et al.*, 1994; Wang *et al.*, 1994; Fernandez-Alnemri *et al.*, 1994)). *Nedd2* (NPC expressed, developmentally down-regulated gene) was cloned by subtraction screening of mouse neural precursor cell (NPC) cDNA library (Kumar *et al.*, 1992). *Nedd2* gene is highly expressed in developing central nervous system and other tissues including kidney (Kumar *et al.*, 1994). In adult mice, expression level of *Nedd2* is lower than that in embryo, but its expression can be detected in all tissues. *Ich-1* (*Ice* and *ced-3* homolog) was isolated from human fetal cDNA library by screening with 3'-portion of *Nedd2* gene as a probe. Structural and functional analysis revealed that *Nedd2* and *Ich-1* are the same gene. Ich-1/Nedd2 protein is similar to both ICE and Ced-3 (approximately 28% identity). Overexpression of the *Ich-1/Nedd2* gene in mammalian cells induces apoptosis and this apoptosis is prevented by *bcl-2*. *Ich-1* induced apoptosis can be only marginally inhibited by *crmA*. This suggests that CrmA is not a general inhibitor for ICE/Ced-3 family. *Ich-1* mRNA is alternatively spliced into two different forms. One mRNA species encodes a protein product of 435 amino acids, named ICH-1$_L$, which is homologous to both the P20 and P10 subunits of ICE as well as the entire Ced-3. The other mRNA encodes a 312 amino acid sequence truncated version of ICH-1$_L$ protein, named ICH-1$_S$, that terminates 21 amino acid residues after the pentapeptide QACRG of ICH-1$_L$. ICH-1$_L$ and ICH-1$_S$ have opposite functions: overexpression of ICH-1$_L$ induces apoptosis, while overexpression of the ICH-1$_S$ suppresses Rat-1 cell death induced by serum deprivation. These results suggest that *Ich-1* may play important roles in both positive and negative regulation of programmed cell death as the *bcl-x* gene does in vertebrates (Boise *et al.*, 1993).

Cytolytic granules of cytotixic T lymphocytes (CTL) and natural killer (NK) cells store multiple serine proteases refered to as granzymes or fragmentins (Bleackley *et al.*, 1988; Jenne *et al.*, 1988; Shi *et al.*, 1992a, 1992b). One of these serine proteases purified from a rat NK cell line RNK-16, fragmentin 2, can induce DNA fragmentation and apoptosis in YAC-1 target cell in the presence of perforin (Shi *et al.*, 1992a, 1992b) and is considered as the rat ortholog of granzyme B. Fragmentin 2, granzyme B and ICE share a unique

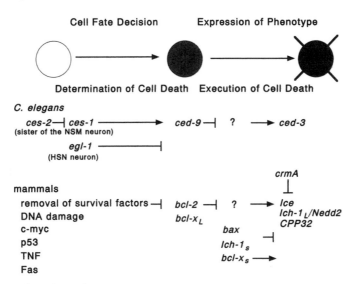

Figure 1. The genetic pathway for programmed cell death in *C. elegans* and mammals. The proposed regulatory interaction between adjacent genes in the pathway are indicated as follows: arrow, positive interaction; bar, negative interaction.

substrate site specificity for aspartic acid. CTL from granzyme B knockout mice have defects in their ability to induce rapid DNA fragmentation and apoptosis in allogeneic target cells (Heusel *et al.*, 1994). Although *in vitro* cleavage studies show that the P45 ICE precursor is not a substrate for granzyme B, ICE substrate can be cleaved by granzyme B (Darmon *et al.*, 1994). These enzymes may share substrates which are crucial for cell survival.

CONCLUSION

Different cell-cell interactions and extracelluar signals may be required for many cells to die in vertebrate animals. In *C. elegans*, the genes that control the determination of programmed cell death may be different in different cell lineages. Many genes may be involved in the determination of programmed cell death. In spite of the differences in the mechanisms of determination of programmed cell death in each cell type, morphological features of programmed cell death in *C. elegans* are quite similar to that of apoptosis in mammals (Wyllie 1981; Robertson and Thomson, 1982; Rotello, *et al.*, 1994; Fernandez, *et al.*, 1994). Our results suggest that mechanisms of *Ice/ced-3* family mediated execution of programmed cell death might be evolutionarily conserved from worm to mammals (Fig.1). Isolation of mammalian genes which regulate ICE family protease activity may allow a better biochemical analysis of the programmed cell death pathways. Such studies in turn may increase our understanding of programmed cell death in *C. elegans*.

ACKNOWLEDGEMENTS

This work is supported in part by grants to J. Y. from Bristol/Myer-Squibb and from National Institute of Aging. M. M. is supported by the Mochida Memorial Foundation for Medical and Pharmaceutical Research and a Fogarty International Fellowship from NIH.

REFERENCES

Black, R.A., Kronheim, S.R., and Sleath, P.R., 1989, Activation of interleukin-1β by a co-induced protease, *FEBS Lett.* 247:386-390.

Bleackley, R.C., Lobe, C.G., Duggan, B., Ehrman, N., Fregeau, C., Meier, M., Letellier, M., Havele, C., Shaw, J., and Paetkau, V., 1988, The isolation and characterization of a family of serine protease genes expressed in activated cytotoxic T-lymphocytes, *Immunol. Rev.* 103:5-19.

Boise, L.H., Gonzalez-Garcia, M., Postema, C.E., Ding, L., Lindsten, T., Turka, L.A., Mao, X., Nunez, G., and Thompson, C.B., 1993, *bcl-x*, a *bcl-2*-related gene that functions as a dominant regulator of apoptotic cell death, *Cell* 74:597-608.

Casciola-Rosen, L. A.,Miller, D. K.,Anhalt, G. J., and Rosen, A., 1994, Specific cleavage of the 70-kDa protein component of the U1 small nuclear ribonucleoprotein is a characteristic biochemical feature of apoptotic cell death, *J. Biol. Chem.* 269:30757-30760.

Cerretti, D.P., Kozlosky, C.J., Mosley, B., Nelson, N., Ness, K.V., Greenstreet, T.A., March, C.J., Kronheim, S.R., Druck, T., Cannizzaro, L.A., Huebner, K., and Black, R.A., 1992, Molecular cloning of the interleukin-1β converting enzyme, *Science* 256:97-100.

Cohen, J.J., Duke, R.C., Fadock, V.A., and Sellins, K.S., 1992, Apoptosis and programmed cell death in immunity, *Annu. Rev. Immunol.* 9:243-269.

Coles, H.S.R., Burne, J.F., and Raff, M.C., 1993, Large-scale normal cell death in the developing rat kidney and its reduction by epidermal growth factor, *Development* 118:777-784.

Cowan, W.M., 1984, Regressive events in neurogenesis, *Science* 225: 1258-1265.

Darmon, A. J.,Ehrman, N.,Caputo, A.,Fujinaga, J., and Bleackley, R. C., 1994, The cytotoxic T cell proteinase granzyme B does not activate interleukin-1ß-converting enzyme, *J. Biol. Chem.*, 269:32043-32046.

Duvall, E., and Wyllie, A.H., 1986, Death and the cell. *Immunol. Today* 7:115-119.

Ellis, H.M., and Horvitz, H.R., 1986, Genetic control of programmed cell death in the nematode *C. elegans*, *Cell* 44:817-829.

Ellis, R.E., Yuan, J., and Horvitz, H.R., 1991, Mechanisms and functions of cell death, *Annu. Rev. Cell Biol.* 7:663-698.

Fernandez, P-A., Rotello, R., Rangini, Z., Doupe, A., Drexler. H. C. A., and Yuan, J., 1994, Expression of a specific marker of avian programmed cell death in both apoptosis and necrosis, *Proc. Natl. Acad. Sci. USA.* 91:8641-8645.

Fernandes-Alnemri, T., Litwack, G., and Alnemri, E. S., 1994, CPP32, a novel human apoptotic protein with homology to *Caenorhabditis elegans* cell death protein *ced-3* and mammalian interleukin-1ß-converting enzyme, *J. Biol. Chem.*, 269:30761-30764.

Fontana, A. F., Kristensen, F., Duds, R., Gemsa, D., and Weber, E., 1982, Production of prostagrandin E and an interleukin-1 like factor by cultured astrocytes and C6 glioma cells, *J. Immunol.* 129:2413-2419.

Forloni, G., Demicheli, F., Bendotti, C., and Angeretti, N., 1992, Expression of amyloid precursor protein mRNAs in endothelial, neuronal and glial cells: modulation by interleukin-1, *Mol. Brain Res.* 16:128-134.

Freiden, M., Bennett, M. V. L., and Kessler, J. A., 1992, Cultured sympathetic neurons synthesize and release the cytokine interleukin 1ß, *Proc. Natl. Acad. Sci. USA.* 89:10440-10443.

Gagliardini, V., Fernandez, P.-A., Lee, R.K.K., Drexler, H.C., Rotello, R.J., Fishman, M.C., and Yuan, J., 1994, Prevention of vertebrate neuronal death by the *crmA* gene, *Science* 263:826-828.

Giulian, D., Baker, T. J., Shih, L. N., and Lachman, L. B., 1986, Interleukin 1 of the central nervous system is produced by ameboid microglia, *J. Exp. Med.* 164:594-604.

Glücksmann, A., 1951, Cell deaths in normal vertebrate ontogeny, *Biol. Rev. Cambridge Philos. Soc.* 26:59-86.

Griffin, W.S.T., Sranley, L.C., Ling, C., White, L., MacLeod, V., Perrot, L.J., White, C.L.I., and Araoz, C., 1989, Brain interleukin 1 and S-100 immunoreactivity are elevated in Down syndrome and Alzheimer disease, *Proc. Natl. Acad. Sci. USA.* 86:7611-7615.

Hamburger, V., and Oppenheim, R.W., 1982, Naturally occuring cell death in vertebrates, *Neurosci. Comment* 1:39-55.

Hengartner, M., Ellis, R.E., and Horvitz, H.R., 1992, *Caenorhabditis elegans* gene *ced-9* protects cells from programmed cell death, *Nature* 356:494-499.

Hengartner, M.O., and Horvitz, H.R., 1994, *C. elegans* cell survival gene *ced-9* encode a functional homolog of the mammalian proto-oncogene *bcl-2*, *Cell* 76: 665-676.

Heusel, J.W., Wesselschmidt, R.L., Shresta, S., Russell, J.H., and Ley, T.J., 1994, Cytotoxic lymphocytes require granzyme B for the rapid induction of DNA fragmentation and apoptosis in allogenetic target cells, *Cell* 76:977-987.

Hinchliffe, J.R., 1981, Cell death in embryogenesis. *In* "Cell death in biology and pathology" (I. D. Brown and R. A. Lockshin, ed.), Chapman and Hall, London.

Hogoquist, K.A., Nett, M.A., Unanue, E.R., and Caplin, D.D., 1991, Interleukin 1 is processed and released during apoptosis, *Proc. Natl. Acad. Sci. USA.* 88:8485-8489.

Horvitz, H.R., Ellis, H.M., and Sternberg, P.W., 1982, Programmed cell death in nematode development, *Neurosci. Comment* 1:56-65.

Ishizaki, Y., Voyvodic, J.T., Burne, J.F., and Raff, M.C., 1993, Control of lens epithelial cell survival, *J. Cell Biol.* 121:899-908.

Ishizaki, Y., Burne, J.F., and Raff, M., 1994, Autocrine signals enable chondrocytes to survive in culture, *J. Cell Biol.* 126:1069-1077.

Jenne, D., and Tschopp, J., 1988, Granzymes, a family of serine proteases released from granules of cytolytic T-lymphocytes upon T-cell receptor stimulation, *Immunol. Rev.* 103:53-71.

Komiyama, T., Ray, C. A., Pickup, D. J., Howard, A. D., Thornberry, N. A., Peterson, E. P., and Salvesen, G.,1994, Inhibition of interleulin-1ß converting enzyme by the cowpox virus serpin crmA: an example of cross-class inhibition, *J. Biol. Chem.* 269:19331-19337.

Koseki, C., Herzlinger, D., and Al-Awqati, Q., 1992, Apoptosis in metanephric development, *J. Cell Biol.* 119:1327-1333.

Kostura, M.J., Tocci, M.J., Limjuco, G., Chin, J., Cameron, P., Hillman, A.G., Chartrain, N.A., and Schmidt, J.A., 1989, Identification of a monocyte specific pre-interleukin-1ß convertase activity, *Proc. Natl. Acad. Sci. USA.* 86:5227-5231.

Kumar, S., Tomooka, Y., and Noda, M., 1992, Identification of a set of genes with developmentally down-regulated expression in the mouse brain, *Biochem. Biophys. Res. Commun.* 185:1155-1161.

Kumar, S., Kinoshita, M., Noda, M., Copeland, N.G., and Jenkins, N.A., 1994, Induction of apoptosis by the mouse *Nedd2* gene, which encodes a protein similar to the product of the *Caenorhabditis elegans* cell death gene *ced-3* and the mammalian IL-1ß-converting enzyme, *Genes Dev.* 8:1613-1626.

Lazebnik, Y.A., Kaufmann, S.H., Desnoyers, S., Poirier, G.G., and Earnshaw, W.C., 1994, Cleavage of poly(ADP-ribose) polymerase by a proteinase with properties like ICE, *Nature* 371:346-347.

Miura, M., Zhu, H., Rotello, R., Hartwieg, E., and Yuan, J., 1993, Induction of apoptosis in fibroblasts by IL-1ß-converting enzyme, a mammalian homolog of the *C. elegans* cell death gene *ced-3*, *Cell* 75:653-660.

Oppenheim, R.W., 1991, Cell death during development of the nervous system, *Annu. Rev. Neurosci.* 14:453-501.

Ray, C.A., Black, R.A., Kronheim, S.R., Greenstreet, T.A., Sleath, P.R., Salvesen, G.S., and Pickup, D.J., 1992, Viral inhibition of inflammation: cowpox virus encodes an inhibitor of the interleukin-1ß converting enzyme, *Cell* 69:597-604.

Reed, J.C., 1994, *Bcl-2* and the regulation of programmed cell death. *J. Cell Biol.* 124:1-6.

Robertson, A.M.G., and Thomson, J.N., 1982, Morphology of programmed cell death in the ventral nerve cord of *Caenorhabditis elegans* larvae, *J. Embryol. Exp. Morph.* 67:89-100.

Rotello, R.J. Fernandez, P.-A., and Yuan, J., 1994, Anti-apogens and anti-engulfens: monoclonal antibodies reveal specific antigenes on apoptotic and engulfment cells during chicken embryonic development, *Development.* 120:1421-1431.

Shi, L., Kraut, R.P., Aebersold, R., and Greenberg, A.H., 1992a, A natural killer cell granule protein that induce DNA fragmentation and apoptosis, *J. Exp. Med.* 175:553-566.

Shi, L., Kam, C.-M., Powers, J.C., Aebersold, R., and Greenberg, A.H., 1992b, Purification of three cytotoxic lymphocyte granule serine proteases that induce apoptosis through distinct substrate and target cell interactions, *J. Exp. Med.* 176: 1521-1529.

Sulston, J.E., and Horvitz, H.R., 1977, Post-embryonic cell lineages of the nematode *Caenorhabditis elegans*, *Dev. Biol.* 56:110-156.

Sulston, J.E., Schierenberg, E., White, J.G., and Thomson, N., 1983, The embryonic cell lineage of the nematode *Caenorhabditis elegans*, *Dev. Biol.* 100:64-119.

Thornberry, N.A., Bull, H.G., Calaycay, J.R., Chapman, K.T., Howard, A.D., Kostura, M.J., Miller, D.K., Molineaux, S.M., Weidner, J.R., Aunins, J., Elliston, K.O., Ayala, J.M., Casano, F.J., Chin, J., Ding, G.J.-F., Egger, L.A., Gaffney, E.P., Limjuco, G., Palyha, O., C,, Raju, S.M., Rolando, A.M., J.P., S., Yamin, T.-T., Lee, T.D., Shively, J.E., MacCross, M., Mumford, R.A., Schmidt, J.A., and Tocci, M.J., 1992, A novel heterodimeric cysteine protease is required for interleukin-1β processing in monocytes, *Nature* 356:768-774.

Vaux, D.L., Weissman, I.L., and Kim, S.K., 1992, Prevention of programmed cell death in *Caenorhabditis elegans* by human *bcl-2*, *Science* 258:1955-1957.

Walker, N.P.C., Talanian, R.V., Brady, K.D., Dang, L.C., Bump, N.J., Ferenz, C.R., Franklin, S., Ghayur, T., Hackett, M.C., Hammill, L.D., Herzog, L., Hugunin, M., Houy, W., Mankovich, J.A., McGuiness, L., Orlewicz, E., Paskind, M., Pratt, C.A., Reis, P., Summani, A., Terranova, M., Welch, J.P., Xiong, L., Moller, A., Tracey, D.E., Kamen, R., and Wong, W.W., 1994, Crystal structure of the cysteine protease interleukin-1ß-converting enzyme: a (p20/p10)$_2$ homodimer, *Cell* 78:343-352.

Wang, L., Miura, M., Bergeron, L., Zhu, H., and Yuan, J., 1994, *Ich-1*, an *Ice/ced-3*-related gene, encodes both positive and negative regulators of programmed cell death, *Cell* 78:739-750.

Wilson, K.P., Black, J-A.F., Thomson, J.A., Kim, E.E., Griffith, J.P., Navia, M.A., Murcko, M.A., Chambers, S.P., Aldape, R.A., Raybuck, S.A., and Livingston, D.J., 1994, Structure and mechanism of interleukin-1ß converting enzyme, *Nature* 370:270-275.

Wyllie, A.H., 1981, Cell death: a new classification separating apoptosis from necrosis. *In* "Cell death in biology and pathology."(I. D. Brown and R. A. Lockshin, ed.), Chapman and Hall, London.

Yuan, J., and Horvitz, H.R.,1990, The *Caenorhabditis elegans* genes *ced-3* and *ced-4* act cell autonomously to cause programmed cell death, *Dev. Biol.* 138:33-41.

Yuan, J., and Horvitz, H.R., 1992, The *Caenorharbditis elegans* cell death gene *ced-4* encodes a novel protein and is expressed during the period of extensive cell programmed cell death, *Development* 116:309-320.

Yuan, J., Shaham, S., Ledoux, S., Ellis, H.M., and Horvitz, H.R., 1993, The *C. elegans* cell death gene *ced-3* encodes a protein similar to mammalian interleukin-1ß converting enzyme, *Cell* 75:641-652.

21

INHIBITION OF CYSTEINE AND SERINE PROTEINASES BY THE COWPOX VIRUS SERPIN CRMA

Tomoko Komiyama, Long T. Quan, and Guy S. Salvesen

Department of Pathology
Duke University Medical Center
Durham, North Carolina 27710

INTRODUCTION

Interleukin 1β converting enzyme (ICE) is the founding member of a family of cysteine proteinases active in the metabolism of intracellular proteins [1]. Other members of this family include Ich-1, CPP32, Ced-3, all of which have been implicated in apoptosis (programmed cell death) [2-4]. Though ICE itself may also be involved in apoptosis, its main role is probably to activate pro-interleukin-1β (proIL-1β) via specific cleavage following Asp residues [1,5].

Processing of proIL-1β by ICE can be inhibited by the product of the cowpox virus serpin CrmA (cytokine response modifier A), which is a tight binding inhibitor of ICE [6,7]. This interaction is unusual since ICE is a cysteine proteinase and CrmA, based on its amino acid sequence, is a member of the serpin family of inhibitors [8] which normally inhibits only serine proteinases [9,10].

SERPINS

Serpins are a wide spread family of proteins found in mammals, plants, insects and viruses [11-13]. The serpins have evolved into a large family of proteins recognizable by sequence homology and a similar structural motif of 9 α-helices and 3 β-sheets [14]. This forms an elaborate substructure which serves to align the reactive site loop (RSL) for proteinase binding in the proteinase-inhibitory members of the family. However, not all serpins are proteinase inhibitors. For example ovalbumin [15], thyroxin binding globulin [16], and angiotensinogen [17] are binding proteins without any apparent inhibitory activity [18-20]. In contrast, the well investigated serpins, α₁proteinase inhibitor (α_1PI), α₁antichymotrypsin (α_1ACT), and plasminogen activator inhibitor 1 (PAI-1) are inhibitory towards their target serine proteinase but can be cleaved by various non-target proteinases [21,22].

Intracellular Protein Catabolism, Edited by Koichi Suzuki and Judith Bond
Plenum Press, New York, 1996

Inhibitory serpins have several defining characteristics. Due to their unstable nature, they are easily denatured by heat or chemical denaturants [23,22]. Thus, they are said to have a stressed (S) conformation which, upon proteinase binding or cleavage of the RSL, can undergo a conformational change into the relaxed (R) form that is very resistant to denaturation. Most non-inhibitory serpins lack the (S) to (R) change upon cleavage of their RSL [24]. For instance, ovalbumin remains in the (R) form while angiotensinogen retains the (S) form even after RSL cleavage [23]. Another feature conserved in inhibitory serpins is the residue equivalent to position 345 in α_1PI at the base of the reactive site loop, which is usually Thr, Val, or Ser. The non-inhibitory serpin ovalbumin has an Arg in this position. In fact, changing this residue from Thr to Arg in several inhibitory serpins has rendered them inactive [25-27].

CRMA-ICE INTERACTION

CrmA is a 38 kDa viral serpin expressed early in the cowpox infection [8]. It is translated without a signal peptide and belongs to a subgroup of intracellular serpins including placenta-type plasminogen activator inhibitor [28], human monocyte/neutrophil elastase inhibitor [29], squamous cell carcinoma antigen [30], and cytoplasmic antiproteinase [31]. ICE is thought to be an intracellular enzyme thus CrmA and ICE most likely interact inside a virally infected cell.

CrmA is a very tight binding inhibitor of ICE [7]. However, to inhibit one molecule of ICE, about three molecules of fully active CrmA, produced by *E. coli* expression system, are required (unpublished result). We define the partition ratio of three for the CrmA-ICE interaction, by using recombinant CrmA. Presumably two out of three CrmA interact with ICE in a substrate-like manner, and the remaining interacts in an inhibitor-like manner. Similar partitioning is reported for the interaction of some serpins with serine proteinases [32-35]. Finally, the CrmA-ICE complex, though stable to 8M urea, is dissociated in SDS-PAGE [7]. This is very unusual, though not unique [35], for serpin reactions with target proteinases.

Since CrmA could inhibit a cysteine proteinase, we asked whether it is a typical inhibitory serpin. We aligned CrmA with other serpins and found that residue 345 (α_1PI numbering) is a conserved Thr as with other inhibitory serpins. Furthermore, we know that CrmA undergoes the (S) to (R) conformational change characteristic of inhibitory serpins [7]. With the above data, we hypothesized that CrmA is not different in its inhibitory mechanism than any other inhibitory serpins. This hypothesis predicted that CrmA should be able to inhibit serine proteinases in addition to the cysteine proteinase ICE.

INTERACTION OF CRMA WITH SERINE PROTEINASES

The P_1 residue of a serpin is the indicator of its primary specificity. CrmA has an Asp at the P_1 position, thus we decided to try several serine proteinases that could cleave after an Asp residue: Granzyme B (GraB), *Streptomyces griseus* Glu specific proteinase (GluSGP), and *Staphylococcus aureaus* V8 (Spase V8). We found that recombinant CrmA could inhibit GraB with an association rate constant (k_{on}) of 2.9×10^5 $M^{-1}s^{-1}$, forming an equimolar SDS stable complex [36]. The fact that both GraB and ICE are specific for Asp in their S_1 pockets and that GraB competed with ICE for CrmA binding implied these two proteinases bound to the same site in the RSL of CrmA [36].

GluSGP and Spase V8 are acidic amino acid specific proteinases from bacterial origin. We found that CrmA could inhibit GluSGP, forming an equimolar SDS stable complex (unpublished result). However, Spase V8 activity was unaffected by CrmA [7].

DISCUSSION

Our results show that CrmA can inhibit a eukaryotic and a prokaryotic serine proteinase in addition to the cysteine proteinase ICE. This confirmed our prediction based on the fact that CrmA has all of the characteristics of a typical inhibitory serpin: it has an (S) to (R) conformational change upon RSL cleavage, it has the conserved Thr345 residue, and it contains an Asp at the P_1 position. CrmA is a cross-class inhibitor that can interact with proteinases from distinct catalytic classes by using the same reactive site.

The fact that CrmA can inhibit two classes of proteinases supports our previous speculation that the substrate binding geometry of ICE might be similar to a serine proteinase [7,37]; indeed, the ICE crystal structure verifies this [38,39]. This similarity supports our interpretation that the interaction between a protein inhibitor and a proteinase is dictated by the substrate binding geometry, rather than the catalytic mechanism. Our hypothesis would predict that an inhibitor should be able to inhibit proteinases containing a similar substrate binding geometry regardless of the mechanism class of the proteinases. From this perspective it would be interesting to test whether inhibitors containing Asp at the P_1 position, but based on distinct frames such as the Kunitz or Kazal folds, could inhibit ICE.

ACKNOWLEDGMENT

We thank Dr. Thornberry for supplying ICE.

REFERENCES

1 Thornberry, N. A. and Molineaux, S. M. (1995) *Protein Science* 4: 3-12.

2 Wang, L., Miura, M., Bergeron, L., Zhu, H. & Yuan, J. (1994) *Cell* 78: 739-750.

3 Fernandes-Alnemri, T., Litwack, G. & Alnemri, E. S. (1994) *J. Biol. Chem.* 269:30761-30764.

4 Yuan, J., Shaham, S., Ledoux, S., Ellis, H. M. & Horvitz, H. R. (1993) *Cell* 75: 653-660.

5 Kuida, K., Lippke, J. A., Ku, G., Harding, M. W., Livingston, D. J., Su, M. S.-S. & Flavell, R. A. (1995) *Science* 267:2000-2003.

6 Ray, C. A., Black, R. A., Kronheim, S. R., Greenstreet, T. A., Salvesen, G. S. & Pickup, D. J. (1992) *Cell* 69: 597-604.

7 Komiyama, T., Ray, C. A., Pickup, D. J., Howard, A. D., Thornberry, N. A., Peterson, E. P. & Salvesen, G. (1994) *J. Biol. Chem.* 269:19331-19337.

8 Pickup, D. J., Ink, B. S., Hu, W., Ray, C. A. & Joklik, W. K. (1986) *Proc. Natl. Acad. Sci.* USA 83:7698-7702.

9 Huber, R. and Carrell, R. W. (1989) *Biochemistry* 28:8951-8971.

10 Potempa, J., Korzus, E. and Travis, J. (1994) *J. Biol. Chem.* 269:15957-15960.

11 Carrell, R. W. and Boswell, D. R. (1986) *in Proteinase Inhibitors* (Barrett, A. J., and Salvesen, G., eds.) pp. 403-420 Elsevier Science Publishers, Amsterdam.

12 Travis, J. and Salvesen, G. S. (1983) *Ann. Rev. Biochem.* 52:655-709.

13 Salvesen, G. and Enghild, J. J. (1993) *in Acute Phase Proteins Molecular biology, biochemistry, and clinical applications* (Mackiewicz, A., Kushner, I., and Baumann, H., eds.) pp. 117-147 CRC Press, Boca Raton.

14 Stein, P. E. and Carrell, R. W. (1995) *Nature Structural Biology* 2:96-113.

15 Hunt, L. T. and Dayhoff, M. O. (1980) *Biochem. Biophys. Res. Comm.* 95:864-871.

16 Flink, I. L., Bailey, T. J., Gustafson, T. A. & Markham, B. E. (1986) *Proc. Natl. Acad. Sci.* USA 83:7708-7712.

17 Doolittle, R. F. (1983) *Science* 222:417-419.

18 Remold-O'Donnell, E. (1993) *FEBS Lett.* 315:105-108.

19 Pemberton, P. A., Stein, P. E., Pepys, M. B., Potter, J. M. & Carrell, R. W. (1988) *Nature* 336:257-258.

20 Stein, P. E., Tewkesbury, D. A. & Carrell, R. W. (1989) *Biochem. J.* 262:103-107.

21 Potempa, J., Fedak, D., Dubin, A., Mast, A. & Travis, J. (1991) *J. Biol. Chem.* 266:21482-21487.

22 Mast, A. E., Enghild, J. J. & Salvesen, G. (1992) *Biochemistry* 31:2720-2728.

23 Mast, A. E., Enghild, J. J., Pizzo, S. V. & Salvesen, G. (1991) *Biochemistry* 30:1723-1730.

24 Carrell, R. W. and Owen, M. C. (1986) *Nature* 322:730-732.

25 Schulze, A. J., Huber, R., Degryse, E., Speck, D. & Bischoff, R. (1991) *Eur. J. Biochem.* 202:1147-1155.

26 Davis, A. E. I., Aulak, K., Parad, R. B., Stecklein, H. P., Eldering, E., Hack, C. E., Kramer, J., Strunk, R. C., Bissler, J. & Rosen, F. S. (1992) *Nature Genetics* 1:354-358.

27 Hood, D. B., Huntington, J. A. & Gettins, P. G. W. (1994) *Biochemistry* 33:8538-8547.

28 Remold-O'Donnell, E., Chin, J. & Alberts, M. (1992) *Proc. Natl. Acad. Sci.* USA 89:5635-5639.

29 Dubin, A., Travis, J., Enghild, J. J. & Potempa, J. (1992) *J. Biol .Chem.* 267:6576-6583.

30 Suminami, Y., Kishi, F., Sekiguchi, K. & Kato, H. (1991) *Biochim. Biophys. Commun.* 181:51-58.

31 Morgenstern, K. A., Henzel, W. J., Baker, J. B., Wong, S. & Pastuszyn, A. (1993) *J. Biol. Chem.* 268:21560-21568.

32 Patston, P. A., Gettins, P., Beechem, J. & Schapira, M. (1991) *Biochemistry* 30:8876-8882.

33 Patston, P. A. (1995) *Inflammation* 19:75-81.

34 Schechter, N. M., Jordan, L. M., James, A. M., Cooperman, B., Wang, Z. M. & Rubin, H. (1993) *J. Biol. Chem.* 268:23626-23633.

35 Enghild, J. J., Valnickova, Z., Thørgersen, I. B., Pizzo, S. V. & Salvesen, G. (1993) *Biochem. J.* 291:933-938.

36 Quan, L. T., Caputo, A., Bleakley, R. C., Pickup, J. P. & Salvesen, G. S. (1995) *J. Biol. Chem.* 270:10377-10379.

37 Salvesen, G. (1994) *in Proteolysis and Protein Turnover* (Bond, J. S. and Barrett, A. J., eds.) pp.57-63 Portland Press, London and Chapel Hill.

38 Wilson, K. P., Black, J.-A. F., Thompson, J. A., Kim, E. E., Griffith, J. P., Navia, M. A., Murcko, M. A., Chambers, S. P., Aldape, R. A., Raybuck, S. A. & Livingston, D. J. (1994) *Nature* 370:270-273.

39 Walker, N. P. C., Talanian, R. V., Brady, K. D., Dang, L. C., Bump, N. J., Ferenz, C. R., Franklin, S., Ghayur, T., Hackett, M. C., Hammill, L. D., Herzog, L., Huguin, M., Houy, W., Mankovich, J. A., McGuiness, L., Oriewicz, E., Paskind, M., Pratt, C. A., Reis, P., Summan, A., Terranova, M., Weich, J. P., Xiong, L., Moller, A., Tracey, D. E., Kamen, R. & Wong, W. W. (1994) *Cell* 78: 343-3522.

22

PARTICIPATION OF CATHEPSINS B, H, AND L IN PERIKARYAL CONDENSATION OF CA1 PYRAMIDAL NEURONS UNDERGOING APOPTOSIS AFTER BRIEF ISCHEMIA

T. Nitatori,[1] N. Sato,[2] E. Kominami,[3] and Y. Uchiyama[2]

[1] Department of Cell Biology and Neuroanatmy
Iwate Medical University School of Medicine
Uchimaru 19-1
Morioka 020, Japan
[2] Department of Anatomy I
Osaka University School of Medicine
Yamadagaoka 2-2
Osaka 565, Japan
[3] Department of Biochemistry
Juntendo University School of Medicine
Hongoh 2-1-1
Bunkyo-ku
Tokyo 113, Japan

INTRODUCTION

The CA1 pyramidal neurons in the hippocampus are known to be selectively vulnerable to brief ischemia of the global cerebrum, resulting in delayed neuronal death several days after ischemic insult (Ito *et al.*, 1975; Pulsinelli *et al.*, 1979; Kirino, 1982; Kirino and Sano, 1984a, b; Petito *et al.*, 1987; Bonnekoh *et al.*, 1990). It remains, however, unknown whether this delayed neuronal death is necrosis or apoptosis. According to the morphological criteria of apoptosis, dying cells show chromatin condensation and cell shrinkage, followed by heterophagocytosis (Kerr *et al.*, 1987; Clarke, 1990). In the course of the study of the delayed death of the CA1 pyramidal neurons after transient ischemia, we noticed that shrinkage of these neurons suddenly occurs 3 or 4 days after ischemic insult. This suggests the possibility that delayed death of the CA1 pyramidal neurons after brief ischemia is not necrotic but apoptotic.

Lysosomes, a membrane-bound cytoplasmic organelle, contain a great variety of hydrolytic enzymes which are capable of breaking down proteins, nucleic acids, complex carbohydrates and lipids. Cathepsins B, H, and L are well characterized cysteine proteinases,

Intracellular Protein Catabolism, Edited by Koichi Suzuki and Judith Bond
Plenum Press, New York, 1996

each being widely distributed in lysosomes of various mammalian cells (Katunuma and Kominami, 1983) .

As for a role of lysosomes in nerve cells, they have been demonstrated to correlate with some sorts of developing nerve cell death (Cataldo *et al.*, 1991). Recently, degenerating nerve cells in Alzheimer's disease have been demonstrated to contain autophagic vacuoles showing histochemical and immunohistochemical reactivities for various lysosomal enzymes including cysteine proteinases, and the intense reactivities for them have also been found in senile plaques with β-amyloid (Cataldo and Nixon 1990; Cataldo *et al.*, 1990; Cataldo *et al.*, 1991). These results suggest that lysosomal cysteine proteinases may be involved in some sorts of neuronal death.

The present study examined cytoplasmic and nuclear alterations in the CA1 pyramidal neurons in gerbil hippocampus after brief ischemia. To detect the cytoplasmic changes, we immunocytochemically examined alterations in lysosomes, while *in situ* nick end labeling of biotinylated dUTP mediated by terminal deoxytransferase (TUNEL reaction) and Southern blot analysis were applied to the tissue to detect nuclear changes.

MATERIALS AND METHODS

Surgical Procedures

The bilateral common carotid arteries of adult male Mongolian gerbils (twelve weeks of age, ca. 70-100 gm) under light ether anesthesia were exposed at the neck and were occluded with aneurysm clips for 5 minutes. The clips were then released to resume the normal blood flow in the forebrain. Gerbils operated on without occluding the carotid arteries were used as sham operated controls.

Antisera

Rabbit antibodies to rat cathepsins B, H, and L, and anti-cystatin b , were purified by affinity chromatography, as previously reported (Bando *et al.*, 1986; Kominami *et al.*, 1984, 1985; Watanabe *et al.*, 1988). These antibodies were immunologically different from each other and showed no crossreactivity (Kominami *et al.*, 1984). Rabbit antibodies against ubiquitin which had been covalently coupled to bovine-γ-globulin were prepared and purified, as previously reported (Haas and Bright, 1985; Ueno and Kominami, 1991). These antibodies prefentially recognized ubiqutin-protein conjugates on Westernblots (Ueno and Kominami, 1991).

Sample for Immunohisto/Cytochemistry and Electron Microscopy

The sham operated controls and ischemia-treated animals at 12 hr, and 1, 2, 3, 4, and 7 days after the resumption of the cerebral blood flow were deeply anesthetized and fixed by cardiac perfusion with 4% paraformaldehyde buffered with 0.1 m cacodylate-hcl buffer containing 4% sucrose, ph 7.2, for light microscopic histochemistry and immunohistochemistry, 4% paraformaldehyde-0.1% glutaraldehyde buffered with the same buffer for electron microscopic immunocytochemistry, or 2% paraformaldehyde-2% glutaraldehyde buffered with the same buffer for ordinal electron microscopy. the brains were then removed from each animal.

Immunohisto/Cytochemistry

Light Microscopy. The samples were rapidly frozen after cryoprotection. Cryosections cut at 10 μm with a cryostat were immunostained using a method fully described elsewhere (Uchiyama *et al.*, 1990). Sections were incubated with the following first antibodies: anti-cathepsin B (5 μg/ml), anti-cathepsin H (7 μg/ml), anti-cathepsin L (3 μg/ml), and anti-cystatin b (9 mg/ml). For semi-thin sections, some samples of the 3 mm thick-brain sections were dehydrated with graded alcohols and embedded in Epon 812. Serial semi-thin sections cut at 1 μm with an ultramicrotome (Ultracut N, Reichert-Nissei: Tokyo, Japan) were immunostained with anti-ubiquitin (6 μg/ml) after removal of the epoxy resine with sodium methoxide.

Electron Microscopy. The samples were dehydrated with graded alcohols and embedded in LR White. Thin sections were cut with an ultramicrotome and mounted on nickel grids. For double immunostaining using anti-cathepsin B (11 μg/ml) and anti-ubiquitin (6 μg/ml), a two face technique according to Bendayan (1982) was performed as described previously (Uchiyama *et al.*, 1990). After the immunoreactions, sections were stained with a saturated aqueous solution of uranyl acetate and lead citrate and observed with a Hitachi H-7100 electron microscope.

Ordinal Electron Microscopy

The samples were postfixed with 2% osmium tetroxide buffered with 0.1 M cacodylate-HCl, pH 7.2, dehydrated with graded alcohols, and embedded in Epon 812. Thin sections of the CA1 pyramidal layer were cut with an ultramicrotome.

Dna Nick End Labeling

The terminal deoxynucleotidyl transferase (TdT)-mediated dUTP-biotin nick end labeling (TUNEL) reaction was applied to the cryosections according to the modified method of Gavrieli et al. (1992). Briefly, the cryosections were incubated with 100 U/ml TdT and 10 nmol/ml biotinylated 16-2'-dUTP at 37°C for 60 min. Further incubation with peroxidase-conjugated streptavidin or Texas red-conjugated avidin was carried out for 60 min at room temperature.

Southern Blotting

Hippocampal tissues including the CA1 zone were dissected from gerbil brains of the sham operated controls and of those 2, 3, and 4 days after ischemic insult. Genomic DNA from each hippocampal tissue was prepared by a modification of the method of Sambrook et al. (1989). Each sample was subjected to electrophoresis and transferred to a nylon membrane. The membrane was hybridized with [32P]-labeled EcoRI-digested gerbil brain DNA. The membrane was then subjected to autoradiography.

RESULTS

Immunohisto/Cytochemistry

Light Microscopy: Cysteine Proteinases and Cystatin B. In the sham operated gerbil brains, immunoreactivity for cathepsins B, H, and L was detected in the CA1, CA2, and

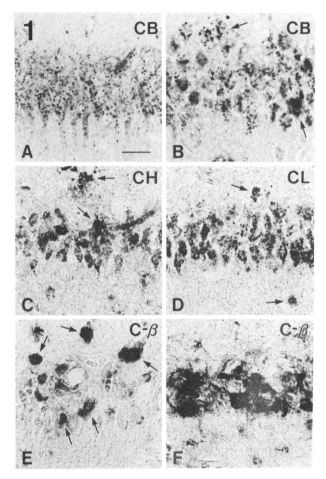

Figure 1. Changes in immunoreactivity for cathepsins B (CB), H (CH), and L (CL), and cystatin b (C-b) in the CA1 pyramidal layer of the hippocampus after 5 min ischemia. The immunoreactivity for cathepsin B in pyramidal neurons after ischemic insult progressively increases, compared with that of the sham operated control (A). Particularly, coarse granular immunodeposits for these three enzymes are densely localized in the neurons 3 d after ischemia (B-D). Phagocytic microgria-like cells showing immunoreactivity for cathepsins B, H, and L (B-D, arrows), and Cystatin b (E, arrows) appear on day 3 , and occupy the pyramidal cell layer on day 7 (F). Scale bar = 20 mm.

CA3 pyramidal layers of the hippocampus, and in the granular layer of the dentate gyrus, although that for cathepsin H was negative or faint (Fig. 1A). Three days after ischemic insult, the immunoreactivity for these enzymes became intense only in the CA1 pyramidal layer (Fig. 1B-D). Granular immuno-deposits for cathepsins B, H, and L gradually increased and became coarse in the CA1 pyramidal neurons after ischemic insult (Fig. 1B-D). Particularly, 3 days after ischemic insult, the CA1 pyramidal neurons, which had shrunk, were densely immunostained (Fig. 1B-D). At this stage, microglial cells showing coarse immunodeposits for cysteine proteinases and diffuse immunoreactivity for cystatin b also appeared close to the CA1 pyramidal neurons (Fig. 1B-E). On day 7, the CA1 pyramidal layer was occupied by microglial cells immunopositive for the cysteine proteinases and cystatin b (Fig. 1F). It is interesting that cystatin b-positive cells of a small size and round shape were also intensely immunostained by anti-cathepsin H.

Ubiquitin. Diffuse immunoreactivity for ubiquitin was localized in the nuclei and the perikarya of pyramidal neurons in the hippocampus and granular cells in the dentate gyrus of the untreated control brains (Fig. 2A). Twelve hr after the ischemic operation, the immunoreactivity disappeared from the nuclei of the CA1 pyramidal neurons and decreased in the perikarya. At this stage fine granular immunodeposits for ubiquitin appeared in the perikarya of the CA1 pyramidal neurons. On day 1 (24 hr), the immunoreactivity further decreased in the perikarya of these neurons, while the fine granular immunodeposits were detected in them (Fig. 2B). On day 3, coarse granular immunodeposits for ubiquitin were distinctly present in the perikarya of the CA1 pyramidal neurons and some neurons at the hilus of the dentate gyrus (Fig. 2C); these immunodeposits in the pyramidal neurons were very similar to those seen in the neuron immunostained by anti-cysteine proteinases (Fig. 1B-D).

Electron Microscopy. By electron microscopy, immunogold particles indicating cathepsin B were co-localized with those showing ubiquitin in large lysosomal structures (Fig. 3), indicating that large lysosomal structures are autolysosomes. Immunogold particles indicating ubiquitin were also detected in the ground cytoplasm.

Tunel Reaction

Positive staining of the TUNEL reaction was weakly detected in nuclei of the CA1 pyramidal neurons 3 days after ischemic insult (Fig. 4B), but not in those obtained in the earlier stages nor in the sham operated control brains (Fig. 4A). Four days after ischemic insult, the positive staining in the neurons was not only localized in their cell bodies, but also in their dendrites forming the molecular layer (Fig. 4C).

Electron Microscopy

Cytoplasmic Alterations. The control CA1 pyramidal neurons possessed well-developed rough endoplasmic reticulum (rER), numerous polysomes, and several dense bodies in the perikarya (data not shown). Three days after ischemic insult, profiles of rER and polysomes decreased from the perikarya of the neurons, whereas numerous dense bodies and membrane-bounded vacuoles containing membranous structures and parts of the cytoplasm, which seemed to be autophagic vacuoles, appeared in them (data not shown). Profiles of mitochondria and the Golgi complexes appeared intact 3 days after ischemic insult. Considering the fact that the cathepsin B- and ubiquitin-immunopositive lysosomes increased in the CA1 pyramidal neurons 3 days after ischemic insult (Fig. 3), these autolysosomes correspond well with membrane-bounded (autophagic) vacuoles.

Nuclear Alterations. The CA1 pyramidal neurons of the control brains had large round nuclei with a homogeneous electron density. Three days after ischemic insult, nuclear profiles of the CA1 neurons were often indented. Four days later, the CA1 neurons contained irregularly-shaped nuclei with dense chromatin masses , while apoptotic bodies containing distinct chromatin condensation were often found near microglial cells located in the CA1 pyramidal layer (Fig. 5). These nuclear changes occurred concomitantly with perikaryal condensation; numerous autophagic vacuoles and vacuolar structures (clear vacuoles) appeared in the shrunken perikarya of the neurons.

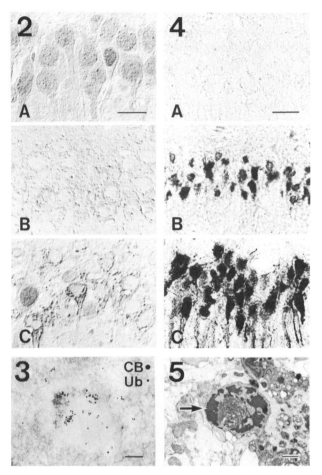

Figure 2. Changes in immunoreactivity for ubiquitin in the CA1 pyramidal layer of the hippocampus after 5 min ischemia. The immunoreactivity appears in both the nuclei and perikarya of the control pyramidal neurons (A). The diffuse immunoreactivity disappears from the nuclei 12 hr and from nuclei and perikarya 24 hr after ischemic insult, whereas granular immunodeposits for ubiquitin appear in the perikarya 24 hr after ischemic insult (B) and become coarse on day 3 (C). Scale bar = 20mm.

Figure 3. Double immuno-staining of cathepsin B (CB) and ubiquitin (Ub) in a CA1 pyramidal neuron 3 d after 5 min ischemia. A vacuolar structure containing amorphous structures in the perikarya of the neuron is colabeled with small and large immunogold particles. Scale bar = 0.1 mm.

Figure 4. TUNEL staining in the CA1 area of the hippocampus obtained from gerbil brains after 5 min ischemia. No staining is detected in the CA1 pyramidal neurons on day 2 (A), whereas the positive staining specifically appears in the CA1 pyramidal neurons 3 d (B) and 4 d (C) after ischemic insult. Scale bar = 20mm.

Figure 5. An electron micrograph of an apoptotic body (arrow) observed in the CA1 pyramidal layer 4d after 5 min ischemia. Scale bar = 1 mm.

Southern Blot Analysis

Since the nuclear changes in the CA1 neurons suddenly occurred 3 or 4 days after ischemic insult, we examined genomic DNA in hippocampal tissue by Southern blotting.

DNA fragmentation into oligonucleosomal fragments appeared only in the tissue 4 days after ischemic impact, but not in tissue at prior stages or in the control hippocampal tissue (data not shown).

DISCUSSION

In the present study, the following cytoplasmic as well as nuclear alterations were demonstrated in the CA1 pyramidal neurons of the gerbil hippocampus after brief ischemia: 1) perikaryal condensation with the formation of autophagic vacuoles (autolysosomes), 2) positive nick-end labeling of the nuclei, 3) dense chromatin masses in the nuclei and apoptotic bodies, 4) laddering of DNA, and 5) heterophagocytosis by cystatin b-immunopositive microglial cells.

Cytoplasmic Alterations

By immunocytochemistry, cathepsins B and L are localized in the lysosomes of nerve cells and astrocytes in the rat CNS, while cathepsin H is only found in perivascular microglia-like cells (Bernstein *et al.*, 1989, 1990; Taniguchi *et al.*, 1993). The present study demonstrated that the immunoreactivity for cathepsin H appeared in the CA1 pyramidal neurons in the early stages after the ischemic insult, corresponding to the increase in the immunoreactivity for cathepsins B and L. It has been shown that pyramidal neurons in the gerbil hippocampus decrease uptake of radiolabeled L-tyrosine immediately after ischemic insult, but the uptake by the neurons mostly recovers within 24 hr after ischemic impact, except for that by the CA1 pyramidal neurons (Thilmann *et al.*, 1986). Lysosomal cysteine proteinases degrade old, unneeded proteins into amino acids within lysosomes of mammalian cells, which are re-used for protein synthesis in the cells (Kirschke *et al.*, 1980, Katunuma and Kominami, 1983). It is interesting that the decrease in the uptake of amino acids in pyramidal neurons of gerbil hippocampus after brief ischemia occurs concomitantly with the increase in the immunoreactivity for lysosomal cysteine proteinases including cathepsin H.

The present study exhibited that cathepsin B-immunopositive lysosomes were colocalized with ubiquitin in the CA1 pyramidal neurons 3 days after ischemic insult, when autophagic vacuoles containing membranous structures and parts of the cytoplasm also increased in the perikarya. These events suggest that the degradation of proteins in lysosomes of the CA1 pyramidal neurons is highly activated after the ischemic insult and that this process proceeds by forming autolysosomes.

Autophagic alternations frequently occur in dying cells, and the role of the autophagy is believed to protect the cells from death (Clarke, 1990). That is, lysosomes in the CA1 pyramidal neurons early after ischemic insult may serve as a source of amino acids by actively degrading proteins. Concerning this point, the increases in the immuno-reactivity for cysteine proteinases in the CA1 pyramidal neurons may play an important role in protecting them from further alterations.

The dying CA1 pyramidal neurons contained numerous autophagic vacuoles and vacuolar structures (clear vacuoles) 4 days after ischemic insult when perikaryal condensation occurred. According to Clarke (1990), dying cells with autophagic vacuoles have been shown to undergo intense endocytosis, which can be thought of as inward blebbing and serves equally well to reduce the area of the plasma membranes. Therefore, it seems likely that the formation of autophagic vacuoles late in the stage after ischemic insult is closely associated with perikaryal condensation.

Nuclear Alterations

The present study showed the positive staining of the TUNEL reaction even in nuclei of the ischemia-damaged CA1 pyramidal neurons. Moreover, this positive reaction occurred in the nuclei prior to the morphological alterations with chromatin condensation. By Southern blot analysis, we also found DNA fragmentation into oligonucleosomal fragments. The decrease in protein synthesis of cultured neonatal sympathetic neurons after nerve growth factor deprivation is insufficient to cause subsequent death, whereas DNA fragmentation of the neurons occurs close to the neuronal death (Deckwerth and Johnson, 1993). This implies that DNA fragmentation in the CA1 pyramidal neurons after brief ischemia is linked to irreversible alterations of the neurons.

Heterophagocytosis by Cystatin B-Immunopositive Microglial Cells

It has recently been shown that several types of specific inhibitors for cysteine proteinase in mammalian tissues (Katunuma and Kominami, 1983). Cystatin b, an endogenous cysteine proteinases exist inhibitor, has been purified and sequenced from rat liver (Takio et al., 1983) and shown to be localized in alveolar macrophages and liver Kupffer cells (Ishii et al., 1991). Until the present study, however, little was known about the precise localization of cystatin b in brain tissue. In the present study, we found that cystatin b was specifically localized in activated microglial cells, suggestting that cystatin b is an excellent marker for activated microglia cells, in the central nervous tissue.

From the present results, we conclud that the delayed neuronal death in the CA1 pyramidal layer after brief ischemia is apoptosis, since the dying cells show morphological and biochemical findings consistent with those of apoptosis (Kerr et al., 1987; Clarke, 1990). Moreover, lysosomal cysteine proteinases, cathepsins B, H, and L participate in the process of perikaryal condensation by forming autophagic vacuoles or autolysosomes.

ACKNOWLEDGMENTS

This paper was supported by a Grant-in-Aid for Scientific Research on Priority Areas, Ministry of Education, Science and Culture, Japan and by a Grant from Keiryoukai Research Foundation (No. 44).

REFERENCES

Bando, Y., Kominami, E., and Katunuma, N., 1986, Purification and tissue distribution of rat cathepsin L, J. Biochem. 100:35-42.

Bendayan, M., 1982, Double immunocytochemical labelling applying the protein A-gold technique, J. Histochem. Cytochem. 30: 81-85.

Bernstein, H.-G., Kirschke, H., Roskoden, T., and Wiederanders, B., 1990, Distribution of cathepsin L in rat brain as revealed by immunohistochemistry, Acta Histochem. Cytochem. 23: 203-207.

Bernstein, H.-G., Sormunen, R., Järvinen, M., Kloss, P., Kirschke, H., and Rinne, A., 1989, Cathepsin B immunoreactive neurons in rat brain. A combined light and electron microscopic study, J. Hirnforsch. 30: 313-317.

Bonnekoh, P., Barbier, A., Oschlies, U., and Hossmann, K.-A., 1990, Selective vulnerability in the gerbil hippocampus: Morphological changes after 5-min ischemia and long survival times, Acta Neuropathol. (Berl.) 80: 18-25.

Cataldo, A.M. and Nixon, R.A., 1990, Enzymatically active lysosomal proteases are associated with amyloid deposits in Alzheimer brain, Proc. Natl. Acad. Sci. USA 87: 3861-3865.

Cataldo, A.M., Paskevich, P.A., Kominami, E. and Nixon R.A., 1991, Lysosomal hydrolases of different classes are abnormally distributed in brains of patients with Alzheimer disease, *Proc. Natl. Acad. Sci. USA* 88: 10998-11002.

Cataldo, A.M., Thayer C.Y., Bird, E.D., Wheelock, T.R. and Nixon, R.A., 1990, Lysosomal proteinase antigens are prominently localized within senile plaques of Alzheimer's disease: evidence for a neuronal origin, *Brain Res.* 513: 181-192.

Clarke, P. G. H., 1990, Developmental cell death: morphological diversity and multiple mechanisms, *Anat. Embryol.* 181: 195-213.

Deckwerth, T. L., and Johnson, E. M. Jr., 1993, Temporal analysis of events associated with programmed cell death (apoptosis) of sympathetic neurons deprived of nerve growth factor, *J. Cell Biol.* 123: 1207-1222.

Gavrieli, Y., Sherman, Y., and Ben-Sasson, A. J., 1992, Identification of programmed cell death *in situ* via specific labeling of nuclear DNA fragmentation, *J. Cell Biol.* 119: 493-501.

Haas, A. and Bright, P.M., 1985, The immunochemical detection and quantitation of intracellular ubiquitin-protein conjugates, *J. Biol. Chem.* 260: 12464-12473.

Ishii, Y., Hashizume, Y., Watanabe, T., Waguri, S., Sato, N., Yamamoto, M., Hasegawa, S., Kominami, E., and Uchiyama, Y., 1991, Cysteine proteinases in bronchoalveolar epithelial cells and lavage fluid of rat lung, *J. Histochem. Cytochem.* 39: 461-468.

Ito, U., Spatz, M., Walker, J. T., and Klatzo, I., 1975, Experimental cerebral ischemia in Mongolian gerbils. I. Light microscopic observations, *Acta Neuropathol. (Berl.)* 32: 209-223.

Katunuma, N., and Kominami, E., 1983, Structures and functions of lysosomal thiol proteinases and their endogenous inhibitors, *Curr. Top. Cell Regul.* 22: 71-101.

Kerr, J. F. R., Searle, J., Harmon, B. V., and Bishop, C. J., 1987, Apoptosis. In: Perspectives on mammalian cell death (Potten CS ed), pp 93-128. Oxford: Oxford University Press.

Kirino, T., 1982, Delayed neuronal death in the gerbil hippocampus following ischemia, *Brain Res.* 239: 57-69.

Kirino, T., and Sano, K., 1984a, Selective vulnerability in the gerbil hippocampus following transient ischemia, *Acta Neuropathol. (Berl.)* 62: 201-208.

Kirino, T., and Sano, K., 1984b, Fine structural nature of delayed neuronal death following ischemia in the gerbil hippocampus, *Acta Neuropathol. (Berl.)* 62: 209-218.

Kirschke, H., Langer, J., Riemann, S., Wiederanders, B., Ansorge, S., and Bohley, P., 1980, Lysosomal cysteine proteinases. In Protein degradation in health and disease (Evered D, Whelan J eds), Ciba Foundation Symposium 75, pp 15-35. Amsterdam: Excerpta Medica.

Kominami, E., Bando, Y., Wakamatsu, N., and Katunuma, N., 1984, Different tissue distributions of two types of thiol proteinase inhibitors from rat liver and epidermis, *J. Biochem.* 96: 1437-1442.

Kominami, E., Tsukahara, T., Bando, Y., and Katunuma, N., 1985, Distribution of cathepsins B and H in rat tissues and peripheral blood cells, *J. Biochem.* 98: 87-93.

Petito, C. K., Feldmann, E., Pulsinelli, W. A., and Plum, F., 1987, Delayed hippocampal damage in human following cardiorespiratory arrest, *Neurology* 37: 1281-1286.

Pulsinelli, W. A., and Brierley, J. B., 1979, A new model of bilateral hemispheric ischemia in the unanesthetized rat, *Stroke* 10: 267-272.

Sambrook, J., Fritsch, E. F., and Maniatis, T., 1989, Molecular cloning: A laboratory manual. 2nd ed. Cold Spring Harbor, NY: Cold Spring Habor Press, B.20,E.2.

Takio, K., Kominami, E., Wakamatsu, N., Katunuma, and Titani, K., 1983, Amino acid sequence of rat liver thiol proteinase inhibitor, *Biochem. Biophys. Res. Commun.* 115: 902-908.

Taniguchi, K., Tomita, M., Kominami, E., and Uchiyama, Y., 1993, Cysteine proteinases in rat dorsal root ganglion and spinal cord, with special reference to the co-localization of these enzymes with calcitonin gene-related peptide (CGRP) in lysosomes, *Brain Res.* 601: 143-153.

Thilmann, R., Xie, Y., Kleihues, P., and Kiessling, M., 1986, Persistent inhibition of protein synthesis precede delayed neuronal death in postischemic gerbil hippocampus, *Acta Neuropathol. (Berl.)* 71: 88-93.

Uchiyama, Y., Nakajima, M., Muno, D., Watanabe, T., Ishii, Y., Waguri, S., Sato, N., and Kominami, E., 1990, Immunocytochemical localization of cathepsins B and H in corticotrophs and melanotrophs of rat pituitary gland, *J. Histochem. Cytochem.* 38: 633-639.

Ueno, T. and Kominami, E., 1991, Mechanism and regulation of lysosomal sequestration and proteolysis, *Biomed. Biochem. Acta* 50: 365-371.

Watanabe, M., Watanabe, T., Ishii, Y., Matsuba, H., Kimura, S., Fujita, T., Kominami, E., Katunuma, N., and Uchiyama, Y., 1988, Immunohistochemical localization of cathepsins B, H, and their endogenous inhibitor, cystatin b, in islet endocrine cells of rat pancreas, *J. Histochem. Cytochem.* 36: 783-791.

23

PROTEIN AND GENE STRUCTURES OF 20S AND 26S PROTEASOMES

Keiji Tanaka, Tomohiro Tamura, Nobuyuki Tanahashi,
and Chizuko Tsurumi

Institute for Enzyme Research
University of Tokushima
Tokushima 770, Japan

ABSTRACT

The two types of proteasomes with apparent sedimentation coefficients of 20S and 26S consist of a number of heterogeneous polypeptides and are unusually large protein complexes of approximately 750 kDa and 2000 kDa, respectively. The 26S proteasome is a cylindrical caterpillar-shaped complex with a symmetrical assembly of a four-layered central 20S proteasome and two terminal 22S regulators each with a V-like structure. The central core and the terminal structures are formed by multiple polypeptides with molecular masses of 21-31 kDa and 28-112 kDa, respectively. We have been studying their detailed structures by protein-chemical and molecular biological techniques. In this review, we summarize the structural features of eukaryotic 20S and 26S proteasomes. We also discuss the possible function(s) of the terminal multi-protein regulator complex based on current information.

INTRODUCTION

The proteasome is a major compoment of a non-lysosomal, ubiquitin-dependent proteolysis system, which is involved in a variety of biologically important processes such as metabolic regulation, cell cycle control and the immune response (1). The proteasome has been recognized as a eukaryotic ATP-dependent protease responsible for selective degradation of various naturally occurring short-lived proteins and accelerated breakdown of abnormal proteins generated by various stresses in cells (2). Recently, the proteasome has been demonstrated to have the ability for not only exhaustive degradation of target proteins, but also for processing, such as in the conversion of a precursor p110 NFkB to the mature p50 form (3) and generation of antigenic peptides from endogenous cellular proteins (4).

Intracellular Protein Catabolism, Edited by Koichi Suzuki and Judith Bond
Plenum Press, New York, 1996

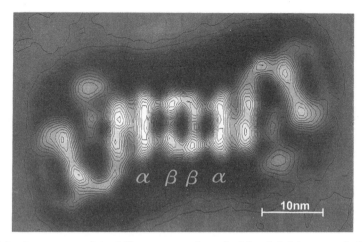

Figure 1. Molecular structure of the 26S proteasome determined by digital image analysis of electron micrographs. It consists of one central 20S proteasome with four rings of αββα structure and two polar 22S regulator complexes with a roughly V-shaped structure. For details, see Yoshimura *et al.*, (11), Peter *et al.*, (12) and Fujinami *et al.* (13).

Thus there is much evidence for the importance of various physiological functions of the proteasome.

Initially proteasomes were found as a 20S complex, but later various cellular components were shown to be associated with the 20S proteasome to form the 26S proteasome (2, 5). To understand the mechanisms for the complex functions of proteasomes, it is essential to clarify the structure-function relationships of their individual constituents. Eukaryotic proteasomes have been classified into two isoforms with apparent sedimentation coefficients of 20S and 26S (1). We have been examining the protein and gene structures of eukaryotic 20S and 26S proteasomes (6-8), and here we review recent information on the structural features of these proteasomes.

PROTEIN STRUCTURES

We have been studing the structures of 20S and 26S proteasomes from rat liver (9-11). Although the latter are referred to as 26S complexes, recent accurate measurements of its sedimentation coefficient indicated a value of approximately 30S. The physicochemical properties of the 20S and 26S proteasomes are listed in Table I. The estimated molecular masses of the 20S and 26S proteasomes were determined to be approximately 750 kDa and 2000 kDa, respectively, by various physical techniques. By electron microscopy, the 20S proteasome appears to be a cylindrical particle with a dimeric structure of two distinct discs. On the other hand, by electron microscopy in conjunction with digital image analysis, the 26S proteasome appears to be a caterpillar-shaped particle with two large V-like or U-like terminal structures in opposite orientations attached to a smaller central 20S proteasome (Fig. 1). The gross structures of 26S proteasomes purified from amphibia (12) mammals (11) and plants (13) are similar, suggesting their high evolutionary conservation possibly related to common functions in eukaryotes.

Table 1. Physicochemical Properties of 20S and 26S Proteasomes from Rat Liver.

Parameter	Method of analysis	20S Proteasome	26S Proteasome
Sedimentation coefficient.	Density gradient centrifugation	20S	26S
	Sedimentation velocity	20S	30S
Diffusion coefficient	Quasi-elastic light scattering	2.50×10^{-7} (cm² s⁻¹)	1.38×10^{-7} (cm² s⁻¹)
Stokes Radius	Quasi-elastic light scattering	8.5 (nm)	15.5 (nm)
	Small-angle X-ray scattering		7.6 (nm)
Gyration Radius	Small-angle X-ray scattering	6.6 (nm)	
	Static light scattering		16.8 (nm)
Partial specific volume	Amino acid composition	0.734 (ml g⁻¹)	
Isoelectric point	Isoelectric focusing	5.0	5.0
Extinction coefficient	Differential refractometry	9.61 ($E^{1\%}$ at 280 nm)	
Molecular ellipticity	Circular dichroism	-12000 (deg cm² dmol⁻¹)	
Molecular weight	Sedimentation velocity and diffusion	722000	2020000
	Sedimentation equilibrium	743000	
	Low-angle laser light scattering	760000	
	Small-angle X-ray scattering	750000	
	Static light scattering		1910000

Data are cited from Tanaka et al. (9, 10); and Yoshimura et al.(11).

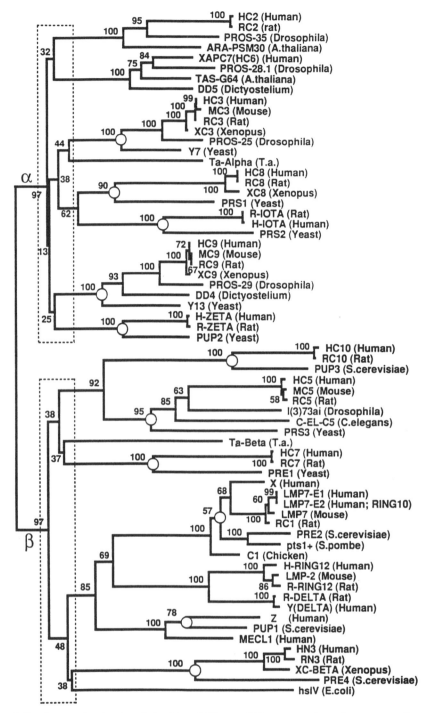

Figure 2. Dendrogram showing the relationships of all known 20S proteasomal protein sequences. Proteasome subunits can be assigned to one of two groups based on their similarity of the Thermoplasma acidophilim a (Ta-Alpha) or b (Ta-Beta) subunit (15). Note that hslV is a eubacterial gene (16). The sequence data are cited from reviews (6,7) and the data bank.

GENE STRUCTURES

The Proteasome Gene Family

Two dimensional electrophoretic analysis showed that the 20S proteasome consists of a set of polypeptides of small size (21-32 kDa) but with strikingly different charges (pI=3-10). We have determined the primary structures of most subunits of mammalian 20S proteasomes by cDNA cloning (6, 7). The protein sequences of all the subunits of the 20S proteasome from budding yeast have also been deduced from the nucleotide sequences of their genes isolated by recombinant DNA techniques (14). All the proteasomal genes examined so far encode previously unidentified proteins. The primary structures of these subunits show considerably high inter-subunit homology in the same species, and high evolutional conservation in various eukaryotes. We named these homologous genes encoding 20S proteasomal subunits the proteasome gene family (7).

The 20S proteasome of the archaebacterium *Thermoplasma acidophilum* consists of two polypeptides named the α- and β-subunits, and has the subunit organization $\alpha_7\beta_7\beta_7\alpha_7$ (15). Eukaryotic proteasome genes can also be classified into two distinct subgroups, α and β, with high similarities to archaebacterial α- and β-subunits, respectively. The α-type and β-type subunits also show considerable similarity, suggesting that they may have originated from a common ancestral gene. Immunoelectron microscopic studies indicated that the 20S eukaryotic proteasome is a complex dimer and that the α-type subunits are located on the outer side of a cylindrical particle consisting of four rings (15). An evolutionary tree of the various eukaryotic proteasomal genes studied so far clearly shows that the α-type and β-type subunits branched into 7 subgroups independently (Fig. 2). Based on these findings, the 20S proteasome is thought to be a dimeric assembly of two symmetrical discs, each consisting of 7 α-type subunits and 7 β-type subunits, having the molecular organization $\alpha n(1-7)\beta n(1-7)\beta n(1-7)\alpha n(1-7)$, where "n" indicates the number of 7 heterogeneous subunits. Interestingly, recently the HslV (*heat shock locus*) gene of *E coli* was identified and found to have high similarity with the β-type subunit (for review, see 16), suggesting that it may be a prototype of the eukaryotic proteasome gene. The HslV gene perhaps forms an operon with another heat shock gene named HslU that has high similarity to members of the 26S proteasomal ATPase family (see below), suggesting that a HslV-HslU protein complex may be a ATP-dependent protease in prokaryotes. Thus proteasome genes are ubiquitous not only in eukaryotes, but also in archaebacteria and eubacteria.

22s Regulator Complex

The 26S proteasome is a dimeric complex, consisting of a central 20S proteasome and two 22S regulators that are attached to the central part with opposite, asymmetric orientations. The terminal multi-protein complex consists of many polypeptides with molecular masses of 28-112 kDa (7). This complex has been referred to by various names, such as the 20S ball, u-particle, 19S cap, 19S ATPase, and PA700, and is here named "22S the regulator complex", since, in contrast to the 19S or 20S copmlex reported previously, when we recently purified the native complex in the presence of ATP, it's sedimentation was considerably different from that of the 20S proteasome and indicated an apparent sedimentation coefficient of "22S" (to be published). Of the components of the 22S multi-protein regulator complex, the primary structures of several members of a ATPase family have been determined, and structural analyses of the other compoments by cDNA cloning are currently in progress. The subunits of the 22S complex identified so far are shown in Fig. 3.

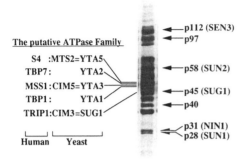

Figure 3. Electrophoretic analysis of the 22S regulator complex from rat liver, which is a regulatory part of the 26S proteasome. The 22S complex purified was analyzed by SDS-PAGE and proteins were stained with Coomassie brilliant blue. Details of the procedures for purification will be reported elsewhere. ATPases detected by immunoblotting are shown on the left. We recently determined primary structures of the subunits shown on the right by cDNA cloning. For details, see text.

ATPase Subunits : The purified 26S proteasome was found to show intrinsic ATPase activity (5). Presumably one role of the ATPase is to supply energy continuously for the selective degradation of ubiquitinated and non-ubiquitinated proteins by the active 26S proteasome (17). In addition, the purified 22S regulator complex was found to show high ATPase activity (11). Perhaps at least five different species of ATPase belonging to the same family are associated with the 26S proteasome, as shown in Fig. 3. In the human 26S proteasome, these 5 ATPases are named S4 (18), MSS1 (19), TBP1 (20), TBP7 (21) and p45 (to be published, see below). Genes homologous to all 5 human ATPases have been cloned from yeast (22). Interestingly, human MSS1 is a modulator of human immuno-deficiency virus HIV Tat-mediated transactivation, TBP1 and TBP7 are HIV Tat-binding proteins, and yeast SUG1 is a factor repressing mutation of the acidic activation domain in the universal transcriptional apparatus GAL4 (23). Moreover, yeast SUG1 was also found to be a component of a mediator complex, consisting of about 20 polypeptides that are associated with core RNA polymerase II to form a holoenzyme (24). We recently isolated a cDNA that encodes a new regulatory subunit, named p45, of the 26S proteasome of human hepatoblastoma HepG2 cells and found, by computer analysis, that p45 belongs to the family of putative ATPases associated with the 26S proteasome (to be published). The overall structure of p45 is homologous with that of yeast Sug1p, which has been identified as a transcripional mediator, but it is most closely similar to Trip1, reported as a functional homolog of Sug1p in human tissues (25). Thus, p45/Trip1 and Sug1p appear to function in the same way in protein degradation and transcriptional regulation in human cells, suggesting dual functions of this putative ATPase.

Another set of 26S proteasomal ATPase genes designated in *S. cerevisiae* as CIM5 and CIM3 (*co-lethal in mitosis*) are temperature sensitive (ts) and their growth is arrested at the G2/M boundary without loss of Cdc28p kinase activity under restrictive conditions (26). CIM5 is a homolog of human MSS1, and CIM3 is identical to SUG1. Moreover, in *S. pombe*, ts-mutation of the MTS2 gene (a homolog of the human S4 gene) also causes defective G2/M transition (27). These results indicate that putative proteasomal ATPases are closely related to cell cycle control. These ATPases show similarity to ATPase domains or subunits of the bacterial ATP-dependent proteases, the ClpA/X (ATPase subunits) - ClpP (common proteolytic subunit) complex and FtsH (a membrane-bound, ATP-requiring metallo-endoprotease found recently). Interestingly, ClpA or ClpX itself is suggested to have chaperone-like functions, because it induced conversion of many aggregated proteins to their native structures as reported for various chaperonins. However, if ClpP is associated with ClpA/X, the complex functions as an ATP-dependent protease, resulting in rapid hydrolysis of denatured proteins (28). Thus an ATP-dependent protease may be defined as the non-covalent assembly of an ATPase subunit with a protease component.

These ATP-binding domains appear to be members of an expanding family of ATPases that have been reported to be involved in not only proteolysis, gene expression and cell cycle regulation, but also vesicle-mediated secretion and peroxisome assembly (5). Thus 200 amino acid-ATPase modules were proposed to be an AAA family (ATPases associated with a variety of cellular activities) by Confalonieri and Duguet (29). Possibly these ATPase modules function as molecular chaperons related with protein-protein interactions, in which function requires metabolic energy.

Non-ATPase Subunits : Recently DeMartino et al. (30) isolated more than 10 components of the 22S regulator as a PA700 complex by HPLC, and partially sequenced them. Some of them are identical with the known ATPases mentioned above. In collaboration with DeMartino's group we recently cloned 6 other subunits named p28, p31, p40, p58, p97 and p112 in addition to the putative ATPase p45, and found that they were structurally unrelated to members of the ATPase family (Fig. 3). Subunit p31 was found to be a homolog of the cell cycle gene NIN1 (nuclear integrity) in S. cerevisiae (31). A mutation of Nin1p caused the accumulation of polyubiquitin-conjugated cellular proteins, indicating that Nin1p plays an important regulatory role in the function of the 26S proteasome (to be published). A temperature-sensitive nin1-1 mutant displayed various subtle phenotypes, such as chromosome instability and sensitivity to UV-irradiation. When the nin1-1 mutant and wild type cultures were first arrested by mating pheromone or hydroxyurea in the late G1 and early S phase respectively, and then released at a restrictive temperature, cell cycle progression was arrested almost completely. Moreover, activation of CDC28p-dependent histone H1 kinase was not observed in nin1-1 mutant cells, unlike in wild type cells, when they were exposed to a restrictive temperature. These results suggest that proteolysis catalyzed by the 26S proteasome containing Nin1p is necessary for activation of Cdc28p kinase and functions in G1/S and G2/M transitions of the cell cycle. The phenotype of the nin1-1 mutant, in which Cdc28p kinase is not activated, is in clear contrast to those of MTS2 or CIM5 ts-mutants that cannot exit from metaphase of the cell cycle due to dysfunction of cyclin B degradation. Why do mutations of these proteasomal components cause different defects in cell cycle progression? These subunits are perhaps involved in the breakdown of key factors functioning in distinct stages of the cell cycle. Therefore, the proteasome perhaps plays an essential role as a surveillance system for quantity control of many key proteins responsible for cell cycle regulation.

Human p28 and p58 were identified as homologs of SUN1 and SUN2 proteins, respectively, which are multicopy suppressors of the nin1-1, ts mutant of NIN1 in yeast, suggesting their physical association (to be published). Moreover, the overall structure of p112 is highly homologous to that of the SEN3 encoded factor affecting tRNA-splicing endonuclease in S. cerevisiae, whose sequence was deposited in the Gene databank by Culbertson et al. (personal communication). Interestingly, yeast SEN3 is a non-essential gene, although it's disruption caused growth-arrest at high temperature. Complementation analysis revealed that human p112 can substitute for SEN3 in yeast, indicating that SEN3 is a homolog of p112. Yeast cells with a defective SEN3 gene showed abnormal accumulation of poly-ubiquitinated proteins, suggesting failure of the 26S proteasome with a dysfunctional Sen3p subunit to degrade ubiquitinated proteins (to be published). Thus, the 26S proteasome may be involved in RNA processing directly or indirectly via degradation of specific proteins. Moreover, the SEN3 gene showed the synthetic lethality with the nin1-1 mutation, suggesting a genetic interaction between SEN3 and NIN1.

PERSPECTIVES

In this review we briefly summarized recent findings by physical analyses and cDNA cloning on the structural features of 20S and 26S proteasomes. However, the primary structures of about half the non-ATPase subunits of the 22S regulator complex have not yet been determined. Recently Deveraux *et al.*, (32) found a 50 kDa subunit, S5, of the 26S complex, that specifically binds to Ub-protein conjugates, confirming the recognition of ubiquitin-ligated proteins for selective breakdown by the 26S proteasome. In addition, the 26S proteasome has been reported to be associated with ubiquitin-isopeptidase, generating the ubiquitin moiety from the multiubiquitin chain for reutilization, but the cDNA of ubiquitin isopeptidase associated with the 26S proteasome has not yet been cloned. Other putative activities, such as oxido-reductase that would reduce disulfide bonds for unfolding of target proteins and protein-unfolding activities that may be required for entrapping target proteins within the 26S proteasome complex, have been suggested to be associated with the 26S proteasome from consideration of the molecular mechanisms of proteolysis mediated by the 26S proteasome (14). But the subunits responsible for these activities have also not yet been identfied. Therefore, studies on the protein sequences of 26S proteasome subunits should provide valuable information on the structure-function relationships of the individual components, the mechanisms of assembly of the 20S and 26S complexes, and the patho-physiological functions of the 26S proteasome.

REFERENCES

1. Goldberg, A.L., & Rock, K.L. (1992) *Nature* 357, 375-379.
2. Ciechanover, A. (1994) *Cell* 79, 13-21.
3. Palombella, V.J., Rando, O.J., Goldberg, A.L., & Maniatis, T. (1994) *Cell* 78, 773-785.
4. Cazynska, M., Rock, K.L., & Goldberg, A.L. (1993) *Enzyme Protein* 47, 354-369.
5. Rechsteiner, M., Hoffman, L., & Dubiel, W. (1993) *J. Biol. Chem.* 268, 6065-6068.
6. Tanaka, K., Tamura, T., Yoshimura, T., & Ichihara, A. (1992) *New Biol.* 4, 173-187.
7. Tanahashi, N., Tsurumi, C., Tamura, T., & Tanaka, K. (1993) *Enzyme Protein* 47, 241-251.
8. Tanaka, K. (1995) *Mol. Biol. Rep.* in press.
9. Tanaka, K., Yoshimura, T., Ichihara, A., Kameyama, K., & Takagi, T. (1986) *J. Biol. Chem.* 261, 15204-15207.
10. Tanaka, K., Yoshimura, T., Ichihara, A., Ikai, A., Nishigai, M., Morimoto, M., Sato, M., Tanaka, N., Katsube, Y., Kameyama, K., & Takagi, T. (1988) *J. Mol. Biol.* 203, 985-996.
11. Yoshimura, T., Kameyama, K., Takagi, T., Ikai, A., Tokunaga, F., Koide, T., Tanahashi, N., Tamura, T., Cejka, Z., Baumeister, W., Tanaka, K., Ichihara, A. (1993) *J. Struct. Biol.* 111, 200-211.
12. Peters, J.-M., Cejka ,Z., Harris, J.R., Kleinschmidt, J.A., & Baumeister, W. (1993) *J. Mol. Biol,* 234, 932-937.
13. Fujinami, K., Tanahashi, N., Tanaka, K., Ichihara, A., Cejka, Z., Baumeister, W., Miyawaki, M., Sato, T., & Nakagawa, H. (1994) *J. Biol. Chem.* 269, 25905-25910.
14. Hilt, W., Heinemeyer, W., & Wolf, D.H. (1993) *Enzyme Protein* 47, 189-201.
15. Lupas, A., Kaster, A.J., & Baumeister, W. (1993) *Enzyme Protein* 47, 252-273.
16. Lupas, A., Zwickl,, P., & Baumeister, W. (1994) *TIBS* 19, 533-534.
17. Murakami, Y., Matsufuji, S., Kameji, T., Hayashi, S., Igarashi, K., Tamura, T., Tanaka, K., Ichihara, A.(1992) *Nature* 360, 597-599.
18. Dubiel, W., Ferrell, K., Pratt, G., & Rechsteiner, M. (1992) *J. Biol. Chem.* 267, 22669-22702.
19. Shibuya, H., Irie, K., Ninomiya-Tsuji, J., Goebl, M., Taniguchi, T., & Matsumoto, K. (1992) *Nature* 357, 700-702.
20. Nelbock, P., Dillon, P.J., Perkins, A., & Rosen, C.A. (1990) *Science* 248, 1650-1653.
21. Ohana, B., Moore, P.A., Ruben, S.M., Southgate, C.D., Green, M.R. & Rosen, C.A. (1993) *Proc. Natl. Acad. Sci. USA* 90, 138-142.
22. Schnall, R., Mannhaupt, G., Stucka, R., Tauer, R., Ehnle, S., Schwarzlose, C., Vetter, I., & Feldmann, H. (1994) *Yeast* 10, 1141-1155.

23. Swaffield, J.C., Bromberg, J.F., & Johnston, S.A. (1992) *Nature* 357, 698-700.

24. Kim, Y.-J., Björklund, S., Li, Y., Sayre, M.H., & Kornberg, R.D. (1994) *Cell* 77, 599-608.

25. Lee, J.W., Ryan, F., Swaffield, J.C., Johnston, S.A., & Moore, D.D. (1995) *Nature* in press.

26. Ghisiain, M., Udvardy, A., & Mann, C. (1993) *Nature* 366, 358-362.

27. Gordon, C., McGurk, G., Dillon, P., Rosen, C. & Hastle, N. (1993) *Nature* 366, 355-357.

28. Craig, E.A., Weissman, J.S., & Horwich, A.L. (1994) *Cell* 78, 365-372.

29. Kunau, W.H., Beyer, A., Franken, T., Götte, K., Marzioch, M., Saidowsky, J., Skaletz-Rorowski, A., & Wiebel, F.F. (1993) *Biochimie*, 75, 209-224.

30. DeMartino, G.N., Moomaw, C.R., Zagniko, O.P., Proske, R.J., Chu-Ping, M., Afendis, S.J., Swaffield, J.C., & Slaughter, C.A. (1994) *J. Biol. Chem.* 269, 20878-20884.

31. Nisogi H, Kominami K, Tanaka K, & Toh-e, A. (1992) *Exp. Cell. Res.* 200: 48-57.

32. Deveraux, Q., Ustrell, V., Pickart, C., & Rechsteiner, M. (1994) *J. Biol. Chem.* 269, 7059-7061.

24

THE PROTEASOME AND PROTEIN DEGRADATION IN YEAST

Wolfgang Hilt, Wolfgang Heinemeyer, and Dieter H. Wolf

Institut für Biochemie
Universität Stuttgart
Pfaffenwaldring 55
70569 Stuttgart, Germany

20S PROTEASOMES: ACTIVITIES, GENES AND PROTEINS

In 1984 a high molecular mass multisubunit protease complex was isolated from *Saccharomyces cerevisiae* [Achstetter *et al.* 1984] which proved to be the yeast homologue of the 20S proteasome complexes found in all eukaryotic cells [Kleinschmidt *et al.* 1988]. The yeast 20S proteasome is able to cleave chromo- and fluorogenic peptides at the carboxyterminus of hydrophobic, basic or acidic amino acids (chymotrypsin-like-, trypsin-like- and peptidyl-glutamyl-peptide hydrolyzing activity, respectively) [Heinemeyer *et al.* 1991]. The yeast 20S proteasome is composed of different subunits, showing a set of protein bands in the SDS-PAGE with molecular masses ranging from 20 to 35 kDa. They can be separated into 14 protein spots after two-dimensional gel electrophoresis [Heinemeyer *et al.* 1991]. Genes named Y7, Y13, PRS1 and PRS2 (independendly cloned as Y8 and SCL1) were cloned and sequenced on the basis of protein sequences of 20S proteasome subunits, genes named PRS3, PUP1, PUP2 and PUP3 were sequenced by chance [for summary see Hilt *et al.* 1993b]. We cloned the β–type genes PRE1, PRE2, PRE3 and PRE4 by complementation of mutants defective in the chymotrypsin-like- (pre1 and pre2 mutants) or the PGPH-activity (pre3 and pre4 mutants) of the proteasome [Heinemeyer *et al.* 1991, Heinemeyer *et al.* 1993, Hilt *et al.* 1993a, Enenkel *et al.* 1994]. Additionally we cloned two α–type genes PRE5 and PRE6 using peptide sequences derived from purified proteasome subunits, extending the number of yeast 20S proteasome subunit genes to 14 [Heinemeyer *et al.* 1994]. The now known 14 proteasomal subunits show structural relationships to each other and to those of other species but are not similar to any other known protein. According to their degree of sequence similarity and their relationship to either the α– or the β–subunit of the *Thermoplasma* proteasome they can be grouped into seven α–type and seven β–type proteins. We propose that the known 14 yeast proteasomal subunits most likely represent the complete set of proteins constituting the 20S core complex in this organism [Heinemeyer *et al.* 1994]: (i) 14 cloned yeast proteasomal genes nicely correspond to the 14 protein spots obtained after two-dimensional separation of the subunits of the purified yeast 20S proteasome [Heinemeyer *et al.* 1991]. (ii) The seven α–type and seven β–type subunits found in yeast can easily be arranged into a structure with a two-fold a7b7 symmetry as found in the ancestral archaebacterial proteasome [Puhler *et al.* 1992]. (iii) Multiple alignment of all known eukaryotic 20S-proteasome sequences yields a dendrogram which clearly shows 14 subgroups (seven α–type- and seven β–type), each of the subgroups containing a single yeast member. According to the structure proposed for proteasomes of higher eukaryotes [Kopp *et al.* 1993, Schauer *et al.* 1993] the yeast 20S proteasome most probably constitutes a complex dimer with C2-symmetry. Within such a structure each of the 14 different subunits is used two times and occupies two defined positions. The calculated molecular masses of the 14 yeast proteasome subunits ranging from 21.2 to 31.6 kDa are in agreement with molecular masses of the protein bands found in SDS-PAGE studies of purified yeast 20S proteasomes [Tanaka *et al.* 1989, Kleinschmidt *et al.* 1988]. Proteasomes most probably represent a completely new type of proteolytic enzymes. Genetic studies in yeast link certain β–type subunits to different proteolytic activities. Intact Pre1 and Pre2 proteins are necessary for the chymotrypsin-like activity [Heinemeyer *et al.* 1991, Heinemeyer *et al.* 1993] whereas intact Pre3 and Pre4 proteins are necessary for the PGPH activity [Enenkel *et al.* 1994, Hilt *et al.* 1993a]. These results support the idea, that proteolytically active sites may be formed by interaction of two side by side arranged subunits. On the other hand active sites may be located at individual subunits within the proteasome. In this case inactivation may also be caused by conformational changes induced by a mutation in a protein located next to the catalytic subunit. So far only β–type subunits have been found to contain mutations which lead to defects of the 20S proteasome's peptide cleaving activities. By analogy to the archaebacterial "urproteasome" with its β–subunits forming the inner two rings of the cylindrical particle [Puhler *et al.* 1992], the sites of proteolysis in eukaryotic proteasomes are most probably located in the central rings made up of the β–type subunits. It is tempting to speculate that degradation of the finally unfolded proteins occurs in the hole of the 20S cylinder particle [Hilt & Wolf 1992].

THE 26S PROTEASOME

In 1986 a 26S protease complex which degrades ubiquitinylated proteins *in vitro* had been purified from rabbit reticulocytes [Hough *et al.* 1986]. Studies in several higher eukaryotes demonstrated that this 26S protease complex consists of the 20S proteasome as a core and additional subunits attached at both ends of the 20S cylinder [Peters *et al.* 1993]. Strong evidence appeared that this larger proteinase complex (now called 26S proteasome) also exists in yeast: (i) Indication for the function of a 26S proteasome in the yeast S. *cerevisae* is provided by the fact that mutants with defects in 20S proteasome typic peptide cleaving activities are also defective in the degradation of ubiquitinylated proteins in vivo, which are specific *in vitro* substrates of the 26S proteasome [Heinemeyer *et al.* 1993, Heinemeyer *et al.* 1991, Hilt *et al.* 1993a, Richter-Ruoff *et al.* 1992, Seufert & Jentsch 1992]. (ii) In recent work a 26S protease complex has been purified from yeast, which exhibits peptide cleaving activities of the 20S proteasome but is in addition able to degrade ubiquitinylated proteins *in vitro* [Fischer *et al.* 1994].

PROTEASOME FUNCTIONS

Proteasomes Are Essential to Life

Proteasomes are essential tools of the yeast cell. In 13 cases chromosomal deletion of one of the 14 different yeast 20S proteasome subunits was lethal. Spores derived from different heterozygous diploids carrying a null mutation of one of several 20S proteasome genes were able to germinate, but in each case cell growth stopped after 2-3 cell divisions [Heinemeyer et al. 1994, Hilt et al. 1993a]. We suggest that chromosomal deletion of one of the 20S proteasome genes abolishes the assembly of the 20S and 26S proteasome complex and therefore leads to complete loss of all proteasome functions.

Proteasomes Act in Stress-Dependent and Ubiquitin-Mediated Proteolytic Pathways

Yeast mutants have been used to demonstrate the *in vivo* function of the 20S proteasome in stress-dependent and ubiquitin-mediated proteolytic pathways. Mutations affecting the 20S proteasomal genes *PRE1* and *PRE2* lead to defects of the chymotrypsin-like activity [Heinemeyer et al. 1991, Heinemeyer et al. 1993], whereas mutants of *PRE3* and *PRE4* are defective in the PGPH activtiy of the complex [Hilt et al. 1993a, Enenkel et al. 1994]. *Pre1-1* single and even more so *pre1-1 pre2-2* as well as *pre1-1 pre4-1* double mutants are sensitive to heat stress [Heinemeyer et al. 1991, Heinemeyer et al. 1993, Hilt et al. 1993a] and also to the arginine analogue canavanine. Under these stress conditions the mutants accumulate ubiquitinylated proteins. It is suggested that heat and canavavine stress lead to the formation of large amounts of abnormal proteins which are ubiquitinylated but remain undegraded in the proteolysis defective 20S proteasome mutants [Heinemeyer et al. 1991, Heinemeyer et al. 1993, Hilt et al. 1993a].

Proteasomes Degrade Short-Lived N-End-Rule Substrates.

The function of proteasomes in the degradation of ubiquitinylated proteins *in vivo* has been confirmed by analyzing the degradation of the short-lived protein substrates of the N-end-rule pathway in 20S proteasome mutants [Richter-Ruoff et al. 1992, Seufert &Jentsch 1992]. The short-lived N-end-rule substrates which constitute artificial protein substrates in yeast, and are ubiquitinylated prior to proteolysis, are clearly stabilized in yeast 20S proteasome mutants *pre1-1* and *pre2-2* with defective chymotrypsin-like activity [Richter-Ruoff et al. 1992, Seufert & Jentsch 1992]. Stabilization is strongly enhanced in *pre1-1 pre2-2* double mutants [Richter-Ruoff 1992]. 20S proteasome mutants also stabilize the short lived Ub-Pro-ß-gal protein [Richter-Ruoff et al. 1992, Seufert & Jentsch 1992] which is ubiquitinylated in a N-end-rule independent pathway [Johnson et al. 1992].

IN VIVO SUBSTRATES OF THE PROTEASOME

Unassembled Proteins: Proteasomes Degrade Free α-Subunits of the Fatty Acid Synthase

The yeast fatty acid synthase complex which is composed of 6α- (Fas2) and 6ß- (Fas1) subunits, is a rather stable protein ($t_{1/2}$=20h) [Egner et al. 1993]. In contrast, in yeast cells containing a chromosomal *fas1* deletion free Fas2 α-subunits are rapidly degraded

($t_{1/2}$=1h). However, in 20S proteasome mutants (*pre1-1*) containing a *fas1* deletion allele the Fas2 protein is clearly stabilized, as compared to *PRE1* wild type cells [Egner *et al.* 1993]. The free α-subunit seems to be recognized as some sort of abnormal protein which is rapidly degraded via the proteasomal pathway.

Proteasomes Degrade Fructose-1,6-Bisphosphatase

Proteasome dependent degradation of a metabolic enzyme has been shown in the case of fructose-1,6-bisphosphatase which is a key enzyme in gluconeogenesis. This enzyme is subject to catabolite inactivation: upon addition of glucose to yeast cells, which have been grown on a non-fermentable carbon source, fructose-1,6-bisposphatase is inactivated by phosphorylation and then rapidly degraded ($t_{1/2}$=1 h). In *pre1-1* 20S proteasome mutants, fructose-1,6-bisphosphatase is strongly stabilized. In *pre1-1 pre2-1* double mutants defective in two subunits of the 20S proteasome, degradation of fructose-1,6-bisposphatase is nearly absent [Schork *et al.* 1994].

Proteasomes Degrade the Yeast MATα2-Repressor

Proteasomes are also needed for degradation of regulatory proteins: This has been demonstrated in yeast for the MATα2-protein which is required for mating type differentiation of this organism. The MATα2-protein is responsible for repression of MATa specific genes in haploid MATα cells and of haploid specific genes in diploid cells. The very short-lived MATα2-repressor ($t_{1/2}$=5min at 30°C) [Hochstrasser & Varshavsky 1990] contains two independent destruction boxes and is ubiquitinylated via the ubiquitin conjugating enzymes Ubc4/Ubc5 and Ubc6/Ubc7 [Chen *et al.* 1993, Hochstrasser &Varshavsky 1990]. The function of proteasomes in degradation of the MATα2-repressor has been demonstrated by the stabilization of this protein in 20S proteasome mutants (*pre1-1* and *pre1-1 pre2-2*) [Richter-Ruoff *et al.* 1994].

PROTEASOMES FUNCTION IN YEAST CELL CYCLE CONTROL

Multiple kinase complexes formed by the association of different types of cyclins with a single kinase subunit (Cdc28) are crucial for cell cycle control in yeast [Nasmyth 1993]. The appearance and disappearance of particular kinase forms is regulated by the synthesis and proteolytic degradation of specific cyclins during different phases of the cell cycle. Certain β–type cyclins (Clb1-4) are important for function and assembly of the spindle apparatus during mitosis. Such β–type cyclins are thought to be degraded by proteasomes via ubiquitin dependent pathways: Studies of rapid mitotic degradation of β–type cyclins revealed an aminoterminally located highly conserved stretch of nine amino acids (designated as the "destruction box ") and ubiquitin modification prior to proteolysis [Glotzer *et al.* 1991]. We now uncovered a link between proteasome and Clb2 function. Yeast cells with a proteolytically stable form of the cyclin Clb2 are arrested in the mitotic telophase [Nasmyth 1993, Surana *et al.* 1993]. Overexpression of a single *CLB2* copy from the *GAL1* promoter is tolerated by yeast cells while overexpression of four copies of *CLB2* is lethal [Surana *et al.* 1993]. However in proteolysis defective 20S proteasome *(pre1-1)* mutants, overexpression of even one copy of *CLB2* leads to a halt in cellular growth [Richter-Ruoff & Wolf 1993] suggesting that proteasome dependent proteolysis is needed for control of the cell cycle.

CONCLUDING REMARKS

Thus using yeast as a model organism of the eukaryotic cell, the necessity of the proteasome for life and its involvement in the ubiquitin pathway of protein degradation and in central cellular functions *in vivo* could be shown for the first time. It can be expected that the action of the proteasome as the counterpart of the ribosome will be found in many more essential cellular processes.

REFERENCES

Achstetter, T., Ehmann, C., Osaki, A. and Wolf, D.H. 1984 Proteinase yscE, a new yeast peptidase *J. Biol. Chem.* 259:13344-13348

Chen, P., Jentsch, S. and Hochstrasser, M. 1993 Multiple ubiquitin-conjugating enzymes *Cell* 74:357-369

Egner, R., Thumm, M., Straub, M., Simeon, A., Schuller, H.J. and Wolf, D.H. 1993 Tracing intracellular proteolytic pathways. Proteolysis of fatty acid synthase and other cytoplasmic proteins in the yeast Saccharomyces cerevisiae *J. Biol. Chem.* 268:27269-27276

Enenkel, C., Lehmann, H., Kipper, J., Guckel, R., Hilt, W. and Wolf, D.H. 1994 *PRE3*, highly homologous to the human major histocompatibility complex-linked *LMP2* (RING12) gene, codes for a yeast proteasome subunit necessary for the peptidylglutamyl-peptide hydrolyzing activity *FEBS Lett.* 341:193-196

Fischer, M., Hilt, W., Richter-Ruoff, B., Gonen, H., Ciechanover, A., and Wolf, D.H. 1994 The 26S Proteasome of the yeast *Saccharomyces cerevisiae FEBS Lett.* 355:69-75

Glotzer, M., Murray, A.W. and Kirschner, M.W. 1991 Cyclin is degraded by the ubiquitin pathway *Nature* 349:132-138

Heinemeyer, W., Tröndle, N., Albrecht, G. and Wolf, D.H. 1994 *PRE5* and *PRE6*, the last missing genes encoding 20S proteasome subunits from yeast? Indication for a set of 14 different subunits in the eukaryotic proteasome core *Biochemistry* 33:12229-12237

Heinemeyer, W., Gruhler, A., Möhrle, V., Mahe, Y. and Wolf, D.H. 1993 *PRE2*, highly homologous to the human major histocompatibility complex-linked RING10 gene, codes for a yeast proteasome subunit necessary for chymotryptic activity and degradation of ubiquitinated proteins *J. Biol. Chem.* 268:5115-5120

Heinemeyer, W., Kleinschmidt, J.A., Saidowsky, J., Escher, C. and Wolf, D.H. 1991 Proteinase yscE, the yeast proteasome/multicatalytic-multifunctional proteinase: mutants unravel its function in stress induced proteolysis and uncover its necessity for cell survival *EMBO J.* 10:555-562

Hilt, W., Enenkel, C., Gruhler, A., Singer, T. and Wolf, D.H. 1993a The *PRE4* gene codes for a subunit of the yeast proteasome necessary for peptidylglutamyl-peptide-hydrolyzing activity - mutations link the proteasome to stress-dependent and ubiquitin-dependent proteolysis *J. Biol. Chem.* 268:3479-3486

Hilt, W., Heinemeyer, W. and Wolf, D.H. 1993b Studies on the yeast proteasome uncover its basic structural features and multiple in vivo functions *Enyzme Protein* 47:189-201

Hilt, W. and Wolf, D.H. 1992 Stress-induced proteolysis in yeast *Molecular Microbiology* 6:2437-2442

Hochstrasser, M. and Varshavsky, A. 1990 In vivo degradation of a transcriptional regulator: the yeast alpha 2 repressor *Cell* 61:697-708

Hough, R., Pratt, G. and Rechsteiner, M. 1986 Ubiquitin-lysozyme conjugates. Identification and characterization of an ATP-dependent protease from rabbit reticulocyte lysates *J. Biol. Chem.* 261:2400-2408

Johnson, E.S., Bartel, B., Seufert, W, and Varshavsky, A. 1992 Ubiquitin as a degradation signal *EMBO J.* 11:497-505

Kleinschmidt, J.A., Escher, C. and Wolf, D.H. 1988 Proteinase yscE of yeast shows homology with the 20 S cylinder particles of *Xenopus laevis FEBS Lett.* 239:35-40

Kopp, F., Dahlmann, B. and Hendil, K.B. 1993 Evidence indicating that the human proteasome is a complex dimer *J. Mol. Biol.* 229:14-19

Nasmyth, K. 1993 Control of the yeast cell cycle by the Cdc28 protein kinase *Curr. Opin. Cell. Biol.* 5:166-179

Peters, J.M., Cejka, Z., Harris, J.R., Kleinschmidt, J.A. and Baumeister, W. 1993 Structural Features of the 26S Proteasome Complex *J. Mol. Biol.* 234:932-937

Pühler, G., Weinkauf, S., Bachmann, L., Müller, S., Engel, A., Hegerl, R. and Baumeister, W. 1992 Subunit stoichiometry and three-dimensional arrangement in proteasomes from *Thermoplasma acidophilum EMBO J.* 11:1607-1616

Richter-Ruoff, B., Wolf, D.H. and Hochstrasser M. 1994 Degradation of yeast MATα2 transcriptional regulator is indicated by the proteasome *FEBS Lett.* 354:50-52

Richter-Ruoff, B.and Wolf, D.H. 1993 Proteasome and cell cycle. Evidence for a regulatory role of the protease on mitotic cyclins in yeast *FEBS Lett..* 336:34-36

Richter-Ruoff, B., Heinemeyer, W. and Wolf D.H. 1992 The proteasome/ multicatalytic-multifunctional proteinase. In vivo function in the ubiquitin-dependent N-end rule pathway of protein degradation in eukaryotes *FEBS Lett.* 302:192-196

Schauer, T.M., Nesper, M., Kehl, M., Lottspeich, F., Mllertaubenberger, A. Gerisch, G. and Baumeister, W. 1993 Proteasomes from *Dictyostelium discoideum* - Characterization of structure and function *J. Struct. Biol.* 111:135-147

Schork, S.M., Bee, G., Thumm, M. and Wolf, D.H. 1994 Site of catabolite inactivation *Nature* 369:283-284

Seufert, W. and Jentsch, S. 1992 In vivo function of the proteasome in the ubiquitin pathway *EMBO J.* 11:3077-3080

Surana, U., Amon, A., Dowzer, C., McGrew, J., Byers, B. and Nasmyth, K. 1993 Destruction of the CDC28/CLB mitotic kinase is not required for the metaphase to anaphase transition in budding yeast *EMBO J.* 12:1969-1978

Tanaka, K., Tamura, T., Kumatori, A., Kwak, T.H., Chung, C.H., Ichihara, A. 1989 Separation of yeast proteasome subunits. Immunoreactivity with antibodies against ATP-dependent protease Ti from *Escherichia coli Biochem. Biophys. Res. Commun.* 164:1253-1261

25

UBIQUITIN C-TERMINAL HYDROLASES IN CHICK SKELETAL MUSCLE

Chin Ha Chung, Seung Kyoon Woo, Jae Il Lee, Il Kyoo Park, Man-Sik Kang, and Doo Bong Ha

Department of Molecular Biology and SRC for Cell Differentiation
College of Natural Sciences
Seoul National University
Seoul 151-742, Korea

INTRODUCTION

Ubiquitin is the highly conserved 76 amino acid protein found in all eukaryotic cells. It exists as a free monomer and covalently ligated to a variety of intracellular proteins through an isopeptide linkage between the C-terminal Gly of ubiquitin and the εNH group of Lys of target proteins (Rechsteiner, 1987). Ubiquitins by themselves or that have already been conjugated to proteins may also be ligated to additional ubiquitin molecules to form a branched multi-ubiquitin chain. This ubiquitination has been implicated in the regulation of a variety of cellular processes, such as selective protein breakdown, cell cycle regulation, and stress response (Hershko and Ciechanover, 1992). In addition, protein ubiquitination *in vivo* is a dynamic, reversible process that is under control responding to external stimuli, such as heat shock and starvation (Carlson and Rechsteiner, 1987; Carlson *et al.*, 1987; Haas, 1988). Therefore, the enzymes that proteolytically remove ubiquitins from ubiquitin-protein conjugates should be of importance in maintaining the steady-state levels of free ubiquitin for a variety of its cellular functions.

Ubiquitins are encoded by two distinct classes of genes. One is a poly-ubiquitin gene which encodes a polyprotein of tandemly repeated ubiquitins (Ozkaynak *et al.*, 1984; Lund *et al.*, 1985). The other encodes a fusion protein of which a single ubiquitin is linked to a ribosomal protein consisting of 52 or 76-80 amino acids. The transient association of ubiquitin with the ribosomal proteins has been suggested to promote their incorporation into ribosomes (Finley *et al.*, 1989). Therefore, proteolysis at the peptide bonds between ubiquitin and the extension proteins is required for generation of ribosomal proteins for ribosome biogenesis as well as of free ubiquitin.

A number of ubiquitin C-terminal hydrolases (UCHs) have been identified from *Saccharomyces cerevisiae*. The YUH1 protease releases ubiquitin from its linear C-terminal conjugates to relatively short peptides (Miller *et al.*, 1989). Three different ubiquitin-specific proteases, UBP1, UBP2 and UBP3, also hydrolyze linear ubiquitin conjugates but irrespec-

Intracellular Protein Catabolism, Edited by Koichi Suzuki and Judith Bond
Plenum Press, New York, 1996

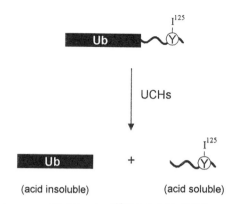

Figure 1. Assay of UCHs using ^{125}I-labeled Ub-PESTc as a substrate.

tive of the sizes of C-terminal extensions (Tobias and Varshavsky, 1991; Baker *et al.,* 1992). An additional deubiquitinating enzyme, called DOA4 (UBP4), releases ubiquitin that is conjugated to proteins by εNH-isopeptide linkage as well as by αNH-peptide bond (Papa and Hochstrasser, 1993). These studies with yeast containing at least 5 different deubiquitinating enzymes imply the existence of a variety of uncharacterized UCHs also in higher eukaryotic organisms.

A family of UCHs from bovine calf thymus, named L1, L2, L3 and H2, has been identified using ubiquitin-O-C_2H_5 as a substrate (Mayer and Wilkinson, 1989). However, only two of the enzymes (L1 and L3) have so far been purified (Wilkinson *et al.,* 1992), partly due to the difficulty and/or insensitivity of the available assay methods. Therefore, a simple and sensitive assay method is necessary for facilitating the purification of UCHs, for studying more systematically the diversity of the enzymes in a single source, and for comparison of the properties of the enzymes from different sources.

When Ub-PESTc is incubated with UCHs, free ubiquitin and PESTc peptide molecules will be generated. Because Ub-PESTc is radio-iodinated almost exclusively in the peptide portion (*i.e.,* most likely on the 17th Tyr residue), the activity of any UCH can be measured simply by counting the radioactivity of the acid-soluble products.

A NEW PROTOCOL FOR ASSAYING UCHS

Rechsteiner and coworkers have constructed ubiquitin-αNH-peptide extensions containing "PEST" sequences (Rogers *et al.,* 1986; Yoo *et al.,* 1989). Of these, ubiquitin-MHISP-PEPESEEEEEHYC (referred to as Ub-PESTc) can be almost exclusively radio-iodinated on the extension peptide under mild labeling conditions (*e.g.,* using Iodo-Beads). In addition, the peptide portion containing 18 amino acids is short enough to be soluble in acid solutions, such as 10% trichloroacetic acid (TCA). Therefore, the activities of UCHs can rapidly be assayed by simple measurement of the radioactivity released into acid-soluble products (Figure 1). For example, when the purified YUH1 was incubated with ^{125}I-labeled Ub-PESTc, production of TCA-soluble radioactivity increased in a time-dependent manner (Figure 2). Of interest is an unusual observation that poly-L-Lys dramatically stimulates the activity of YUH1. However, the mechanism by which the highly basic polypeptide activates the enzyme is not known, particularly because it shows little or no effect on the hydrolysis of ubiquitin-carboxyl extension protein of 80 amino acids (Ub-CEP80) by YUH1 or the partially purified UCHs from muscle (see below). Since YUH1 is known to specifically

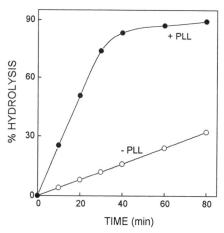

Figure 2. Hydrolysis of ^{125}I-labeled Ub-PESTc by the purified YUH1. Reaction mixtures containing YUH1 and the radiolabeled substrate were incubated for various periods in the absence (-) and presence (+) of poly-L-Lys (PLL). The mixtures were then added to TCA and centrifuged. The resulting supernatant fractions were counted for their radioactivity.

hydrolyze the carboxyl side of the C-terminal RGG sequence of ubiquitin, the acid-soluble products should represent the PESTc peptides released from the ubiquitin-peptide extension.

SEPARATION AND PROPERTIES OF MUSCLE UCHS

Using the newly developed assay protocol, at least 10 different UCHs active against ^{125}I-labeled Ub-PESTc have been identified and partially purified from the extracts of chick skeletal muscle. These enzymes are named tentatively as UCH-1 to -10. In addition, the muscle UCHs, except UCH-1, are also markedly activated by poly-L-Lys. The summary of the chromatographic procedures for the separation of UCHs are shown in Figure 3. All the muscle UCHs are capable of specifically hydrolyzing the carboxyl side of the C-terminal RGG sequence of ubiquitin as revealed by polyacrylamide gel electrophoresis of the cleavage products in the presence of SDS and by determination of the N-terminal sequence of the PESTc peptide released as acid-soluble products.

The sizes of the muscle UCHs range from 35 kDa to 810 kDa as determined under nondenaturing conditions using Superose-6 and -12 columns. Among the enzymes, UCH-1 has the smallest molecular mass of 35 kDa while UCH-7 has the largest size of about 810 kDa. All the UCHs are sensitive to inhibition by sulfhydryl blocking agents, such as N-ethylmaleimide, but not by the inhibitors of serine proteases including diisopropyl fluorophosphate or phenylmethylsulfonyl fluoride. Therefore, the muscle UCHs appears to contain a sulfhydryl residue(s) necessary for its catalytic activity, similar to the previously identified UCHs (e.g., YUH1 and L3). In addition, the activities of the muscle enzymes, except that of UCH-1, are strongly inhibited by high salt concentrations, such as 0.3 M NaCl, although they can be fully recovered upon dialysis to remove the salt. Therefore, the ten newly identified UCHs from the extracts of chick skeletal muscle appear distinct from each other at least in their chromatographic behaviors and sizes.

The molarities shown in the flow-chart indicate the NaCl concentrations where the peaks of Ub-PESTc-cleaving activities were eluted. The numerals in the parenthesis represent the molecular masses of UCHs determined by gel filtration under nondenaturing conditions. The fractions eluted at about 0.2 M NaCl from the HiLoad-Q column (UCH-5s) still contained multiple peaks of the UCH activity but could not be further separated due to their instability. Therefore, their sizes were not determined (ND[*]).

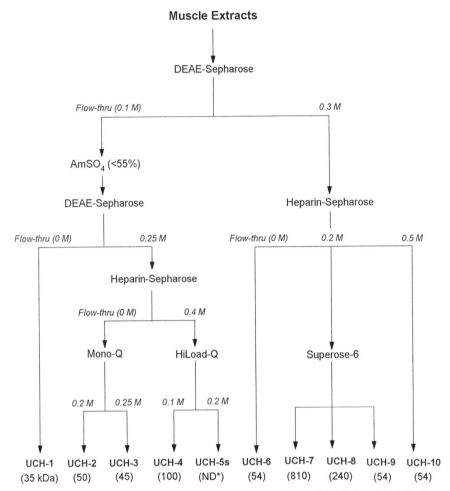

Figure 3. Summary of the chromatographic procedures for separation of UCHs in chick skeletal muscle.

SUBSTRATE SPECIFICITY OF MUSCLE UCHS

All the muscle UCHs are also capable of releasing free ubiquitin from Ub-CEP80 and Ub-dihydrofolate reductase (Ub-DHFR) (Table 1). In the absence of poly-L-Lys, Ub-CEP80 is hydrolyzed by the individual UCH, except UCH-6, more-or-less at the same rate as Ub-PESTc. On the other hand, Ub-DHFR is hydrolyzed by all the muscle UCHs much less efficiently than Ub-PESTc or Ub-CEP80. In addition, UCH-6 cleaves Ub-PESTc much faster than Ub-CEP80. Thus, the size and/or the tertiary structure of extension proteins appear to influence the susceptibility of ubiquitin-extension proteins to the muscle enzymes. On the other hand, only five of the enzymes (UCH-1 to -5) can cleave polyHis-tagged di-ubiquitin. Therefore, the muscle UCHs may play a role in the production of both the ribosomal protein and free ubiquitin and the five of them may also participate in the generation of ubiquitin from the product of the poly-ubiquitin gene in muscle cells.

In addition to the ubiquitin-extension proteins, UCH-1 and -7 can also hydrolyze the isopeptide linkage of mono-ubiquitinated Ub(A)-αNH-PARRKWQKTGHAVRAIGRLSS

Table 1. Hydrolysis of various ubiquitin conjugates by muscle UCHs. The partially purified muscle UCHs were incubated with the purified substrates at 37 °C for appropriate periods. The samples were then heated at 55 - 85 °C for 10 min and centrifuged. The supernatant fractions were then subjected to polyacrylamide gel electrophoresis in the presence of SDS using Tris-Tricine buffer, as described by Schagger and von Jagow (1987), and the appearance of free ubiquitin band or disappearance of mono-ubiquitinated Ub(A)-P21 band was monitored.

Substrates	Hydrolysis by UCH-									
	1	2	3	4	5s	6	7	8	9	10
Ub-CEP80	+	+	+	+	+	+	+	+	+	+
Ub-DHFR	+	+	+	+	+	+	+	+	+	+
His-tagged diUb	+	+	+	+	+	−	−	−	−	−
Mono-ubiquitinated Ub(A)-P21	+	−	−	−	−	−	+	−	−	−

(Ub(A)-P21), of which the C-terminal Gly of ubiquitin is replaced by Ala for preventing the action of UCHs on the αNH-peptide bond between the ubiquitin variant and the peptide extension. However, none of these enzymes and other muscle UCHs are capable of releasing ubiquitin from di- or tri-ubiquitinated Ub(A)-P21. The isopeptidase T has been shown to remove ubiquitin from high-molecular-weight, multi-ubiquitinated protein conjugates but not from low-molecular weight forms (Hadari et al., 1992). It has also been demonstrated that the 26S proteasome has inherent deubiquitinating activity against adducts of which a single ubiquitin is linked to εNH-Lys group of protein as well as against conjugates containing multiple ubiquitins (Eytan et al., 1993). Therefore, UCH-1 and -7 together with the 26S proteasome may be involved in the release of ubiquitin from Lys residue of the end products generated by the action of the isopeptidase T against the poly-ubiquitinated protein conjugates for complete recycling of ubiquitin molecules.

CONCLUSION

The present work provides a newly developed assay method for UCH using ^{125}I-labeled Ub-PESTc and summarizes a protocol for partial purification of at least ten UCHs from the extracts of chick skeletal muscle. These enzymes appear distinct from each other in their chromatographic behavior, size and substrate specificity. Possible roles of the muscle UCHs include processing of newly synthesized polyubiquitin and ubiquitin fusion proteins, regulation of ubiquitin-dependent proteolytic system, recycling of ubiquitin from side products with thiols and amines and reversal of nondegradative protein modification by ubiquitination (Hershko and Ciechanover, 1992). Future studies must focus on complete purification of the muscle UCHs and isolation of their cDNAs for detailed characterization of the enzymes, for clarification of their identity with the previously isolated UCHs from other sources, and ultimately for their functional analysis.

ACKNOWLEDGEMENTS

We are grateful to Drs. Keiji Tanaka (Institute for Enzyme Research, Japan) and Yung Joon Yoo (Lucky Biotech., Korea) for their critical comments and supply of many strains and plasmids. This work was supported by grants from Korea Science and Engineering Foundation through SRC for Cell Differentiation and Ministry of Education.

REFERENCES

Baker, R. T., Tobias, J. W., and Varshavsky, A., 1992, Ubiquitin-specific proteases of *Saccharomyces cerevisiae:* Cloning of UBP2 and UBP3, and functional analysis of the UBP gene family. *J. Biol. Chem.* 267, 23364-23375.

Carlson, N., and Rechsteiner, M., 1987, Microinjection of ubiquitin: Intracellular distribution of and metabolism in HeLa cells maintained under normal physiological conditions. *J. Cell Biol.* 104, 537-546.

Carlson, N., Rogers, S., and Rechsteiner, M., 1987, Microinjection of ubiquitin: Changes in protein degradation in HeLa cells subjected to heat-shock. *J. Cell Biol.* 104, 547-555.

Eytan, E., Armon, T., Heller, H., Beck, S., and Hershko, A., 1993, Ubiquitin C-terminal hydrolase activity associated with the 26S protease complex. *J. Biol. Chem.* 268, 4668-4674.

Finley, D., Bartel, B., and Varshavsky, A., 1989, The tails of ubiquitin precursors are ribosomal proteins whose fusion to ubiquitin facilitates ribosome biogenesis. *Nature* 338, 394-401.

Haas, A. L., 1988, Immunochemical probes of Ub pool dynamics. In *Ubiquitin* (Rechsteiner, M., ed.) pp. 173-206, Plenum Press, New York.

Hadari, T., Warms, J. V. B., Rose, I. A., and Hershko, A., 1992, A ubiquitin C-terminal isopeptidase that acts on polyubiquitin chain. *J. Biol. Chem.* 267, 719-727.

Hershko, A., and Ciechanover, A. , 1992, The ubiquitin system for protein degradation. *Annu. Rev. Biochem.* 61, 761-807.

Lund, P. K., Moasts-Staats, B. M., Simmons, J. G., Hoyt, E., D'Ercole, A. J., Martin, F., and Van Wyk, J. J., 1985, Nucleotide sequence analysis of a cDNA encoding human ubiquitin reveals that ubiquitin is synthesized as a precursor. *J. Biol. Chem.* 260, 7609-7613.

Mayer, A. N., and Wilkinson, K. D., 1989, Detection, resolution, and nomenclature of multiple ubiquitin carboxyl-terminal esterases from bovine calf thymus. *Biochemistry* 28, 166-172.

Miller, H. I., Henzel, W. J., Ridgway, J. B., Kuang, W. J., Chisholm, V., and Liu, C. C., 1989, Cloning and expression of a yeast ubiquitin-protein cleaving activity in *E. coli. Bio/Technology* 7, 698-704.

Ozkaynak, E., Finley, D., and Varshavsky, A., 1984, The yeast ubiquitin gene: Head-to-tail repeats encoding a polyubiquitin precursor protein. *Nature* 312, 663-666.

Papa, F. R., and Hochstrasser, M., 1993, The yeast *DOA4* gene encodes a deubiquitinating enzyme related to a product of the human *tre-2* oncogene. *Nature* 366, 313-319.

Rechsteiner, M., 1987, Ubiquitin-mediated pathways for intracellular proteolysis. *Annu. Rev. Cell Biol.* 3, 1-30.

Rogers, S., Wells, R., and Rechsteiner, M., 1986, Amino acid sequences common to rapidly degraded proteins. *Science* 234, 364-368.

Schagger, H., and von Jagow, G., 1987, Tricine-sodiumdodecylsulfate polyacrylamide gel electrophoresis for the separation of proteins in the range from 1 to 100 kDa. *Anal. Biochem.* 166, 368-379.

Tobias, J. W., and Varshavsky, A., 1991, Cloning and functional analysis of the ubiquitin-specific protease UBP1 of *Saccharomyces cerevisiae. J. Biol. Chem.* 266, 12021-12028.

Wilkinson, K. D., Deshpande, S., and Larsen, C. N., 1992, Comparisons of neuronal (PGP 9.5) and non-neuronal ubiquitin C-terminal hydrolases. *Biochem. Soc. Trans.* 20, 631-636.

Yoo, Y., Rote, K., and Rechsteiner, M., 1989, Synthesis of peptides as cloned ubiquitin extensions. *J. Biol. Chem. 264, 17078-17083.*

26

PROTEIN SYNTHESIS ELONGATION FACTOR EF-1α IS AN ISOPEPTIDASE ESSENTIAL FOR UBIQUITIN-DEPENDENT DEGRADATION OF CERTAIN PROTEOLYTIC SUBSTRATES

Hedva Gonen,[1] Dalia Dickman,[1] Alan L. Schwartz,[2]
and Aaron Ciechanover[1]

[1] Department of Biochemistry and
The Rappaport Institute for Research in the Medical Sciences
Faculty of Medicine
Technion-Israel Institute of Technology
P.O.Box 9649
Haifa 31096, Israel
[2] The Edward Mallincrodt Departments of Pediatrics and
Pharmacology Division of Hematology-Oncology
Children's Hospital and Washington University School of Medicine
#1 Children's Place
St. Louis, Missouri 63110

ABSTRACT

Targeting of different cellular proteins for conjugation and subsequent degradation via the ubiquitin pathway involves diverse recognition signals and distinct enzymatic factors. A few proteins are recognized via their N-terminal amino acid residue and conjugated by a ubiquitin-protein ligase that recognizes this residue. However, most substrates, including N-α-acetylated proteins that constitute the vast majority of cellular proteins, are targeted by different signals and are recognized by yet unknown ligases. In addition to the ligases, other factors may also be specific for the recognition of this subset of proteins. We have previously shown that degradation of N-terminally blocked proteins requires a specific factor, designated FH, and that the factor acts along with the 26S protease complex to degrade ubiquitin-conjugated proteins (Gonen et al., 1991). Further studies have shown that FH is identical to the protein synthesis elongation factor EF-1α, and that it can be substituted by the bacterial elongation factor EF-Tu (Gonen et al., 1994). This, rather surprising, finding raises two important and interesting problems. The first involves the mechanism of action of the factor

Intracellular Protein Catabolism, Edited by Koichi Suzuki and Judith Bond
Plenum Press, New York, 1996

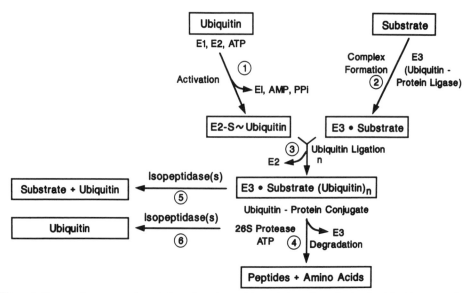

Figure 1. Proposed sequence of events in conjugation and degradation of proteins by the ubiquitin system. Proposed sequence of events in conjugation and degradation of proteins by the ubiquitin system. 1, activation of ubiquitin by the ubiquitin-activating enzyme, E1, and by the ubiquitin carrier protein (ubiquitin-conjugating enzyme), E2. 2, formation of E3 (ubiquitin-protein ligase)-protein substrate complex. 3, conjugation of multiple molecules of ubiquitin to the protein substrate. 4, degradation of conjugates by the 26S proteasome complex. 5, isopeptidase(s)-mediated release of free ubiquitin from "incorrectly" or "mistakenly" conjugated protein substrate. 6, release (recycling) of free ubiquitin from terminal proteolytic products by isopeptidase.

and the second the possibility that protein synthesis and degradation may be regulated by a commonly shared factor. Here, we demonstrate that EF-1α is a ubiquitin C-terminal hydrolase (isopeptidase) that is probably involved in trimming the conjugates to lower molecular weight forms recognized by the 26S proteasome complex. Additional findings demonstrate that its activity is inhibited specifically by tRNA. This finding raises the possibility that under anabolic conditions, when the factor is associated with AA•tRNA and GTP, it is active in protein synthesis but inactive in proteolysis. Under catabolic conditions, when the factor is predominantly found in its apo form, it is active in proteolysis.

INTRODUCTION

Degradation of cellular proteins via the ubiquitin pathway is initiated by the covalent conjugation of the protein substrate with multiple molecules of ubiquitin. The targeted protein is then degraded by a 26S protease complex (Figure 1).

Little is known about the structural signals that target proteins for conjugation and subsequent degradation. A few proteins appear to be targeted via their N-terminal amino acid residues (Varshavsky, 1992; Hershko and Ciechanover, 1992). However, a compelling body of evidence indicates that most cellular proteins are recognized via different signals that reside, most probably, in the "body" of the protein downstream from the N-terminal residue (reviewed in Ciechanover, 1994): **(a)** Approximately 80% of the cellular proteins are N-α-acetylated. As for the remaining free N-termini proteins, the rules that govern removal of the initiator Met by methionine aminopeptidase suggest that in most cases, this residue is cleaved only if the penultimate residue is a "stabilizing" amino acid. Thus, proteins with exposed "destabilizing" N-termini appear to be sparse. **(b)** The ubiquitin system degrades N-α-acetylated proteins in a process that does not require removal of the modifying group and exposure of a free N-terminal residue. **(c)** The recognition of certain proteins with free N-termini (either "destabilizing" or "stabilizing") is not dependent on the identity of their N-terminal residue. **(d)** Most convincing, mutational inactivation of the N-terminal-recognizing ubiquitin-protein ligase (UBR1; E3α) in yeast is not lethal and does not result in a characteristic phenotype. The nature of the targeting signals as well as the identity of the ligating enzymes involved in recognizing non "N-end rule" protein substrates have not been elucidated.

We have recently shown that the degradation of certain N-α-acetylated proteins requires a specific factor that is not required for the breakdown of free N-termini proteins. The factor, designated Factor Hedva (FH), is required for the proteolysis of the core nucleosomal histone H2A, the cytoskeletal protein actin, and the lens protein α-crystallin (Gonen et al, 1991). Recent data indicate that it is also involved in the degradation of the hepatic enzyme, tyrosine aminotransferase (TAT). FH is a homodimer with a subunit molecular mass of 46 kDa. Initial analysis of the mechanism of action of FH revealed that it is not involved in the conjugation process. Rather, it acts along with the 26S protease complex and stimulates degradation of the conjugated substrate. The effect appears to be specific to this group of proteins, as the factor is not required for the degradation of conjugates of several free N-termini proteins, such as oxidized RNase A and lysozyme. Further analysis demonstrated that FH probably interacts with the conjugates prior to their degradation: incubation of conjugates in the presence of purified FH and the protease revealed a short, however significant, time lag that preceded initiation of degradation. The lag was completely abolished when FH was preincubated with the conjugates prior to the addition of the protease. These findings demonstrate that recognition of certain proteins and their targeting for degradation involves both conjugation of ubiquitin and degradation of the adducts by the 26S protease complex. Further analysis revealed that FH is identical to the protein synthesis elongation factor EF-1α (Gonen et al., 1994): **(a)** Partial sequence analysis reveals 100% identity to EF-1α. **(b)** Like EF-1α, FH binds to immobilized GTP (or GDP) and can be purified in one step using the corresponding nucleotide for elution. **(c)** Guanosine nucleotides that bind to EF-1α protect the ubiquitin system-related activity of FH from heat inactivation: nucleotides that do not bind, do not exert this effect. **(d)** EF-Tu, the homologous bacterial elongation factor, can substitute for FH/EF-1α in the proteolytic system. This last finding is of particular interest since the ubiquitin system has not been identified in prokaryotes. The activities of both EF-1α and EF-Tu are strongly and specifically inhibited by ubiquitin-aldehyde, a specific inhibitor of certain ubiquitin C-terminal hydrolases (isopeptidases; Hershko and Rose, 1987). Initial direct findings indicate that EF-1α may be indeed a ubiquitin C-terminal hydrolase, involved in trimming polyubiquitin chains to a form that is susceptible to the activity of the 26S proteasome complex. Thus, it appears that FH/EF-1α is a protein with two functions. In protein synthesis, it generates a ternary complex with GTP and AA•tRNA, a complex that serves as a donor of the activated amino acid residue to the elongating polypeptide chain. In protein degradation it probably serves as an isopeptidase specific to a subset of protein substrates. It is interesting to note that EF-1α has been also implicated in severing stable microtubules which is involved in cytoskeletal rearrangements ocurring during the cell cycle (Shiina et al., 1994). Thus, it appears that this interesting protein may have several distibnct roles. Whether all these functions are coordinated or are carried out independently is a problem of broad biological interest.

Peptide 1

```
                     ↓
EF-1α:     KK-171-IGYNPDTVAFVPISG-185
FH    :           -XXYNPDXVAFVPIXG-
```

Peptide 2

```
                     ↓
EF-1α:     GR-257-VETGVLKPGMVVTFA-271
FH    :           -VETGVLKPGMVVTFA-
```

Peptide 3

```
                     ↓
EF-1α:     VK-281-SVEMHHEALSQALPGDNV-298
FH    :           -SVEMHHEALSQALPGDNV-
```

Peptide 4

```
                     ↓
EF-1α:     LK-383-SGDAAIVDMVPGKP-396
FH    :           -XGXAAIVDMVPGKP-
```

Figure 2. Sequence analysis of four internal tryptic peptides of FH displays identity to EF-1α. Arrows denote trypsin cleavage sites and numbers denote amino acid residues in EF-1α. Underlined X's denote ambiguous residues.

RESULTS

Sequence Analysis Reveals Identity Between FH and Protein Synthesis Elongation Factor EF-1a: Initial sequencing attempt revealed that FH is blocked at the N-terminal residue. To obtain internal sequence information, the protein was partially digested with trypsin and peptide fragments were obtained using HPLC. As can be seen in Figure 2, sequence analysis of four internal peptides derived from FH demonstrates 100% identity to rabbit reticulocytes EF-1α.

Yeast and Rabbit Reticulocytes EF-1a's Stimulate Degradation of Ubiquitinated Histone H2A: To further corroborate that FH is indeed EF-1α, we tested the activity of purified elongation factors from reticulocytes and yeast. As can be seen in Figure 3, the two elongation factors stimulate the degradation of conjugated H2A, and their specific activity is similar to that of purified FH.

FH is a GTP-Binding Protein: Since the remote possibility still existed that FH constitutes a minor contamination that co-purifies with EF-1α in all preparations and

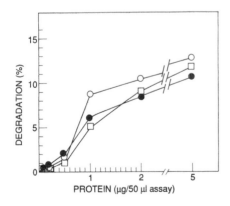

Figure 3. Stimulation of degradation of ubiquitinated-H2A by FH and different EF-1α's. (O), FH; purified EF-1α from yeast (□), and rabbit reticulocytes (●).

Figure 4. Affinity chromatography-purified FH migrates identically to "conventionally" purified EF-1α in SDS-PAGE. A crude preparation of FH was loaded onto immobilized GDP, and the specifically bound protein was eluted using a buffer that contains GDP. Lane 1, molecular weight markers (97.4, phosphorylase B; 66.2, BSA; 45, ovalbumin; 31, carbonic anhydrase). Lane 2, ovalbumin. Lane 3, FH preparation prior to loading on the GDP column. Lane 4, unabsorbed proteins eluted with low salt. Lane 5, unabsorbed proteins eluted with high salt. Lane 6, GDP-eluted fraction. Lane 7, second elution with GDP. Lane 8, "conventionally" purified EF-1α from rabbit reticulocytes.

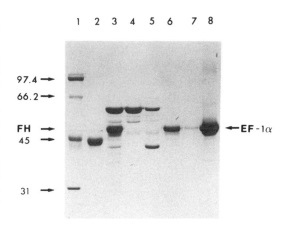

co-migrates with it electrophoretically, it was necessary to demonstrate that the two proteins share common biochemical properties. Mostly, we made use of the GTP-binding property of EF-1α.

As EF-1α is a GTP-binding protein, we determined whether following affinity binding and specific elution from immobilized GDP, EF-1α demonstrates the same ubiquitin system-related activity as the enzyme purified via "conventional" methods. Indeed, as can be seen in Figure 4, the material eluted from immobilized GDP migrates on SDS-PAGE identically to EF-1α that was purified via successive traditional steps, involving ammonium sulfate precipitation, gel filtration and ion exchange chromatography. Furthermore, the purified eluted protein stimulates degradation of ubiquitinated H2A in the presence of the 26S protease (Table 1), and its specific activity is similar to the activities of "conventionally" purified FH and EF-1α. Analogous results were obtained using GTP-eluted EF-1α (not shown).

To further demonstrate that FH has a GTP-binding site, we preincubated the purified factor at 48°C for different periods of time in the presence of various nucleotides. Only GTP, GDP, and the non-cleavable analog, GMP-PNP, the three nucleotides that can bind to the GTP-binding site of EF-1α, protected FH from heat inactivation. GMP, ATP, CTP, and UTP that do not bind, did not protect (not shown).

EF-Tu, the Bacterial Elongation Factor, Can Substitute for EF-1a in the Proteolytic System: To finally confirm that FH is indeed EF-1α, we expressed the EF-1α cDNA (Merrick *et al.*, 1990) in bacteria. As eubacteria do not have any **known** active component of the ubiquitin system (see however in "Discussion"), demonstration of FH activity in bacterial extracts following expression of EF-1α cDNA would provide a strong support to the hypothesis that the two proteins are indeed identical. Surprisingly, similar activity, albeit reduced, was also detected in extracts of bacteria that were not transformed. When this activity was purified, it co-migrated **throughout** the purification process with the bacterial elongation factor EF-Tu. The experiment described in Figure 5 (Panels A and B) demonstrates the last step in the partial purification process of EF-Tu, gel filtration chromatography over Superdex 200 FPLC column. As can be clearly seen, the EF-Tu protein co-migrates along with two activities: GTP-binding activity and stimulation of degradation of ubiquitinated H2A in the presence of purified 26S proteasome complex.

Most importantly, homogeneously purified EF-Tu (Louie *et al.*, 1984) also stimulated degradation of conjugated H2A in the presence of the purified, although the preparation was only 40% as active as the mammalian factor (Table 1).

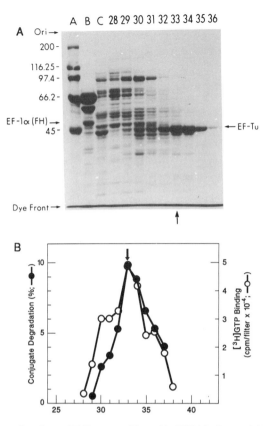

Figure 5. Bacterial elongation factor EF-Tu co-purifies with GTP-binding activity and FH-like conjugate degradation-stimulating activity. **Panel A.** SDS-PAGE-Coomassie Blue staining analysis of Superdex 200 FPLC gel filtration chromatography-derived fractions. EF-Tu was purified via several steps from crude bacterial cytosolic fraction. Shown is the pre-final gel filtration chromatography step. Fractions were resolved via SDS-PAGE and stained by Coomassie Blue. Numbers denote fraction numbers. Arrow denotes peak fraction (No. 33). Lane A, molecular weight markers: 200, myosin; 116.25, β-galactosidase. All other markers are as described in the Legend to Figure 3. Lane B, partially purified FH from rabbit reticulocyte lysate. Lane C, pooled material prior to resolution. **Panel B.** GTP-binding activity (O) and conjugate degradation-stimulating activity (●) of EF-Tu resolved via gel filtration chromatography. EF-Tu was resolved as described in the Legend to Panel A. Conjugate degradation and GTP-binding activity were monitored.

Ubiquitin-Aldehyde, a Specific Inhibitor of Certain Ubiquitin C-Terminal Hydrolases, Inhibits FH/EF-1a-Dependent Degradation of Ubiquitinated H2A: Initial analysis of the mechanism(s) involved in the function of EF-1α in the degradation of ubiquitinated H2A revealed that an activity of a specific ubiquitin C-terminal hydrolase (isopeptidase) plays a role in the process. Ubiquitin-aldehyde, a specific inhibitor of certain isopeptidases (Hershko and Rose, 1987), blocks almost completely the activity of both the mammalian and the bacterial factors (Table 1). Whereas FH/EF-1α can be a specific isopeptidase, it can also serve as a factor that is necessary for the activity of isopeptidase(s) contained in the 26S protease complex (see below).

EF-1α is Probably a Ubiquitin C-Terminal Hydrolase: To test the hypothesis that EF-1α is a specific isopeptidase, we purified the protein to homogeneity and followed its isopeptidatic activity and its ability to stimulate proteolysis of histone H2A conjugates in the presence of the 26S proteasome complex. Throughout the purification procedure, the two activities co-migrated, along with the protein that was detected by Coomassie Blue staining. Figure 6 shows a profile of

Table 1. FH, conventionally purified EF-1a, GDP eluted EF-1a, and EF-Tu
stimulate the degradation of ubiquitinated H2A:
Their activity is blocked by ubiquitin-aldehyde.

Factor added	Degradation of conjugates (% of total radioactivity in conjugates) (numbers in parentheses indicate % inhibition)	Ubiquitin-aldehyde added
FH	13.4	–
FH	2.6 (81)	+
EF-1α	12.5	–
EF-1α	3.9 (69)	+
GDP eluted EF-1α	10.2	–
GDP eluted EF-1α	2.7 (74)	+
EF-Tu	4.9	–
EF-Tu	1.6 (68)	+

Reaction mixture contained purified 26S proteasome, ubiquitin-^{125}I-histone H2A conjugates, ATP, buffer, Mg^{++} and DTT, the indicated factor, and ubiquitin aldehyde (when indicated). Degradation value in the absence of added factor (2.4%) was subtracted from all results. This activity is not sensitive to ubiquitin aldehyde.

a gradient over a cation exchange resin which was used in one of the last purification steps. It is clear that EF-1α along with its two activities, the isopeptidatic activity and the stimulation of degradation of conjugated H2A, attain their peak in Fraction 7.

As can be seen in Figure 7, homogeneously purified EF-1α also stimulates disassembly of high molecular mass adducts to smaller products.

tRNA Specifically Inhibits EF-1a: An intriguing problem concerns the dual role of EF-1α in apparently two opposing processes, protein synthesis and degradation. This raises the interesting possibility that protein synthesis and degradation may be regulated by modulation of a commonly shared factor. It is known that under conditions of rapid cell growth and division, protein synthesis increases, while proteolysis slows down dramatically. The two processes change direction under stressed conditions such as starvation, exposure to heat, and muscle denervation or immobilization. Experimental evidence indicates that, at least in some of these cases, accelerated proteolysis is carried out by the ubiquitin system (Wing and Goldberg, 1993). The activity of the factor can be regulated by the level of charged tRNA or GTP, two of the reactants with which it forms the ternary complex necessary for protein synthesis, or by post-translational modification such as phosphorylation. Indeed, the level of charged tRNA has been implicated as a possible modulator of protein synthesis and degradation in eukaryotic cells (Tischler *et al.*, 1983). Our initial experiments indicate that formation of a ternary complex may inhibit the activity of EF-1α in the ubiquitin system. Addition of RNase A or micrococcal nuclease to isolated complex stimulated its activity several-fold. Inhibition of the nucleases prior to their addition to the mixture, abolished their stimulatory effect. Also, addition of tRNA, but not 28S, 18S, or 5S RNA (the different RNAs extracted from crude reticulocyte Fraction II are shown in Figure 8) to the reaction mixture inhibited the stimulatory effect of EF-1α (not shown).

DISCUSSION

Using several rigorous criteria, we have demonstrated that FH, a protein factor that is required for the degradation of several N-α-acetylated proteins, is identical to the protein

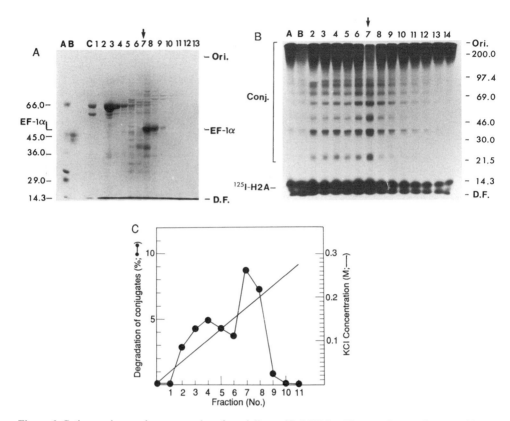

Figure 6. Cation exchange chromatography of partially purified EF-1α: The protein co-migrates with an isopeptidase activity and conjugate degradation stimulating activity. A partially purified preparation of EF-1α was subjected to cation exchange chromatography over Carboxymethyl cellulose. **Panel A,** Coomassie blue staining. Lane A, molecular weight markers. Lane B, ovalbumin. Lane C, unadsorbed, flow through material. Lanes 1-13, gradient fractions. Arrow denotes fraction 7 with highest concentration of EF-1α. EF-1α denotes the elongation factor. Molecular weight markers are: 66.0, BSA; 45.0, ovalbumin; 36, glyceraldehyde-3-phosphate dehydrogenase; 29, carbonic anhydrase; 14.3, lysozyme. Ori. and D.F. denote origin of gel and dye front, respectively. **Panel B,** isopeptidase activity. The different fractions were incubated in the presence of high molecular mass ubiquitin-^{125}I-Histone H2A conjugates, and appearance of low molecular weight products was monitored. Lane A, untreated conjugates. Lane B, conjugates incubated in the presence of the crude fraction prior to chromatography. Molecular weight markers are: 200.0, myosin; 97.4, phosphorylase b; 69.0, BSA; 46, ovalbumin; 30.0, carbonic anhydrase; 21.5, soybean trypsin inhibitor; 14.3, lysozyme. Conj. denotes conjugates. Ori. And D.F. denote origin of gel and dye front, respectively. ^{125}I-H2A denote free labeled histone H2A. **Panel C,** stimulation of degradation of ubiquitinated histone H2A in the presence of purified 26S proteasome complex.

synthesis elongation factor EF-1α (Gonen *et al.*, 1994). Upon analyzing the function of FH/EF-1α in the ubiquitin pathway, one faces two equally interesting and related questions: 1. What is the mechanism of action of the protein in the proteolytic process? and 2. Why does a single protein function in two apparently opposing processes? The factor may function as an enzyme. For example, it can be a ubiquitin C-terminal hydrolase that cleaves ubiquitin moieties from certain polyubiquitin chains, thus rendering the remaining adduct more susceptible to the action of the 26S protease. Alternatively, it may serve as a chaperone that binds to the adduct and folds it in such a manner that renders it accessible to the active sites

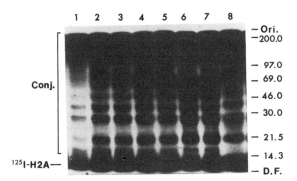

Figure 7. Purified EF-1α is an isopeptidase. Preparations derived from the various steps of purification of EF-1α were incubated in the presence of high molecular mass ubiquitin-^{125}I-histone H2A conjugates, and appearance of low molecular weight products was monitored. In each step, the extract added was derived from the fraction that contained the highest concentration of EF-1α as determined by Coomassie blue staining. Lane 1, free unreacted adducts. Lanes 2 and 3, 55% $(NH_4)_2SO_4$ precipitate of crude reticulocyte Fraction I. Lanes 4 and 5, Sephadex G-100 gel filtration chromatography of the 55% $(NH_4)_2SO_4$ precipitate. Lanes 6 and 7, homogeneously purified EF-1α. Lane 8, conjugates incubated in the presence of crude reticulocyte Fraction I. All symbols and signs are as described in the legend to Figure 6, Panel B.

of the 26S proteasome. Here we demonstrate that FH/EF-1α may be an isopeptidase. It is not clear yet whether it is a specific enzyme, as its isopeptidatic activity has not been tested against ubiquitin conjugates of free N-termini proteins. Also, we have not demonstrated yet that EF-Tu also contains a ubiquitin C-terminal hydrolytic activity. As for the problem of regulation, it is possible that by utilizing a single shared component, the cell can regulate protein synthesis and degradation in a rather tight manner. It is known that during rapid cell growth and division when protein synthesis proceeds at increased rates, protein degradation is slowed down. In contrast, during stressed conditions such as starvation, protein degradation accelerates, whereas proteolysis slows down. Such a regulation of the two processes can be the result of the predominant utilization in one process of a factor that is common to the two pathways. Mechanistically, the activity in one pathway or another can be mediated

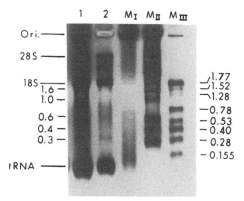

Figure 8. Different RNA species extracted from crude reticulocyte Fraction II. Lanes 1 and 2 are two different concentrations of the extracted RNAs. M_I, M_{II}, and M_{III} denote different RNA molecular mass markers. Numbers denote kb or sedimentation coefficients. tRNA is also marked.

by modulation of the factor, for example, by the level of its association with AA•tRNA. During phases of rapid protein synthesis, the factor is predominantly associated with charged tRNA and GTP and the complex may be inactive in degradation. As the level of AA•tRNA decreases under stress, a large proportion of EF-1α exists in its apo form that may be active in protein degradation but inactive in synthesis. In an initial attempt to test this hypothesis, we have shown that the activity of the factor in the ubiquitin system can be inhibited specifically by tRNA, but not by other species of RNA. However, here too, one has to be cautious. It is necessary to demonstrate that only charged tRNA can inhibit the activity while the uncharged molecule is inactive. Also, one has to demonstrate rigorously that the AA•tRNA generates indeed a complex with the protein, and that GTP is also associated with the two components. Thus, while exciting, these results are only preliminary.

What may be the physiological significance of the activity of EF-Tu? Using rigorous biochemical and molecular approaches, investigators have not been able to identify components or activity of the ubiquitin system in eubacteria. However, three lines of evidence indicate the existence of a "ubiquitin-like" system in prokaryotes. In one study, Tobias and colleagues (1991) have demonstrated that a subset of genetically engineered species of β-galactosidase expressed in *E. coli* are targeted for degradation by the ATP-dependent Clp (Ti) protease following recognition of their N-terminal amino acid residue. Jap and colleagues (1993) have demonstrated a 20S proteasome complex-like structure in the archaebacterium *Thermoplasma acidophilum* (the 20S protease is the core proteolytic subunit of the 26S protease complex). Recently, Wolf and colleagues (1993) have identified ubiquitin in *Thermoplasma*. Thus, it may well be that EF-Tu is an integral part of such a "ubiquitin-like" system. As bacterial proteins are not modified at their N-terminal residues, the elongation factor may be involved in these organisms in recognition of other motifs. Alternatively, bacteria do not have a "ubiquitin-like" system and the structural component of EF-Tu that functions in the eukaryotic ubiquitin system is not functional in bacteria. It evolved late in evolution to become part of the ubiquitin system. It is clear however that the involvement of EF-1α and EF-Tu in the ubiquitin system cannot be attributed to some general properties of the two proteins, their pI for example. The two proteins are different from one another structurally and functionally. EF-Tu cannot substitute for the mammalian factor in cell free translation systems. In addition, EF-1α is a basic protein and does not adsorb to anion exchange resins even at relatively high pH. In contrast, EF-Tu has a pI of ~ 5 and adsorbs to anion exchange resins at neutral pH.

ACKNOWLEDGMENTS

The work was supported by the Council for Tobacco Research-U.S.A., Inc. (CTR), the German-Israeli Foundation For Scientific Research and Development (G.I.F.), the Israeli Academy of Sciences and Humanities, Monsanto, Inc., the Foundation for Promotion of Research in the Technion, the Henri Gutwirth Fund, and a Research Fund administered by the Vice President of the Technion for Research.

REFERENCES

Ciechanover, A. (1994). The ubiquitin-proteasome pathway. *Cell 79*, 13-21.
Gonen, H., Schwartz, A.L., and Ciechanover, A. (1991). Purification and characterization of a novel protein that is required for the degradation of N-α- acetylated poteins by the uiquitin sstem. *J. Biol. Chem.* 266, 19221-19231.

Gonen, H., Smith, C. E., Siegel, N. R., Merrick, W. C., Kahana, C., Chakraburtty, K., Schwartz, A. L., and Ciechanover, A. (1994). Protein Synthesis elongation factor EF-1α is essential for ubiquitin-dependent degradation of N- α-acetylated proteins and may be substituted for by the bacterial elongation factor EF-Tu. *Proc. Natl. Acad. Sci. USA* 91, 7648-7652.

Hershko, A. and Ciechanover, A. (1992). The ubiquitin-mediated proteolytic Pathway. *Annu. Rev. Biochem.* 61, 761-807.

Hershko, A. and Rose, I. A. (1987). Ubiquitin-aldehyde: A general inhibitor of ubiquitin recycling processes. *Proc. Natl. Acad. Sci. USA* 84, 1829-1833

Jap, B., Puhler, G., Lucke, H., Typke, D., Lowe, J., Stock, D., Huber, R., and Baumeister, W. (1993). Preliminary X-ray crystallographic study of the proteasome from Thermoplasma acidophilum. *J. Mol. Biol.* 234, 881-884.

Louie, A., Ribeiro, N. S., Reid, R. B., and Jurnak, F. (1984). Relative affinities of all Escherichia coli aminoacyl-tRNAs for elongation factor Tu-GTP. *J. Biol. Chem.* 259, 5010-5016.

Merrick, W. C., Dever, T. E., Kinzy, T. G., Conroy, S. C., Cavallius, J., and Owens, C. L. (1990). Characterization of protein synthesis factors from rabbit reticulocytes. *Biochim. Biophys. Acta 1050*, 235-240.

Shiina, N., Gotoh, Y., Kubomura, N., Iwamatsu, A., and Nishida, E. (1994). Microtubule severeing by elongation factor 1α. *Science 266*, 282-285.

Tischler, M.E., DeSautels, M., and Goldberg, A.L. (1982). Does leucine, leucyl-tRNA, or some metabolite of leucine regulate protein synthesis and degradation in skeletal and cardiac muscle? *J. Biol. Chem.* 257, 1613-1621.

Tobias, J. W., Shrader, T. E., Rocap, G., and Varshavsky, A. (1991). The N-end rule in bacteria. *Science* 254, 1374-1377.

Varshavsky, A. 1992. The N-end rule. Cell 69, 725-735.

Wing, S.S. and Goldberg, A.L. (1993). Glucocorticoids activate the ATP- ubiquitin-dependent proteolytic system in skeletal muscle during fasting. *Am. J. Physiol.* 264, E668-E676.

Wolf, S., Lottspeich, and Baumeister, W. (1993). Ubiquitin found in the archaebacterium Themoplasma acidophilum. *FEBS Lett.* 326, 42-44.

27

MECHANISMS AND REGULATION OF UBIQUITIN-MEDIATED CYCLIN DEGRADATION

Avram Hershko

Unit of Biochemistry
B. Rappaport Faculty of Medicine and
The Rappaport Institute for Research in the Medical Sciences
Technion-Israel Institute of Technology
Haifa 31096, Israel

We have been studying the biochemical mechanisms of intracellular protein degradation in an ATP-dependent proteolytic system from reticulocytes. Our studies led to the identification of a pathway mediated by ubiquitin (Ub), a highly conserved 76-amino acid residue polypeptide. We found that proteins are committed to degradation by their covalent ligation to Ub. We have also identified several enzymatic reactions in the formation and breakdown of Ub-protein conjugates. Currently available information on the enzymatic reactions of the Ub proteolytic pathway is reviewed in refs. 1-3 and is summarized in Fig. 1. The initial reaction is the activation of the C-terminal Gly residue of Ub by a specific enzyme, E_1 (Step 1). Next, activated Ub is further transferred by transacylation to thiol groups of a family of Ub-carrier proteins, E_2s (Step 2). Multiple species of E_2 were observed in reticulocytes (4) and yeasts (5). Some species of E_2 may transfer Ub directly to certain proteins, such as histones (Step 3). In most cases leading to protein degradation, the ligation of Ub to specific proteins requires a third enzyme, E_3. The Ub-protein ligase E_3 binds suitable substrates (ref. 6, Step 4) and an E_2 (7), and thus allows the transfer of Ub from E_2 to the protein substrate. Two species of E_3, called $E_3\alpha$ and $E_3\beta$, have been isolated from reticulocytes. $E_3\alpha$ binds proteins with basic or hydrophobic N-terminal amino acid residues (8, 9), while $E_3\beta$ mainly acts on proteins with small, uncharged amino acid residues at the N-terminal position (10). In the ligation process, isopeptide bonds are formed between the C-terminal Gly residue of Ub end ϵ-amino groups of Lys residues of proteins. In all cases of E_3-dependent ligation and in some E_3-independent reactions, multiple Ub units are linked to proteins. Some of these are arranged in polyUb chains (11), in which a major site of linkage was shown to be at Lys^{48} of Ub (12). Proteins ligated to multiple Ub units are degraded by a 26S protease complex (13). The 26S complex is formed by the assembly of three components, designated conjugate-degrading factors 1-3 (CF-1 to CF-3), in a ATP-de-

Intracellular Protein Catabolism, Edited by Koichi Suzuki and Judith Bond
Plenum Press, New York, 1996

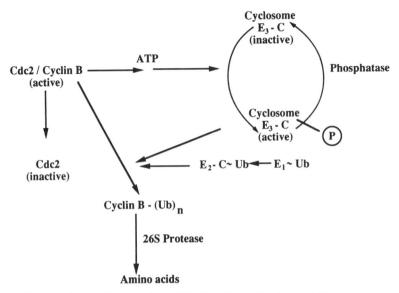

Figure 1. Intermediary reactions of the ubiquitin-mediated proteolytic system.

pendent process (Step 6, ref 14). CF-3 was identified as the 20S "multicatatalytic" protease complex, a particle previously observed in many eukaryotic cells (15). The assembled 26S protease complex degrades the protein moiety of Ub-protein conjugates in a reaction that requires the hydrolysis of ATP (Step 7, ref 17) . Following proteolysis, free and reusable Ub is liberated, a process carried out by Ub- C-terminal hydrolases or isopeptidases. We have recently characterized two different isopeptidases that are involved in different aspects of Ub recycling at the final stages of protein degradation.

The above studies on the basic biochemical mechanisms of the Ub-mediated pro-teolytic pathway were carried out in the reticulocyte system, using artificial (extracellular or denatured) protein substrates. While many gaps remain in our understanding of the basic mechanisms of this pathway, it seemed most important at this stage to address the problem of how the Ub system carries out the selective and regulated degradation of specific cellular proteins. Recent work from several laboratories indicates the involvement of the Ub system in the degradation of many cellular proteins, playing important roles in a variety of cellular processes. Examples are the red light - induced degradation of the plant photoreceptor phytochrome (18), the rapid degradation of yeast MATα2 repressor (19), the cytokine - regulated proteolytic processing required for the activation of mammalian transcription factor NF-KB (20) and the processing of antigens for presentation by MHC class I molecules (21). The Ub system is also involved in the rapid degradation of tumor-suppressor protein p53 (22) and of oncoproteins or proto-oncogene products such as c-myc, c-fos and E1A (23) c-jun (24) and c-mos (25). Particularly noteworthy is the role of Ub-mediated proteolysis in oscillations of the levels of regulators of temporally controlled processes, such as the cell cycle. Presently known examples are the various cyclins and at least some inhibitors of cyclin-dependent protein kinases (refs. 26-28 and see below).

While the widespread involvement of the Ub system in the selective degradation of a variety of specific cellular proteins is now well established, it is not clear in many cases how such proteins are recognized by the Ub system at such remarkably high selectivity. Our lack of understanding of the selectivity of protein degradation is mainly due to insufficient information on specific E_3 enzymes, ligases that recognize specific proteins and target them to ligation with Ub and to degradation. The two previously known E_3 species, $E_3\alpha$ and $E_3\beta$, act on some extracellular proteins and artificial "N-end rule" substrates (29), but they do not seem to be involved in the degradation of most cellular proteins (30). It seems reasonable to assume that a large variety of different E_3 enzymes exist, which recognize different (not N-end rule) signals in different cellular proteins. However, the only other well-characterized E_3-like enzyme is that participating in the Ub-dependent degradation of p53 in reticulocyte lysates (22). Ub-p53 ligation requires the human papillomavirus E6 protein and a cellular protein of 100 kDa, termed E6-associated protein (E6-AP). E6 binds both p53 and E6-AP, and thus promotes the formation of a ternary complex required for the ligation of Ub to p53. E6-AP apparently acts as an E_3 enzyme, since it promotes the ligation of Ub to some cellular proteins in the absence of E6 (22). It is not clear which other proteins, besides p53, are substrates for E6-AP.

I have begun to approach the problems of the selectively and regulation of protein degradation by studying the mode of the degradation of mitotic cyclins. I became attracted to the cyclin degradation system because of its high specificity, intricate regulation and the feasibility to investigate this process in a cell-free system by direct biochemical methods. Cyclins are positive regulatory subunits of protein kinases of the cdc2 family (also called CDKs, or cyclin-dependent kinases, reviewed in refs 31, 32). Several types of cyclins exist, including A-type and B-type mitotic cyclins, S-phase and G1 cyclins, that from complexes with different CDKs. The different cyclins are synthesized and degraded at different stages of the cell cycle, advancing the cell irreversibly to the next stage. The best characterized complex is that between cyclin B and cdc2, also called M-phase promoting factor (MPF). Active cdc2 induces entry into mitosis, presumably by the phosphorylation of some key target proteins such as nuclear lamin or the retinoblastoma growth-suppressor protein. The rise in activity of protein kinase cdc2 at the onset of mitosis requires the prior synthesis of cyclin B (and some phosphorylations and dephosphorylations of cdc2), while the subsequent inactivation of cdc2 is caused by the degradation of cyclin. It has been shown that cyclin B degradation and cdc2 inactivation are required for exit from mitosis, as indicated by observations that non-degradable derivatives of cyclin B keep cdc2 stably hyperactivated and maintain chromosomes in a condensed mitotic state (33).

The molecular mechanisms involved in cyclin degradation and their relationship to the cell cycle regulatory network can now be studied by direct biochemical methods, due to the development of cell-free systems that reproduce events of the cell cycle, including regulated cyclin degradation. Two such systems are available, one from *Xenopus* eggs (33) and another from oocytes of the clam *Spisula* (34) . Studies in both systems indicated that the cyclin degradation machinery is turned on, after a time lag, by a process triggered by active protein kinase cdc2, thus providing a negative feedback mechanism that inactivates cdc2 and terminates mitosis (33, 35, 36). As to the nature of the cyclin degradation machinery, evidence from both systems indicates that cyclin degradation is carried out by the Ub pathway (26,27). Using the *Spisula* cell-free system we found that a derivative of Ub in which all amino groups were blocked by reductive methylation (a modification that prevents the formation of polyUb chains), inhibits the degradation of both cyclin A and cyclin B. The specificity of the action of this agent was indicated by the observation that excess Ub completely overcame the inhibitory effect of methylated Ub on cyclin degradation (27). With extracts from *Xenopus* eggs, Glotzer *et al* (26) showed a direct relationship between cyclin degradation and its ligation to Ub. In M-phase extracts, which degrade cyclin rapidly, levels of cyclin-Ub conjugates are much higher than in interphase extracts, in which cyclins are stable. These investigators also showed that a highly conserved sequence near the N-terminus of mitotic cyclins, the "destruction box", is required for their ubiquitination and degradation in M-phase extracts (26). It was not clear how is this "destruction box" signal recognized the cyclin degradation machinery.

To identify the enzyme components responsible for specific and regulated cyclin-Ub ligation, we have begun a fractionation-reconstitution study of this system from extracts of clam oocytes (37). We preferred the clam oocyte system to the *Xenopus* system because of the availability of large amounts of material, more suitable for biochemical work. Initially, M-phase extracts were separated on DEAE- cellulose into two fractions: Fraction 1, the flow-through, and Fraction 2, that consisted of proteins bound to the resin and eluted with high salt. Neither fraction promoted cyclin-Ub ligation by itself, but such activity could be reconstituted by the combination of the two fractions. We found that the component provided by Fraction 2 was the ubiquitin-activating enzyme, E_1. On the other hand, Fraction 1 contained at least two novel components which could be separated by high-speed centrifugation. The soluble part of Fraction 1 was found to contain a previously unknown species of ubiquitin-carrier protein, called E_2-C, which is apparently specific for the cyclin-Ub ligation system. The third component is associated with particulate material. Cell cycle regulation was retained following fractionation, as shown by the observation that cyclin-Ub ligation was promoted only by Fraction 1 from M-phase extracts, but not by Fraction 1 from interphase extracts. The combination of particulate fraction from M-phase with E_2-C from either M-phase or interphase was active in cyclin-Ub ligation. On the other hand, the combination of the particulate fraction from interphase extracts with E_2-C from either M-phase or interphase did not carry out cyclin-Ub ligation. These findings indicated that the particulate component is regulated in the cell cycle whereas the activity of E_2-C is not regulated. We furthermore found that the particulate component from interphase cells could be activated by purified protein kinase cdc2, after a lag of about 30 min (37).

More recently, we have dissociated the cdc2-regulated component from particles by extraction with high salt. The enzyme was partially purified by ammonium sulfate fractionation and glycerol density gradient centrifugation. It has a very high molecular mass, $M_r\sim$ 1,500 kDa. It appears to be cyclin-selective, since it can be separated from other Ub-protein ligases (which ligate Ub to most other endogenous oocyte proteins) that are of lower molecular size. We have therefore called this enzyme cyclin-Ub ligase, or E_3-C. E_3-C ligates Ub to both cyclin A and cyclin B (in the presence of E_1 and E_2-C) and requires the presence of wild type N-terminal destruction box motifs in each cyclin. As observed previously for the particle-associated activity, the soluble enzyme from interphase cells could be activated *in vitro* by protein kinase cdc2, but only following a time lag (38). It was concluded that the activation of E_3-C, triggered by protein kinase cdc2, causes the sudden increase of the ligation of Ub to cyclin B at the end of mitosis. Ubiquitinated cyclin B is then presumably degraded rapidly by a constitutively active 26S protease complex. The large complex that contains E_3-C was termed the cyclosome, since the regulation of E_3-C associated with this complex is apparently responsible for setting cyclin levels in the embryonic cell cycles (38).

Very recently, we have obtained evidence indicating that the activity of E_3-C is regulated by reversible phosphorylation of the cyclosome (Lahav-Baratz, S., Sudakin, V., Ruderman, J.V., and Hershko, A., unpublished observations). Since the lag kinetics of E_3-C activation by cdc2 indicate a complex mechanism, we thought to approach this problem initially "through the rear door", by asking the question how is E_3-C inactivated in the interphase. It had been reported previously that the inactivation of cyclin degradation activity in cycling extracts of *Xenopus* eggs can be prevented by okadaic acid (39), an inhibitor of protein phosphatases 1 and 2A (40). We have confirmed these findings for cyclin-Ub ligation activity in crude extracts of M-phase clam oocytes. These results indicate that the action of an okadaic acid-sensitive phosphatase is involved in the inactivation of the cyclin-Ub ligation system, but the mechanisms of its action could be rather indirect. For example, the phosphatase may act by the dephosphorylation of Thr-161 of cdc2, a process required for cdc2

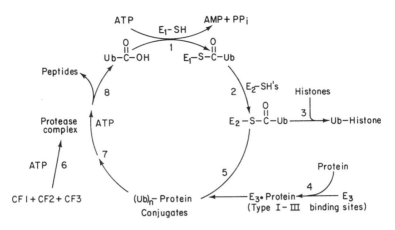

Figure 2. Proposed model for the regulation of cyclin degradation.

inactivation. To examine whether the phosphatase acts on E_3-C directly, we have partially purified an okadaic acid-sensitive phosphatase from extracts of clam oocytes, by gel filtration chromatography followed by ion exchange FPLC and hydrophobic chromatography. We have also further purified active E_3-C and separated it completely from protein kinase cdc2. This last result indicates that the continued presence of protein kinase cdc2 is not required to maintain E_3-C in the active form. The incubation of active E_3-C with partially purified phosphatase markedly inhibited cyclin-Ub ligation activity. This action of the phosphatase was completely prevented by okadaic acid, indicating that the inhibition was not caused by some contamination in the partially purified preparation of the phosphatase. It thus seems that the activity of E_3-C is regulated by reversible phosphorylation of the cyclosome (Lahav-Baratz, S., Sudakin, V., Ruderman, J.V., and Hershko, A., unpublished observations). At present, we do not know whether the E_3-C moiety of the cyclosome is phosphorylated, or whether the phosphorylation of another region of this complex turns on E_3-C activity.

Fig. 2 summarizes our present working hypothesis on the mechanisms of the degradation of cyclin B and its regulation in the cell cycle. The active cdc2/cyclin B complex initiates a process that converts cyclosome-associated E_3-C to an active, phosphorylated form. The activation of E_3-C is presumably carried out by an indirect mechanism, since activation is preceded by a prolonged lag that cannot be overcome by increased levels of protein kinase cdc2. Thus, ATP may be required in E_3-C activation not only for the action of protein kinase cdc2, but also for that of a downstream kinase that phosphorylates E_3-C. It is proposed that E_3-C is a ubiquitin ligase selective for mitotic cyclins, and that it is the only (or main) component of the cyclin degradation system which is regulated in the cell cycle. Other components, such as the cyclin-selective Ub-carrier protein E_2-C, or the nonselective E_1 and 26S protease, are constitutively active. Activated E_3-C carries out cyclin-Ub ligation at the end of mitosis, in concert with E_3-C and E_1. Polyubiquitinated cyclin B is assumed to be rapidly degraded by the 26S protease and cdc2 is thus inactivated. We propose that the cyclin destruction system is subsequently turned off by the action of a phosphatase that dephosphorylates the cyclosome- E_3-C complex.

While the above model is consistent with all available information, much of this information is preliminary or fragmentary, and large gaps remain in our understanding of the molecular mechanisms of the selectivity and regulation of cyclin degradation. The enzymes involved in cyclin-Ub ligation, E_2-C and E_3-C, have been only partially purified, and results obtained with partially purified enzymes can usually yield only tentative conclusions. Much further purification, cloning and expression of E_2-C and E_3-C has to be achieved for the understanding of the specificity and regulation of their action. The subunit composition of the cyclosome complex, the mode of its association with cyclins and with E_2-C, and the possible roles of some of its other subunits in the regulation of E_3-C activity, remain to be addressed. The sequence of events initiated by cdc2 and culminated by the activation of E_3-C, and the delay mechanism which prevents the premature activation of E_3-C, are of obvious interest. It also remains to be seen whether the activity of the phosphatase that turns off cyclin degradation is also subject to cell cycle regulation.

REFERENCES

1. Hershko, A., and Ciechanover, A. (1992). The Ubiquitin System for Protein Degradation. *Annu. Rev. Biochem.*61, 761-807.
2. Finley, D. and Chau, V. (1991). Ubiquitination. *Annu. Rev. Cell Biol.7*, 25-69.
3. Rechsteiner, M. (1993). The Multicatalytic and 26S Proteases. *J. Biol. Chem.* 268, 6065-6068.
4. Pickart, C.M. and Rose, I.A. (1985). Functional Heterogeneity of Ubiquitin Carrier Proteins. *J. Biol. Chem. 260*, 1573-1581.
5. Jentsch, S. (1992). The Ubiquitin-Conjugation System. *Annu. Rev. Genet.* 26, 179-207.
6. Hershko, A., Heller, H. Eytan, E. and Reiss, Y. (1986). The Protein Substrate Binding Site of the Ubiquitin-Protein Ligase System. *J. Biol. Chem. 261*, 11992-11999.
7. Reiss, Y., Heller, H. and Hershko, A., (1989). Binding Sites of Ubiquitin-Protein Conjugates and of Ubiquitin-Carrier Protein. *J. Biol. Chem.* 264, 10378-10383.
8. Reiss, Y., Kaim, D. and Hershko, A., (1988). Specificity of Binding of NH_2-terminal Residue of Proteins to Ubiquitin-Protein Ligase: Use of Amino Acid Derivatives to Characterize Specific Binding Sites. *J. Biol. Chem.* 263, 2693-2698.
9. Reiss, Y. and Hershko, A. (1990). Affinity Purification of Ubiquitin-Protein Ligase on Immobilized Protein Substrates: Evidence for the Existence of Separate NH_2-Terminal Binding Sites on a Single Enzyme. *J. Biol. Chem.* 265, 3685-3690.
10. Heller, H. and Hershko, A., (1990). A Ubiquitin-Protein Ligase Specific for Type III Protein Substrates. *J. Biol. Chem.* 265, 6532-6535.
11. Hershko, A., and Heller, H. (1985). Occurrence of a Polyubiquitin Structure in Ubiquitin-Protein Conjugates. *Biochem. Biophys. Res. Commun.*128, 1079-1086.
12. Chau, V., Tobias, J. W., Bachmair, A., Mariott, D., Ecker, D., Gonda, D.K. and Varshavsky, A. (1989). A Multiubiquitin Chain is Confined to a Specific Lysine in a Targeted Short-lived Protein. *Science*, 243, 1576-1583.
13. Hough, R., Pratt, G. and Rechsteiner, M. (1987). Purification of Two High Molecular Weight Proteases from Rabbit Reticulocyte Lysate. *J. Biol. Chem.* 262. 8308-8313.
14. Ganoth, D., Leshinsky, E., Eytan, E. and Hershko, A. (1988). A Multicomponent System that Degrades Proteins Conjugated to Ubiquitin: Resolution of Factors and Evidence for ATP-Dependent Complex Formation. *J. Biol. Chem.* 263, 12412-12419.
15. Eytan, E., Ganoth, D., Armon, T. and Hershko, A., (1989). ATP-Dependent Incorporation of 20S Protease Into the 26S Complex that Degrades Proteins Conjugated to Ubiquitin. *Proc. Natl. Acad Sci. USA.* 86, 7751-7755.
16. Deleted.
17. Armon, T., Ganoth, D. and Hershko, A., (1990). Assembly of the 26S Complex That Degrades Proteins Ligated to Ubiquitin is Accompanied by the Formation of ATPase Activity. *J. Biol. Chem.* 265, 20723-20726.
18. Shanklin, J., Jabben, M., and Vierstra, R.D. (1987). Red Light-Induced Formation of Ubiquitin-Phytochrome Conjugates: Identification of Possible Intermediates of Phytochrome Degradation. *Proc. Natl. Acad Sci. USA* 84, 359-363.
19. Hochstrasser, M., Ellison, M.J., Chau, V. and Varshavsky, A. (1991). The Short-lived MATα2 Transcriptional Regulator is Ubiquitinated *In Vivo. Proc. Natl. Acad Sci. USA*, 88, 4606-4610.
20. Palombella, V.J., Rando, O.J., Goldberg, A. L. and Maniatis, T. (1994). The Ubiquitin-Proteasome Pathway is Required for the Processing of NF-κB Precursor Protein and the Activation of NF-κB. *Cell*. 78, 773-785.

21. Michalek, M., Grant, E., Gramm, C., Goldberg, A.L. and Rock, K. (1993). A Role for Ubiquitin-dependent Proteolytic Pathway in MHC class-I-restricted Antigen Presentation. *Nature.* 363, 552-554.
22. Scheffner, M., Huigbretze, J.M., Vierstra, R.D. and Howley, P. (1993). The HPV-16 E6 and E6-AP Complex Functions as a Ubiquitin-Protein Ligase in the Ubiquitination of p53. *Cell*,75, 495-505.
23. Ciechanover, A., Di Giuseppe, J.A., Bercovitch, B., Orian, A., Richter, J.D., Schwartz, A.L. and Brodner, G.M. (1991). Degradation of Oncoproteins by the Ubiquitin System *In Vitro. Proc. Natl. Acad Sci. USA* 88, 139-143.
24. Treier, M., Staszerwski, L.M. and Bohmann, D. (1994) Ubiquitin-Dependent c-jun Degradation *In Vivo* is Mediated by the δ Domain. *Cell.* 78, 787-798.
25. Nishisawa, M., Furuno, N., Okazaki, K., Tanaka, H., Ogawa, Y. and Sagata, N. (1993). Degradation of Mos by the N-terminal Proline (Pro^2)-dependent Ubiquitin Pathway on Fertilization of *Xenopus* Eggs: Possible Significance of Natural Selection for Pro^2 in Mos. *EMBO J.* 12, 4021-4027.
26. Glotzer, M., Murray, A.W. and Kirschner M.W. (1991). Cyclin is Degraded by the Ubiquitin Pathway. *Nature,* 349, 132-138.
27. Hershko, A., Ganoth, D., Pehrson, J., Palazzo, R.E. and Cohen, L.H. (1991). Methylated Ubiquitin Inhibits Cyclin Degradation in Clam Embryo Extracts. *J. Biol. Chem.* 266 16376-116379.
28. Schwob, E., B_hm, T., Mendenhall, M.D. and Nasmyth K. (1994). The B-type cyclin Kinase Inhibitor p40 ^{sicl} Controls the G1 to S Transition in *S cerevisiae. Cell.* 79, 233-244.
29. Bachmair, A., Finley, D., and Varshavsky, A. (1986). *In Vivo* Half-Life of a Protein is a Function of its Amino-terminal Residue. *Science.* 234, 179-186.
30. Barbel, B., Wunning, I. and Varshavsky, A. (1990). The Recognition Component of the N-End Rule Pathway. *EMBO J.*9, 3179-3189.
31. Norbury C. and Nurse, P. (1992). Animal Cell Cycles and their Control. *Annu. Rev. Biochem.* 61, 441-470.
32. King, W.R., Jackson, P.K. and Kirschner, M.W. (1994). Mitosis in Transition. Cell, 79, 563-571.
33. Murry, A.W., Solomon, M.J. and Kirschner, M.W. (1989). The Role of Cyclin Synthesis and Degradation in the Control of Maturation Promoting Activity. *Nature,* 339, 280-286.
34. Luca, F.C. and Ruderman, J.V. (1989). Control of Programmed Cyclin Destruction in a Cell-Free System. *J. Biol. Chem.* 109, 1895-1909.
35. Luca, F.C., Shibuya, E.K., Dohrmann, C.D. and Ruderman, J.V. (1991). Both Cyclin AΔ60 and BΔ97 are Stable and Arrest Cells at the M-phase, but only Cyclin BΔ97 Turns on Cyclin Destruction. *EMBO J.*10, 4311-4320.
36. Felix, M.A., Labbe, J.C., Doree, M., Hunt, T. and Karsenti, E. (1990). Triggering of Cyclin Degradation in Interphase Extracts of Amphibian Eggs by Cdc2 Kinase. *Nature,* 346, 379-382.
37. Hershko, A., Ganoth, D., Sudakin, V., Cohen, L.H., Luca, F.C., Ruderman, J.V. and Eytan, E. (1994). Components of a System that Ligates Cyclin to Ubiquitin and Their Regulation by Protein Kinase Cdc2. *J. Biol. Chem.* 269, 4940-4946.
38. Sudakin, V., Ganoth, D., Dahan, A., Heller, H., Hershko, J., Luca, F.C., Ruderman, J.V. and Hershko, A. (1995). The Cyclosome, a Large Complex Containing Cyclin-Selective Ubiquitin Ligase Activity, Targets Cyclins for Destruction at the End of Mitosis. *Mol. Biol..Cell,* in press.
39. Lorca, T., Fesquet, D., Zindy, F., LeBouffant, F., Cerutti, M., Brechot, C., Devauchelle, G. and Doree, M. (1991). An Okadaic Acid-sensitive Phosphatase Negatively Controls the Cyclin Degradation Pathway in Amphibian Eggs. *Mol. Cell Biol.* 11, 1171-1175.
40. Cohen, P., Holmes, C.F.B. and Tsukitani, Y. (1990). Okadaic acid: a New Probe for the Study of Cellular Regulation. *Trends Biochem. Sci.* 15, 98-102.

28

PHYSIOLOGICAL FUNCTIONS OF PROTEASOMES IN ASCIDIAN FERTILIZATION AND EMBRYONIC CELL CYCLE

Hitoshi Sawada, Hiroyuki Kawahara, Yoshiko Saitoh, and Hideyoshi Yokosawa

Department of Biochemistry
Faculty of Pharmaceutical Sciences
Hokkaido University
Sapporo 060, Japan

INTRODUCTION

Fertilization is initiated by sperm binding to the vitelline coat, a glycoprotein coat, of the eggs. After the sperm binding, sperm undergoes the acrosome reaction, an exocytosis of the acrosome. By this acrosome reaction, sperm proteases or lytic agents in the acrosome which allows sperm to penetrate through the vitelline coat are exposed and released. Subsequently, the sperm-egg membrane fusion occurs, which triggers a transient increase in intracellular calcium ions, leading to the exocytosis and the resumption of the meiotic division cycle, *i.e.,* egg activation.

Ascidians (prochordate) occupy a phylogenetic position between invertebrates and "true" vertebrates. While this animal is a hermaphrodite, the fertilization is self-sterile in the solitary asicidian, *Halocynthia roretzi,* because of strict self and non-self recognition between gametes (2). Easiness of fertilization experiments and good synchrony of cell division cycle in *H. roretzi* enabled us to investigate the roles of sperm proteases in a process of sperm penetration through the vitelline coat (a lysin system) and those of egg proteases in meiotic and mitotic cell cycles, using this animal.

ROLES OF SPERM PROTEASOMES IN FERTILIZATION

We have previously proposed that trypsin-lilke and chymotrypsin-like proteases are essential for sperm penetration of the vitelline coat on the basis of our results of the effects of various protease inhibitors and substrates on fertilization of naked and intact eggs of *H. roretzi* (1-4). At first, we purified two trypsin-like proteases, desingated acrosin and spermosin, from *H. roretzi* sperm to homogeneity. It has been proved that both of these proteases are involved in

Intracellular Protein Catabolism, Edited by Koichi Suzuki and Judith Bond
Plenum Press, New York, 1996

Table 1. Effects of protease inhibitors on ascidian sperm proteasomes

Inhibitor	Concentration (mM)	Inhibition (%) 20 S proteasome	930 kDa proteasome
Chymostatin	0.1	95	0
PCMB	0.1	98	0
Propioxatin A	0.1	0	98
EDTA	0.1	0	54

PCMB, *p*-chloromercuribenzonte

fertilization, especially in the sperm penetration process of the coat, by comparing the effects of a variety of leupeptin analogs on the purified enzyme activities and fertilization (5, 6). These enzymes, however, were found to be incapable of digesting the vitelline coat.

Next, we attempted to purify the chymotrypsin-like enzyme from sperm by using Suc-Leu-Leu-Val-Tyr-MCA as a substrate, and two isoforms of high molecular weight multicatalytic proteases were isolated (7). One is a typical 20 S proteasome on the basis of its molecular mass and subunit composition, while the other appears to be a higher molecular mass (930 kDa) isoform of the 20 S proteasome, because the antiserum raised against the ascidian egg 20 S proteasome cross-reacted to this isoform. The 20 S proteasome is strongly inhibited by chymostatin and PCMB but not by propioxatin A (an enkephalinase inhibitor) and EDTA. In contrast, the 930 kDa proteasome is strongly inhibited by propioxatin A and EDTA but not by chymostatin and PCMB (Table 1).

Ascidian sperm undergoes the sperm reaction (a similar reaction to the acrosome reaction in other animals) by treatment with slightly alkaline seawater at pH 9.0. The resultant supernatant fractions (a sperm exudate) was concentrated and subjected to Superose 6 FPLC. The vitelline coat digesting activity was detected in the Superose 6 fractions containing proteasomes, which were revealed by dot blot analysis with anti-20 S proteasome antibody. A fraction which contained the 930 kDa proteasome showed stronger vitelline coat-digesting activity than that containing the 20 S proteasome. The fact that the vitelline coat-digesting activity in the sperm exudate is strongly inhibited by propioxatin A fits well with the result that propioxatin A is a strong inhibitor of ascidian fertilization (Fig. 1). Taking into account that proteasomes are released upon the sperm reaction and that they have a vitelline coat digesting activity, the 930 kDa proteasome (and probably 20 S proteasome) is thought to be responsible for sperm penetration through the vitelline coat in the ascidian.

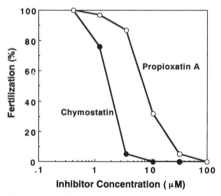

Figure 1. Inhibition of ascidian fertilization by propioxatin A and chymostatin.

It was found that chymostatin, a strong inhibitor of the ascidian fertilization (see Fig 1), significantly inhibits the binding of sperm to the vitelline coat (8). In addition, inhibition patterns of chymostatin and several peptidyl-argininals against the sperm binding to the vitelline coat are well coincident with their inhibition patterns against the chymotrypsin-like activity of the 20 S proteasome. These results indicate that the 20 S proteasome is involved in a process of sperm binding to the vitelline coat during ascidian fertilization (8).

In conclusion, sperm proteasomes play essential roles in ascidian fertilization processes, especially in sperm binding to the vitelline coat and in sperm penetration through the vitelline coat.

ROLE OF PROTEASOME IN EMBRYONIC CELL CYCLE

The 20 S proteasome was isolated from eggs of the ascidian, *H. roretzi* (9), which is the first report on the proteasome from gamete origin. By immunocytochemical analysis using monoclonal antibodies raised against the purified 20 S proteasome, we found that the proteasome undergoes a cell-cycle dependent change of distribution during the meiotic cell cycle and the subsequent cleavage cycle (10). In accordance with the change of distribution of the proteasome, it was found that an ATP-stimulated 26 S proteasome activity is activated at two points, prophase and metaphase, during the egg cleavage cycle of *H. roretzi* (11).

Next, we investigated whether a transient activation of proteasomes occurs during the meiotic division cycle triggered by intracellular calcium mobilization, and analyzed the mechanisms of activation of proteasome (12).

In the egg activation process, a proteasome activity toward Suc-Leu-Leu-Val-Tyr-MCA in a high-molecular weight protein fraction (precipitates obtained by ultracentrifugation at 100,000 x g for 5 h) is transiently increased at 5 min after the addition of 1 μM Ca^{2+} ionophore A23187, the timing of which corresponds to the metaphase-anaphase transition, as analyzed by microscopic observation. In contrast, total amounts of the 20 S and 26 S proteasomes remain constant throughout the meiotic division cycle. The transient activation of proteasome induced by treatment with A23187 is abolished by pretreatment of eggs with BAPTA-AM, a cell-permeable calcium cheleting reagent, indicating that this activation is elicited by intracellular calcium mobilization.

The high-molecular weight protein fractions obtained at the respective times after the addition of A23187 were subjected to Superose 6 FPLC. The results showed that the activity of the 26 S proteasome-containing fraction, which is immunoprecipitable with anti-20 S proteasome antibody, is temporarily increased at 5 min after the addition of the ionophore, and then decreased. A similar result was also obtained when Z-Leu-Leu-Glu-βNA

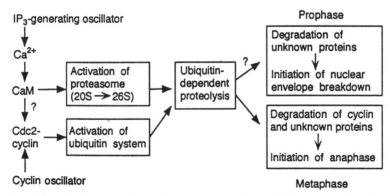

Figure 2. A possible activation mechanism of the proteasome and its role in the progression of the cell cycle.

was used as a substrate. In unfertilized eggs, both 20 S and 26 S proteasomes are present, though the amount of the 20 S proteasome is much higher than that of 26 S proteasome. At 5 min after the initiation of egg activation, the amount of the 26 S proteasome markedly increases, and that of the 20 S proteasome decreases, as revealed by Western blotting. Subsequently, the amount of the 26 S proteasome decreases and that of the 20 S proteasome increases in an inverse manner. These results clearly indicate that the interconversion between the 20 S and the 26 S proteasomes is induced by intracellular calcium mobilization. The transient increase of the ATP-stimulated 26 S proteasome activity, which is due to its reconstitution from the 20 S proteasome, is prevented by pretreatment of eggs with W-7, a calmodulin antagonist, suggesting the involvement of calmodulin in the reconstitution of the 26 S proteasome. Furthermore, a similar proteasome activation coupled to meiotic division cycle I and II triggered by insemination was also demonstrated.

In conclusion, the interconversion between the 20 S and the 26 S proteasomes is regulated by intracellular calcium mobilization, and the 26 S proteasome, reconstituted from the 20 S proteasomes, functions in the ubiquitin-dependent proteolysis of cell cycle regulators in the progression of cell cycle (see Fig. 2).

ACKNOWLEDGEMENT

This study was supported in part by Grants-in-aid for Scientific Research from the Ministry of Education, Science, and Culture of Japan.

REFERENCES

1. Hoshi, M., Numakunai, T., and Sawada, H., 1981, Evidence for participation of sperm proteinases in fertilization of the solitary ascidian, *Halocynthia roretzi:* Effects of protease inhibitors, *Dev. Biol.* 86: 117-121.
2. Sawada, H., Yokosawa, H., Hoshi, M., and Ishii, S., 1982, Evidence for acrosin-like enzyme in sperm extract and its involvement in fertilization of the ascidian, *Halocynhtia roretzi, Gamete Res.* 5: 291-301.
3. Sawada, H., Yokosawa, H., Hoshi, M., and Ishii, S., 1983, Ascidian sperm chymotrypsin-like enzyme; participation in fertilization, *Experientia* 39: 377-378.
4. Yokosawa, H., Numakunai, T., Murao, S., and Ishii, S., 1987, Sperm chymotryspin-like enzyme of different inhibitor-susceptibility as lysins in ascidians, *Experientia* 43: 925-927.
5. Sawada, H., Yokosawa, H., and Ishii, S., 1984, Purification and characterization of two types of trypsin-like enzymes from sperm of the ascidian (Prochordata), *Halocynthia roretzi.* Evidence for the presence of spermosin, a novel acrosin-like enzyme, *J. Biol. Chem.* 259: 2900-2904.
6. Sawada, H., Yokosawa, H., Someno, T., Saino, T., and Ishii, S., 1984, Evidence for the participation of two sperm proteases, spermosin and acrosin, in fertilization of the ascidian, *Halocynthia roretzi:* Inhibitory effects of leupeptin analogs on enzyme activities and fertilization, *Dev. Biol.* 105: 246-249.
7. Saitoh, Y., Sawada, H., and Yokosawa, H, 1993, High-molecular-weight protease complexes (proteasomes) of sperm of the ascidian, *Halocynthia roretzi:* Isolation, characterization, and physiological roles in fertilization, *Dev. Biol.* 158: 238-244.
8. Takizawa, S., Sawada, H., Someno, T., Saitoh, Y., Yokosawa, H., and Hoshi, M., 1993, Effects of protease inhibitors on binding of sperm to the vitelline coat of ascidian eggs: implications for participation of a proteasome (multicatalytic proteinase complex), *J. Exp. Zool.* 267: 86-91.
9. Saitoh, Y., Yokosawa, H., Takahashi, K., and Ishii, S., 1989, Purification and characterization of multicatalytic proteinase from eggs of the ascidian, *Halocynthia roretzi, J. Biochem.* 105: 254-260.
10. Kawahara, H., and Yokosawa, H. , 1992, Cell cycle-dependent change of proteasome distribution during embryonic development of the ascidian, *Halocynthia roretzi, Dev. Biol.* 151: 27-33.
11. Kawahara, H., Sawada, H., and Yokosawa, H., 1992, The 26 S proteasome is activated at two points in the ascidian cell cycle, *FEBS Lett.* 310: 119-122.
12. Kawahara, H., and Yokosawa, H., 1994, Intracellular calcium mobilization regulates the activity of 26 S proteasome during the metaphase-anaphase transition in the ascidian meiotic cell cycle, *Dev. Biol.* 166: 623-633.

29

CELLULAR PROTEASES INVOLVED IN THE PATHOGENICITY OF HUMAN IMMUNODEFICIENCY AND INFLUENZA VIRUSES

Hiroshi Kido, Takae Towatari, Yasuharu Niwa, Yuushi Okumura, and Yoshihito Beppu

Division of Enzyme Chemistry
Institute for Enzyme Research
The University of Tokushima
Tokushima 770, Japan

INTRODUCTION

Current understanding of the interactions between animal viruses and host cellular factors, such as virus receptors in the membrane and intracellular factors for viral replication, which determine the pathogenicity and infectious tropism of viruses, has considerably progressed. Current knowledge of membrane receptors for animal viruses are summarized in Table I. Viruses recognize and specifically bind to adhesion molecules in the membrane as receptors, many of which are members of the immunoglobulin (IgG) superfamily. These include the 70K IgG family of proteins for poliovirus, ICAM-1 for rhinovirus and CD4 for human immunodeficiency virus type-1 (HIV-1). Other virus receptors are receptors for hormones and proteins, such as the EGF receptor for vaccinia virus and β-adrenergic receptor for reovirus, complement receptor for EB virus and sialic acid for influenza and Sendai viruses.

However, membrane receptors for several enveloped viruses are not sufficient alone for viruses to enter host cells. Intracellular or extracellular proteases are required for proteolytic activation of the viral envelope fusion glycoproteins and also for activation of cellular membrane fusion machinery. Examples of viruses that require posttranslational proteolytic cleavage of viral envelope glycoproteins by cellular proteases for fusion activities, are paramyxoviruses, such as Sendai virus and Newcastle disease virus, and orthomyxoviruses, such as influenza viruses(1-4). In addition, binding of the gp120 envelope glycoprotein of HIV-1 to the CD4 receptor of permissive cells is not sufficient to allow virus entry. Additional components in the surface of host cells are required for cell infection as cofactors (5,6). We first described that one of these cofactors could be a cell surface protease (7,8), that proteolytically cleaves the V3 loop region of HIV-1 gp120 (9), leading to a

Intracellular Protein Catabolism, Edited by Koichi Suzuki and Judith Bond
Plenum Press, New York, 1996

Table 1. Molecular basis of virus receptors

Virus	Receptor
DNA virus	
poliomavirus	2.5, 50 and 95K proteins
adenovirus	42K glycoprotein
EB virus	complement receptor
vaccinia virus	EGF receptor
RNA virus	
poliovirus	70K IgG family protein
rhinovirus	ICAM-1
echovirus	VLA2
coronavirus	aminopeptidase N
LDH virus	Ia antigen
Sendai virus	sialic acid
influenza vairus	sialic acid
reovirus	β-adrenergic receptor
murine leukemia virus	100K glycoprotein
HIV	CD4

conformational change in the gp120-gp41 protein complex on the viral surface following fusion of the lipid bilayer of the virus with that of the target cells. The present discussion describes current knowledge of the cellular protease tryptase Clara, which proteolytically activates the fusion glycoproteins of Sendai and influenza viruses and tryptase TL2 from the membrane of human CD4$^+$ T cells as a cofactor for HIV-1 entry.

PNEUMOPATHOGENICITY OF SENDAI AND INFLUENZA VIRUSES IS DETERMINED BY TRYPTASE CLARA FROM BRONCHIOLAR EPITHELIAL SECRETORY CELLS

Sendai and influenza viruses are exclusively pneumotropic and the target of the viral infection is restricted to the bronchial epithelial cells of rodents and humans, respectively, although the virus receptor sialic acid is widely distributed among various cell types in the lungs and other organs (1-4, 10). It has been postulated that the pneumotropism of the viruses is determined by the presence of a specific trypsin-like serine protease(s) in the respiratory tract, which cleaves precursors of envelope fusion glycoprotein of the progeny viruses to introduce fusion activity, thereby enabling the viruses to undergo multiple cycles of replication (10,11).

We isolated a novel trypsin-like serine protease, designated tryptase Clara, from the rat lung (12). This protease is localized only in non-ciliated secretory cells (Clara cells) of the bronchial and bronchiolar epithelia of rats, and is secreted into the airway lumen. Figure 1 shows that tryptase Clara converts F0 of Sendai virus to the subunits F1 and F2, in a similar manner to trypsin (13). Infectivity in nonactive wild-type Sendai viruses grown in LLC-MK2 cells was activated in a dose-dependent manner (Fig. 1a). Although the activation of infectivity by the protease was less efficient than that induced by trypsin, it should be noted that in contrast to trypsin, tryptase Clara at higher concentrations did not decrease viral infectivity. SDS-PAGE of [^{3}H]glucosamine-labeled wild-type virus activated by tryptase Clara shows that tryptase Clara cleaved the precursor polypeptide F0 into subunits F1 and F2, similar to trypsin (Fig. 1b). Direct amino acid sequencing of the NH$_2$ terminus of the F1

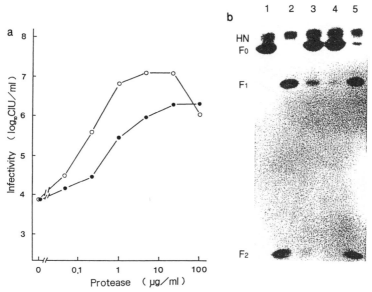

Figure 1. Proteolytic activation of wild-type Sendai virus *in vitro* by tryptase Clara. (a) Activation of virus infectivity. Sendai virus propagated in LLC-MK2 cells were suspended in PBS (pH 7.2) and digested with tryptase Clara for 15 min (●) or with trypsin for 10 min (○) at 37°C. The infectivity of activated virus was assayed by immunofluorescent cell-counting, using LLC-MK2 cells in the absence of trypsin, and is shown as cell-infecting units (CIU). (b) SDS-PAGE of the viral glycoproteins. Virus propagated in LLC-MK2 cells were labeled with [^{3}H] glucosamine (lane 1), then digested with trypsin at 5 µg/ml for 10 min (lane 2) orwith tryptase Clara at 50 (lane 3) and 10 (lane 4) µg/ml for 15 min or at 50 µg/ml for 30 min (lane 5). Viral glycoproteins were analyzed by SDS-PAGE under reducing conditions. HN, hemagglutinin-neuraminidase.

subunit produced by tryptase Clara revealed the residues Phe-Phe-Gly-Ala-Val-Gly-, indicating that the cleavage site of wild-type F0 by tryptase Clara was between R^{116} and F^{117} (13). These results suggested that tryptase Clara cleaves specifically at the activation cleavage site of the F0 protein and not at other sites that may be cleaved by trypsin at high concentrations. The effect of tryptase Clara on hemagglutinin (HA) of influenza virus was similar (12). Tryptase Clara also specifically cleaved the peptide bond between R^{325} and G^{326} of HA of influenza A/Aichi/2/68 (H3N2) to form HA1 and HA2. These viruses have a consensus cleavage motif Q/E-X-R in the precursors of the envelope fusion glycoproteins, which may be specifically recognized and cleaved by tryptase Clara, although amino acid sequences of other regions of the glycoproteins are highly variable among viruses.

The physiological role of tryptase Clara in the respiratory tract is unknown, but the activity of the enzyme, like those of many proteases, may be strictly regulated by an endogenous inhibitor(s). We found that pulmonary surfactant, which is secreted by alveolar type II cells and by bronchiolar epithelial Clara cells, is a specific endogenous inhibitor of tryptase Clara (14). The inhibition was non-competitive and the *Ki* value of the purified surfactant was 0.13 µM. The inhibition was specific for tryptase Clara, and the other trypsin-type proteases, such as trypsin, factor Xa, plasmin and rat mast cell tryptase, were not inhibited. Figure 2 shows the effect of surfactant on the proteolytic cleavage of F0 of Sendai virus by tryptase Clara. SDS-PAGE revealed that surfactant inhibited the cleavage of F0 to its subunits F1 and F2 by tryptase Clara dose dependently, but did not affect trypsin cleavage (14). The proteolytic cleavage of HA of influenza

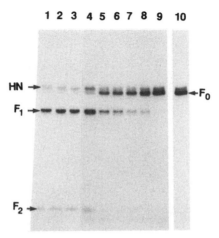

Figure 2. Effect of surfactant on proteolytic cleavage of F0 of Sendai virus by tryptase Clara. Trypsin (10 μg/ml, lanes 1-3) and tryptase Clara (50 μg/ml, lanes 4-9) were incubated without surfactant (lanes 1 and 4) or with 0.26 μg/ml (lanes 2 and 5), 0.52 μg/ml (lane 6), 0.78 μg/ml (lane 7), 1 mg/ml (lanes 3 and 8), 3.0 mg/ml (lane 9) of surfactant in 20 μl of 100 mM Tris-HCl buffer, pH 7.2, at 0°C for 5 min. Thereafter, [³H] glucosamine-labeled inactive Sendai virus (lane 10) was incubated with the reaction mixtures at 37°C for 10 min. The viral polypeptides were separated by SDS-PAGE under reducing conditions followed by fluorography. The concentrations of surfactant are expressed as those of phospholipids in the samples.

A/Aichi virus by tryptase Clara was also inhibited by pulmonary surfactant (data not shown).

A summary of the role of tryptase Clara and its specific inhibitor, pulmonary surfactant, upon influenza and Sendai virus activation in the respiratory tract, is described in Fig. 3.

THE V3 DOMAIN OF HIV-1 GP120 IS CLEAVED BY TRYPTASE TL2 IN THE MEMBRANE OF HUMAN CD4 LYMPHOCYTES

Although proteolytic processing of the precursor gp160 is required for HIV-1 to infect cells (15-17), the mechanism by which the HIV-1 envelope glycoproteins trigger membrane fusion is poorly understood. Several reports have shown that the V3 loop region of gp120 is essential for HIV-1 infection and membrane fusion because (i) mutations introduced into the V3 loop inhibit infectivity and syncytium formation (18), (ii) antibodies against the principal neutralizing domain in the V3 loop region inhibit HIV-1 infection of cells without interfering with the binding of HIV-1 gp120 to its cellular receptor, the CD4 molecule (19) and (iii) the V3 loop region is the target for a serine protease(s) on the host cell surface (7-9). It has therefore been proposed that after gp120 present on the viral surface binds to the CD4 receptor on the target cell membrane, the V3 loop region in gp120 is proteolytically cleaved by a cell surface protease, leading to a conformational change in the gp120-gp41 protein complex on the viral surface (7,20). Such a change is thought to be responsible for exposing the fusogenic domain of the transmembrane protein gp41, resulting in the fusion of the viral-particle membrane with the cell membrane.

We showed that a serine esterase, named tryptase TL2 in the membrane of human CD4⁺ lymphocytes binds specifically to the gp120 of HIV-1, by interacting with its V3 loop (8,9). This binding was selectively blocked by inhibitors of tryptase TL2 with a GPCR

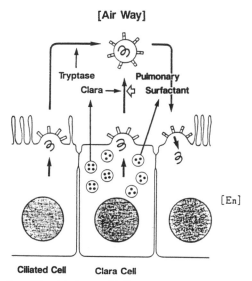

Figure 3. Proteolytic activation of Sendai virus and influenza virus in the respiratory tract and its inhibition by pulmonary surfactant.

sequence in their reactive site, synthetic peptides corresponding with the sequences of the V3 domains of various HIV-1 strains with the GPGR sequence, and an antibody against tryptase TL2, or a neutralizing antibody against the V3 domain. Kunitz type protease inhibitor of tryptase TL2, trypstatin (21) and an antibody against tryptase TL2 inhibited multi nucleated cell-to-cell fusion (syncytia) induced by HIV-1 (7). These results suggest that cleavage of V3 domain of gp120 by cellular protease, such as tryptase TL2, could be important for HIV infection.

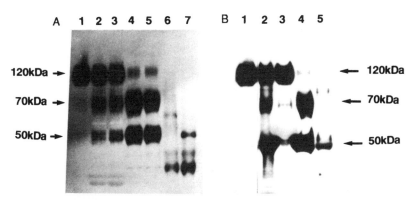

Figure 4. Processing of HIV-1 gp120 by tryptase TL2, thrombin and cathepsin G and its inhibition by the neutralizing antibody against the V3 loop of gp120. (A) HIV-1$_{SF2}$ gp120 (3 μg)(lane 1) was incubated with 1.0 μg of tryptase TL2 (lanes 2 and 3), 0.2 μg of thrombin (lanes 4 and 5) and 0.2 μg of cathepsin G (lanes 6 and 7) with (lanes 3, 5 and 7) or without (lanes 1, 2, 4 and 6) 1 μg of rsCD4 for 8h at 37°C and analyzed by immunoblotting against anti-gp120 antibodies. (B) gp120 (3 μg) (lane 1) was incubated with 2 μg of non-immunized IgG (lane 2) or 2 μg of the neutralizing antibody against the V3 loop of gp120 (lanes 3, 4 and 5) for 1h at 37°C, then with 1.0 μg of tryptase TL2 (lanes 2 and 3), 0.2 μg of thrombin (lane 4), or 0.2 μg of cathepsin G (lane 5) for 8h at 37°C. Proteolytic products were analyzed by immunoblotting.

It has been reported that the outer envelope protein gp120 of highly purified virus preparations was cleaved, most likely at the V3 loop, relative to the CD4 concentration and the incubation time, and neutralizing antibody inhibited the processing and HIV-1 infection (22). The results are explained by copurification, or the incorporation of a cellular protease into the viral membrane during budding. To study the potential role of tryptase TL2 in the membrane, which selectively binds to the V3 loop of gp120 in HIV-1 infection, we analyzed the proteolytic processing of gp120 by this enzyme. In our studies, purified glycosylated recombinant wild type gp120 expressed in CHO cells was incubated with tryptase TL2 and other serine proteases with or without the neutralizing antibody against the V3 domain for 8h at 37°C .

As shown in Fig. 4, tryptase TL2 partly cleaved full-length gp120 into two protein species of approximately 70 and 50kDa, and the cleavage was suppressed by the neutralizing antibody which binds to the center tip of the V3 domain (IRIQRGPGR). The cleavage was highly specific, since there was no additional cleavage of the breakdown products. Processing of gp120 by thrombin was also examined as a positive control, because thrombin has been reported to selectively cleave the gp120 at the V3 loop and forms 70 and 50kDa products *in vitro* system, although thrombin is not a cellular protease in the T cell membranes (23). However, the neutralizing antibody did not block the processing by thrombin. Cathepsin G, which has also been reported to interact with the V3 loop of HIV-1 gp120 in macrophages (24), converted gp120 to various small fragments and the neutralizing antibody did not inhibit this processing. The molecular masses of the proteolytic products generated by tryptase TL2 were similar to those released by thrombin (23) and those from a purified HIV-1 preparation incubated with soluble CD4 without exogenous proteases (22). In our system however, soluble CD4 had no effect on the cleavage. The results suggested that a conformational modification of virion-associated gp120 by soluble CD4 is essential for the proteolytic processing by virion-associated cellular protease(s), whereas CD4 is not required for the processing of soluble gp120 by tryptase TL2 *in vitro*. We thus proposed that the V3 loop of gp120 interacts with a tryptase TL2 but not with a cathepsin G and a thrombin in the membrane or in the virion that would have to cleave it as a prerequisite for viral entry. The selective cleavage of the V3 loop leads to a conformational change of gp120 resulting in the dissociation of gp120 from gp41, in exposing the fusogenic domain of the transmembrane protein gp41 following virus-host cell fusion. Our model of the role of tryptase TL2 in the initial step of HIV-1 entry is summarized in Fig. 5.

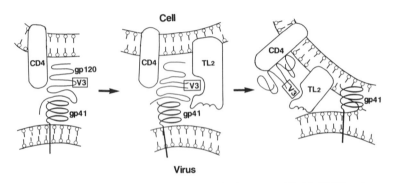

Figure 5. Proposed role of V3 loop processing by tryptase TL2 in HIV-1 entry.

CONCLUSION

Several membrane receptors that bind enveloped viruses are not sufficient by themselves to allow viruses to entry host cells. Intracellular or extracellular proteases are required for the proteolytic activation of the viral envelope glycoproteins and probably also for activation of cellular membrane fusion machinery.

We isolated a novel trypsin-like protease, designated tryptase Clara which is secreted by the Clara cells of the bronchial epithelium. The enzyme specifically recognizes the concensus cleavage motif Q/E-X-R of influenza and Sendai viruses and proteolytically activates the envelope fusion glycoproteins of the progeny viruses extracellularly in air way lumen. Pulmonary surfactant in the lungs and bronchus inhibited the enzyme activity and suppressed multiple cycles of viral replication in the respiratory tract.

In HIV-1 infection, binding of the gp120 envelope glycoprotein to the CD4 receptor is not sufficient in itself to allow virus entry, and an additional component(s) in the membrane is required for cell infection as a cofactor. We isolated a serine esterase, named tryptase TL2, in the membrane of CD4$^+$ lymphocytes, which specifically binds to the V3 loop of HIV-1 gp120 as a cofactor. After binding, tryptase TL2 proteolytically processed gp120 into two protein species of 70 and 50kDa and the cleavage was suppressed by a neutralizing antibody against the V3 loop. The neutralizing antibody and the antibody against the enzyme inhibited membrane fusion induced by HIV-1. We therefore proposed that the selective cleavage of the V3 loop by tryptase TL2 leads to a conformational change of gp120, resulting in the dissociation of gp120 from gp41, exposing the fusogenic domain of the transmembrane protein gp41 following virus-host cell fusion.

ACKNOWLEDGEMENTS

This work was supported in part by Grants-in-Aid 04670278, 05557014 and 06255209 from the Ministry of Education, Science and Culture of Japan. We thank E. Inai and M. Shiota for expert secretarial assistance.

REFERENCES

1. Homma, M., and Ohuchi, M., 1973, Trypsin action on the growth of Sendai virus in tissue culture cell, *J. Virol.* 12: 1457-1465.
2. Klenk, H.-D., and Rott, R., 1988, The molecular biology of influenza virus pathogenicity, *Adv. Virus Res.* 34: 247-281.
3. Rott, R., 1979, Molecular basis of infectivity and pathogenicity of myxoviruses, *Arch. Virol.* 59: 285-298.
4. Scheid, A., and Choppin, P.W., 1974, Identification of biological activities of paramyxovirus glycoproteins. Activation of cell fusion, hemolysis, and infectivity by proteolytic cleavage of an inactive precursor protein of Sendai virus, *Virology* 57: 475-490.
5. Maddon, P.J., Dalgleish, A.G., and McDougal, J.S., 1986, The T4 geme encodes the AIDS vius receptor and is expressed in the immune system and the brain, *Cell* 47: 333-348.
6. Clapham, P.R., Blanc, D., and Weiss, R.A., 1991, Specific cell-surface requirements for infection of CD4-positive cells by human immunodeficiency virus type 1, type 2 and simian imunodeficiency virus, *Virology* 181: 703-715.
7. Hattori, T., Koito, A., Takatsuki, K., Kido, H., and Katunuma, N., 1989, Involvement of tryptase-related cellular protease(s) in human immunodeficiency virus type 1 infection, *FEBS Lett.* 248: 48-52.
8. Kido, H., Fukutomi, A., and Katunuma, N., 1990, A novel membrane-bound serine esterae in human T4$^+$ lymphocytes immunologically rective with antibody inhibiting syncytia induced by HIV-1. Purification and charcterization, *J. Biol. Chem.* 265: 21979-21985.

9. Kido, H., Fukutomi, A., and Katunuma, N., 1991, Tryptase TL2 in the membrane of human T4⁺ lymphocytes is a novel binding protein of the V3 domain of HIV-1 envelope glycoprotein gp120, *FEBS Lett.* 286: 233-236.

10. Tashiro, M., and Homma, M., 1983, Pneumotropism of Sendai virus in relation to protease-mediated activation in mouse lungs, *Infect. Immun.* 39: 879-888.

11. Tashoro, M., Yamakawa, M., Tobita, K., Klenk, H.-D., Rott, R., and Seto, J.T., 1990, Organ tropism of Sendai virus in mice: proteolytic activation of the fusion glycoprotein in mouse organs and budding site at the bronchial epithelium, *J. Virol.* 64: 3627-3634.

12. Kido, H., Yokoigoshi, Y., Sakai, K., Tashiro, M., Kishino, Y., Fukutomi, A., and Katunuma, N., 1992, Isolation and characterization of a novel trypsin-like protease found in rat bronchiolar epithelial Clara cells. A possible activator of the viral fusion glycoprotein, *J. Biol. Chem.* 267: 13573-13579.

13. Tashiro, M., Yokogoshi, Y., Tobita, K., Seto, J.T., Rott, R., and Kido, H., 1992, Tryptase Clara, an activating protease for Sendai virus in rat lungs, is involved in pneumopathogenicity, *J. Virology* 66: 7211-7216.

14. Kido, H., Sakai, K., Kishino, Y., and Tashiro, M., 1993, Pulmonary surfactant is a potential endogenous inhibitor of proteolytic activation of Sendai virus and influenza A virus, *FEBS Lett.*, 322: 115-119.

15. Hu, S.H., Kosowski, S.G., and Dalrymple, J.M., 1986, Expression of AIDS virus envelope gene in recombinant vaccinia viruses, *Nature* 320: 537-540.

16. Travis, M.M., Dykers, T.T., Hu, S.-L., and Lewis, J.B., 1992, Functional role of the V3 hypervariable regions of HIV-1 gp160 in the processing of gp160 and in the formation of syncytia in CD4⁺ cells, *Virology* 186: 313-317.

17. Kido, H., Kamoshita, K., Fukutomi, A., and Katunuma, N., 1993, Processing protease for gp160 human immunodeficiency virus type 1 envelope glycoprotein precursor in human T4⁺ lymphocytes. Purification and characterization, *J. Biol. Chem.* 268: 13406-13413.

18. Ivanoff, L.A., Looney, D.J., McDanal, C., Morris, J.F., Wong-Stal, F., Langlois, A.J., Petteway, S.R., and Matthews, T.J., 1991, Alteration of HIV-1 infectivity and neutralization by a single amino acid replacement in the V3 loop domain. *AIDS Res. Hum. Retroviruses* 7: 595-603.

19. Javaherian, K., Langlois, A.J., McDanal, C., Ross, K.L., Eckler, L.L., Jellis, C.L., Profy, A.T., Rusche, J.R., Bolognesi, D.P., Putney, S.D., and Matthews, T.J., 1989, Principal neutralizing domain on the human immunodeficiency virus type 1 envelopoe protein, *Proc. Natl. Acad. Sci. USA* 86: 6768-6772.

20. Putney, S., 1992, How antibodies block HIV infection: paths to an AIDS vaccine, *TIBS* 17: 191-196.

21. Kido, H., Yokogoshi, Y., and Katunuma, N., 1988, Kunitz type protease inhibitor found in rat mast cell: purification, properties and amino acid sequence, *J. Biol. Chem.* 263: 18104-18107.

22. Werner, A., and Levy, J.A., 1993, Human immunodeficiency virus type 1 envelope gp120 is cleaved after incubation with recombinant soluble CD4, *J. Virology* 67: 2566-2574.

23. Clements, G.J., Price-Jones, M.J., Stephens, P.E., Sutton, C., Schulz, T.F., Clapham, P.R., Mckeating, J.A., McClure, M.O., Thomson, S., Marsh, M., Kay, J., Weiss, R.A., and Moore, J.P., 1991, The V3 loops of the HIV-1 and HIV-2 surface glycoproteins contain proteolytic clevage sites: a possible function in viral fusion, *AIDS Research and Human Retroviruses* 7:3-16.

24. Avril, L-E., Martino-Ferrer, M.D., Pignede, G., Seman, M., and Gauthier, F., 1994, Identification of the U-937 membrane-associated proteinase interacting with the V34 loop of HIV- gp120 as cathepsin G, *FEBS Lett.* 345: 81-86.

30

HIV PROTEASE MUTATIONS LEADING TO REDUCED INHIBITOR SUSCEPTIBILITY

B. Korant, Z. Lu, P. Strack, and C. Rizzo

Molecular Biology Division
DuPont Merck Pharmaceutical Co
500 South Ridgeway Ave
Glenolden, Pennsyvania 19036

INTRODUCTION

Human immunodeficiency viruses have infected millions of people world-wide, and infection almost invariably leads, over a several year time frame, to progression to AIDS with high mortality.

Current therapies based on inhibition of the viral reverse transcriptase, are widely used, but also have intrinsic side effects and after a course of therapy, may lead to emergence of drug-resistant viruses. For the nucleoside-based therapeutics, resistance is characterized by specific mutations in the viral gene coding for reverse transcriptase (1).

The virus codes for a second enzyme essential for replication; namely the viral aspartic protease. The enzyme plays a key role in several steps of replication, and high resolution crystal structures have been reported with several complexed inhibitors. This has helped to drive an intense effort by multiple industrial and academic laboratories to develop potent, selective inhibitors of the protease, and several of these inhibitors are showing encouraging results in treatment of HIV-infected people, although none has yet been approved for clinical use.

A concern of many is that, as observed with drugs directed at reverse transcriptase, the virus, which has a powerful ability to mutate rapidly, may be able to overcome anti-protease effects of small molecule inhibitors, and there have been several early reports of this in HIV-infected cell cultures (2,3,4).

In this paper, we examine the ability of HIV to alter its protease by mutation, and consequently lose susceptibility to protease inhibitors containing a cyclic urea core (5). We also show that the loss of inhibitor sensitivity is accompanied by reduction in enzyme efficiency.

Intracellular Protein Catabolism, Edited by Koichi Suzuki and Judith Bond
Plenum Press, New York, 1996

Table 1. Amino acids in the HIV-1 protease
involved in inhibitor complexing

Amino Acid No.	Residue
8	arginine
23	leucine
25	aspartic
27	glycine
28	alanine
29	aspartic
30	aspartic
32	valine
47	isoleucine
48	glycine
49	glycine
50	isoleucine
53	phenylalanine
76	leucine
81	proline
82	valine
84	isoleucine

RESULTS

Review of the current literature reporting structural analysis of inhibitors bound to HIV protease finds that of the 99 amino acids in the protease subunit, 17 different amino acids may be involved in inhibitor complex formation (Table 1).

Interestingly, a survey (6) of the amino acid variations observed in scores of HIV isolates indicated that there were no reports of changes in any of those 17 amino acids, although other residues of the protease were subject to spontaneous (no protease inhibitor present) mutations.

This encouraging sign was, however, replaced by new data, as several laboratories reported that HIV variants could be selected, by exposing infected cell cultures to limiting inhibitor concentrations for extended periods of time, and allowing viruses with reduced sensitivity to slowly become dominant. A summary of some of the reported changes is given in Table 2. It is clear from the results that changes in protease amino acids which are involved in substrate/inhibitor binding of competitive inhibitors are possible, and do not necessarily preclude virus replication. The observation is also confirmed that for each class of protease inhibitor, there is a characteristic pattern to the mutations associated with reduced sensitivity.

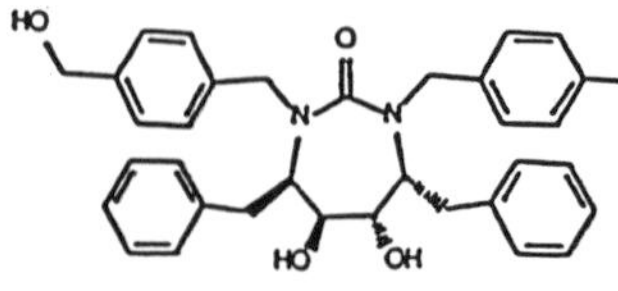

DMP 323

Figure 1. Structure of DMP323.

Table 2. Altered amino acids in the HIV-1 protease:
Reported mutations associated with
loss of susceptibility to protease inhibitors.

Amino Acid No.	Reported Change
8	arg -> gln or lys
10	leu -> phe
32	val -> ile
33	leu -> phe
36	met -> ile
45	lys -> thr
46	met -> leu, phe, ile
47	ile -> val
48	gly -> val
50	ile -> val
53	phe -> leu
54	ile -> thr
63	leu -> pro
71	ala -> val, thr
77	val -> ile
82	val -> ile, thr, ala, phe
84	ile -> val, ala
90	leu -> met
91	thr -> ala
97	leu -> val

We are presently concentrating efforts on a novel classs of HIV protease inhibitors containing a cyclic urea core, exemplified by DMP323 (Fig. 1). When selections were carried out in virus-infected cultures with DMP323 present, viruses emerged which had alterations in protease amino acids valine 82 and/or isoleucine 84 (7,8). The greatest degree of resistance was associated with a change at position 84 from isoleucine to valine, yielding about a 10-fold reduced sensitivity.

The enzyme mutant containing the I84V change was constructed and expressed in *E. coli* (9) using the T7 expression system of Studier (10). Expression and purification of the recombinant enzyme permitted direct assessment of fold-resistance to DMP323 as well as other classes of HIV protease inhibitors.

As shown in Table 3, the enzyme activity was measured on a protein substrate, namely luciferase. Luciferase is efficiently cleaved by the viral protease, rapidly leading to a large loss in the ability of the luciferase to act on its coelenterazine substrate and emit light (manuscript in preparation). Using this or peptide-based assays, it was determined that the I84V mutant enzyme had lost 8-12 fold sensitivity to the drug, and that the effect was specific to cyclic urea-type inhibitors. Also to be noted, however, is that catalytic efficiency of the protease, as measured by autocatalytic processing in *E. coli* or Km/Kcat is depressed by about 80%.

Table 3. Comparative activities of wild-type HIV-1 protease and the I84V
mutant on a luciferase substrate.

Enzyme	Km/Kcat (rel.)	Ki (DMP323)	Ki (Ro-31-8959)
Wild-type	1	1.7 nM	0.18 nM
I84V	0.25	20 nM	0.21 nM

Figure 1. Cleavage pattern of firefly luciferase by wildtype HIV-1 protease and the I84V mutant.

This is in general agreement with an independent measure of enzyme efficiency as measured by virus infectivity. In the report of El-Farrash et al, infectivity of the I84V HIV mutant was decreased more than 90% in cultured human lymphocytes (3).

We looked further at whether the substrate preferences for the I84V mutant were changed vs the wild-type. As shown in Figure 2, using firefly luciferase as a protein substrate, the cleavage patterns caused by wild-type enzyme and the I84V mutant were strikingly similar. Several minor variations in the processing pattern were not reproducible. Cleavage patterns for viral proteins were also virtually indistinguishable for wild-type and the I84V mutant (not shown).

CONCLUSIONS

Hypothetical concerns about genetic hypervariability of HIV interfering with prevention/treatment strategies have been well-founded. There has been little progress toward a vaccine, due in large part to multiple variations of the external glycoprotein of the virus. More recently, clinical observations have been made showing emergence of strains resistant to various classes of drugs directed at essential viral enzymes, including the DNA polymerase as well as the protease.

Competitive inhibitors directed at the protease active site are thought to be substrate mimics. Therefore, it has been more difficult for the virus to circumvent the action of the compounds. Some of the side chains of the enzyme which bind substrate/inhibitors can tolerate changes, but these changes have been relatively subtle, leading to only small (and perhaps clinically acceptable) losses of potency of the potential drug. However, it should be recalled that clinical resistance to the polymerase inhibitor AZT is usally a step-wise process, taking several months to occur, and comprised of several altered amino acids in the reverse transcriptase. In principle, something similar could occur to a protease inhibitor in a clinical setting, which is a more complex environment than a virus-infected cell culture. There is apparently more viral variation possible in persons infected with HIV than in cultured cells (see ref. 11 for a review).

Nevertheless, progress is being steadily made toward developing protease inhibitors as drugs for AIDS. Small, tight-binding inhibitors lacking peptide bonds are showing reasonable pharmacology in animals and some clinical benefit in patients, as very potent antiviral agents.

It is encouraging also that protease inhibitor-resistant viruses which have emerged have done so only slowly, and with minor chemical modifications to the protease structure. Furthermore, the small fold-resistance observed is often compensated by a reduction in enzyme efficiency and lower infectivity or temperature sensitivity of the virus in cell culture models. Obviously, it is not possible to completely determine the virulence profile of the viral protease mutants, since disease is only caused in man. As various protease inhibitors proceed in clinical trials, it will be important to monitor emergence of resistant strains, and attempt to correlate genotypic changes in the protease, with any associated changes in the viral cleavage sites and in the course of disease progression.

REFERENCES

1. Larder, B., Darby, G. and Richman, D., 1989, *Science* 243:1731.
2. Otto, M., Garber, S., Winslow, D., Reid, C., Aldrich, P., Jadhav P., Patterson, C., Hodge, C. and Cheng, Y., 1993, *Proc. Nat'l. Acad. Sci. USA* 90: 7543.
3. El-Farrash, M., Kuroda, M., Kitazaki, T., Masuda, T., Kato, K., Hatanaka, M., and Harada, S. J., 1994, *Virol.* 68:233.
4. Kaplan, A., Michael, S., Wehbie, R., Knigge, M., Paul, D.,Everitt, L., Kempt, D., Norbeck, D., Erickson, J., and Swanstrom, R., 1994, *Proc. Nat'l Acad. Sci. USA* 91:5597.
5. Lam, P., Jadhav, P., Eyermann, C., Hodge, C., Ru, Y., Bacheler, L., Meek, J., Otto, M., Rayner, M., Wong, N., Chang, C., Weber, P., Jackson, D., Sharpe, T., and Erickson-Viitanen, S., 1994, *Science* 263:380.
6. Human Retroviruses and AIDS Database (G. Meyers, ed.) Los Alamos National Lab., (1993).
7. Korant, B., in press, *Aspartic Proteinases* 5.
8. Winslow, D.L., Anton E.D., Horlick, R.A., Zagursky, R.J., Tritch, R.J., Scarnati, H., Ackerman, K., and Bacheler, L.T., 1994, *Biochem. Biophys. Res. Comm.* 205: 1651-1657.
9. Rizzo, C. and Korant, B., 1994, *Meth. Enz.* 241:16.
10. Studier, F., Rosenberg, A., Dunn, J. and Dubendorff, J., 1990, *Meth. Enz.* 185:60.
11. Wain-Hobson, S., 1995, *Nature* 373:102.

31

ASPARTIC PROTEINASES FROM PARASITES

John Kay, Lorraine Tyas, Michelle J. Humphreys, Jeff Hill,
Ben M. Dunn,[1] and Colin Berry

School of Molecular and Medical Biosciences
University of Wales
P.O. Box 911
Cardiff, CF1 3US
Wales, United Kingdom
[1] Department of Biochemistry and Molecular Biology
J. Hillis Miller Health Center
University of Florida
Gainesville, Florida 32610

Malaria is endemic throughout much of the tropics, causing in excess of 100 million clinical cases and 1 million child deaths in Africa per year. With increasing resistance of the parasites to currently available anti-malarials, paralleled by the increasing resistance of vector mosquitoes to pesticides, malaria is resurgent and represents the most serious threat to human health in the developing world.

The crucial importance of an aspartic proteinase in the metabolism of the malarial parasite *Plasmodium falciparum* (the most deadly form of malaria) has been demonstrated (5). During the erythrocytic stage of its life cycle, the malarial parasite takes up host cell cytoplasm and transports it to a specialised, lysosome-like organelle, the food vacuole. Within this vacuole, host cell haemoglobin is broken down sequentially to provide the parasite's predominant source of amino acids. The entire digestive pathway can be inhibited at its first step by pepstatin, indicating the crucial role of an aspartic proteinase at this point (5).

Two highly homologous aspartic proteinases (designated plasmepsins I and II; 72% identical) have been identified within the parasite food vacuole (4) and the genes encoding these proteins have been cloned and sequenced (2,3). The proteinases encoded by these genes display the active site sequence motifs that are commonly found in aspartic proteinases except that the second Asp-Thr-Gly sequence is replaced by Asp-Ser-Gly (as has also been observed in plant aspartic proteinases (11)). The plasmepsins show approximately 30% identity to the human aspartic proteinases (Table 1) and structures have been modelled for them on the basis of the known crystal structure of human cathepsin D (2). The plasmepsins

Intracellular Protein Catabolism, Edited by Koichi Suzuki and Judith Bond
Plenum Press, New York, 1996

Table 1. Amino acid identities between the mature regions of the two
plasmepins from *p. falciparum* and the five know human aspartic proteinases

	IDENTITY %	
	PLASMEPSIN I	PLASMEPSIN II
CATHEPSIN D	35	36
CATHEPSIN E	31	36
RENIN	33	33
PEPSIN	29	31
GASTRICSIN	28	27

are thus able to adopt the characteristic structure of enzymes belonging to the aspartic proteinase family.

The proplasmepsin zymogens (Fig 1) are atypical however, in having unusually long pro-regions (124 amino acids compared to the approximately 50 residues found in the precursors of mammalian aspartic proteinases). The functional significance of these extended pro-regions is not yet understood but hydrophobic sequences within this segment of the zymogen might allow association with membranes.

Attempts have been made to express in *E. coli* genes encoding each plasmepsin in constructs containing different lengths of the propart segment (Fig. 1). When full-length propart was included, the yields of proplasmepsins were low. Consequently, three quarters (76 residues) of each propart were deleted from the N-terminus of each zymogen to leave a truncated pro-region of only 48 amino acids (producing so-called semi-pro-plasmepsin II (Fig 1)) that was more comparable in size to the pro-segments of aspartic proteinases from other species (8). Expression of these constructs in *E. coli* resulted in high levels of protein accumulation (7). However, like other aspartic proteinases, the semi-pro-plasmepsins were produced in an insoluble form in *E. coli*. Solubilisation in 6M urea and refolding by rapid dilution were achieved by the method of Chen *et al.* (1) as modified by Hill *et al.* (6) and the semi-pro-plasmepsin II thus produced was further purified by DEAE ion exchange chromatography. Upon incubation at pH4.7 (Fig 1) this purified zymogen was able to

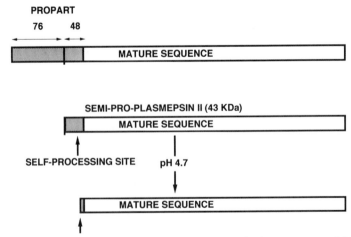

Figure 1. Top: The precursor of proplasmepsin II has 124 residues in the pro-segment. Middle: The gene encoding a truncated form of the zymogen with only 48 residues of the pro-segment was constructed for expression in *E. coli*. The resultant protein was refolded and purified and, upon exposure to acid pH, Bottom: a form of mature plasmepsin II was generated as a result of autoactivation.

Table 2. Kinetic parameters for the hydrolysis of peptide substrates at pH 4.7 by plasmepsin II. * marks the position of the scissile peptide bond.

SUBSTRATE	Km (μM)	kcat (sec-1)
Lys-Pro-Phe-Glu-Phe*Nph-Arg-Leu	20	1.5
Lys-Pro-Ile-Glu-Phe*Nph-Arg-Leu	30	0.9
Lys-Pro-Ile-Val-Phe*Nph-Arg-Leu	10	1.6

undergo auto-activation to generate a form of mature plasmepsin II (7). Crystallisation trials are currently being pursued with both the purified mature enzyme and the zymogen.

Analysis of the subsite preferences of mature plasmepsin II with synthetic peptide substrates (Table 2) indicate that plasmepsin II is not restricted to cleavage of the peptide bond (Phe33-Leu34) of its natural substrate alpha-globin, but rather has a broad specificity with a preference for hydrophobic residues, particularly in the P2 and P3 positions.

Given the homology between plasmepsin II and human renin (33% identical at the amino acid level), the potential of a series of known renin inhibitors to inhibit plasmepsin II was evaluated but found to be ineffective. Isovaleryl, acetyl and lactoyl-pepstatins (13) were also assessed for inhibitory potency. Whereas lactoyl-pepstatin exhibited little ability to inhibit plasmepsin II, the isovaleryl and acetyl-variants of pepstatin were both potent (sub-nanomolar) inhibitors of the enzyme, with isovaleryl-pepstatin apparently being the better inhibitor of the two. These findings are reflective of those observed previously (13) with fungal aspartic proteinases and in contrast to the situation described for HIV-proteinases (10) against which acetyl -pepstatin is the more potent form.

Genes encoding proplasmepsins have also been identified in *Plasmodium vivax* (the most common human malaria) and in *P. berghei*, a rodent form of malaria which may provide an animal model for evaluation of designed inhibitors *in vivo*. Whilst the primary aim of this research is to target *P. falciparum* which causes most malaria fatalities, an ideal inhibitor would also destroy other malaria parasites without causing adverse effects in the host. The design of plasmepsin-specific inhibitors as lead compounds for novel anti-malarial drugs is now underway.

In addition to the two plasmepsins in *P. falciparum*, a third related gene encoding a novel type of proteinase (designated proteinase B) with approximately 60% identity to plasmepsins I and II has been discovered but its function and location are as yet unknown (except that unlike the plasmepsins, it does not appear to be present in the food vacuole). Expression, characterisation and localisation of this enzyme will indicate its role in such processes as invasion and release of the parasite from red blood cells.

The fundamental importance of aspartic proteinases in one class of parasites (causing malaria) has thus been clearly established. The presence of this type of enzyme in other parasites also seems likely and similar strategies to strike at diseases by inhibiting crucial parasite proteinases may also be appropriate. In this light, the protozoan parasites *Eimeria tenella* and *E. acervulina* which cause great financial loss in the poultry industry, appear to be strategic targets. Aspartic proteinases are produced by these organisms as major antigenic proteins and genes encoding these enzymes have been cloned (9). Whilst the sequences are generally homologous to typical aspartic proteinases, the *Eimeria* enzymes have unusual, extended C-terminal sequences (approximately 35 residues longer than typical aspartic proteinases). These extra sequences are very rich in serine and alanine residues and their function is presently unknown. In addition, the *Eimeria* sequences contain a putative ATP binding site motif 21-14 residues upstream from the second Asp-Thr-Gly motif. Whereas it has been shown previously that human cathepsin E is stabilised by ATP (12), this human enzyme does NOT contain such an ATP binding motif. Experiments to express and characterise these *Eimeria* enzymes are in progress to determine structure/activity relationships, especially with regard to the novel features of these proteins. This work will attempt to exploit the *Eimeria* aspartic proteinases as therapeutic targets for both specific inhibition and (as important antigenic parasite proteins), as potential components of new vaccines.

ACKNOWLEDGMENTS

The authors would like to thank Dr John B. Dame, Dr Robert Ridley, Dr Richard Moon and Dr Paul Dunn for their help in providing DNA libraries, clones, interesting discussions and other assistance for the studies described. This investigation received financial support from the UNDP/World Bank/WHO Special Programme for Research and Training in Tropical Diseases (TDR) (for J.H and M.J.H.) and by awards from The Royal Society (for C.B.), and the Science & Engineering Research Council (for L.T.).

REFERENCES

1. Chen, Z., Koelsch, G., Han, H.-P., Wang., X.-J., Lin, X.-L., Hartsuck, J.A., and Tang, J., 1991, Recombinant rhizopuspepsinogen, *J. Biol. Chem.* 266:11718-11725.
2. Dame, J.B., Reddy, G.R., Yowell, C.A., Dunn, B.M., Kay, J., and Berry, C., 1994, Sequence, expression and modeled structure of an aspartic proteinase from the human malaria parasite *Plasmodium falciparum*, *Mol. Biochem. Parasitol.* 64:177-190.
3. Francis, S.E., Gluzman, I.Y., Oksman, A., Knickerbocker, A., Mueller, R., Bryant., M.L., Sherman, D.R., Russell, D.G., and Goldberg, D.E., 1994, Moleculaar characterisation and inhibition of a *Plasmodium falciparum* aspartic hemoglobinase, *EMBO J.* 13:306-317.
4. Gluzman, I.Y., Francis, S.E., Oksman, A., Smith, C.E., Duffin, K.L., and Goldberg, D.E., 1994, Order and specificity of the *Plasmodium falciparum* hemoglobin degradation pathway, *J. Clin. Invest.* 93:1602-1608.
5. Goldberg, D.E., Slater, A.F.G., Beavis, R., Chait, B., Cerami, A., and Henderson, G.B., 1991, Hemoglobin degradation in the human malaria pathogen *Plasmodium falciparum*: a catabolic pathway initiated by a specific aspartic protease, *J. Exp. Med*, 173:961-969.
6. Hill, J., Montgomery, D.S., and Kay, J., 1993, Human cathepsin E produced in *E. coli*, *FEBS Lett.* 326:101-104.
7. Hill, J, Tyas, L, Phylip, L.H, Kay, J., Dunn, B.M., and Berry, C., 1994, High level expression and characterisation of plasmepsin II, an aspartic proteinase from *Plasmodium falciparum*, *FEBS Let.* 352:155-158.
8. Koelsch, G., Mares, M., Metcalf, P., and Fusek, M., 1994, Multiple functions of pro-parts of aspartic proteinase zymogens, *FEBS Lett.* 343:6-10.
9. Laurent, F., Bourdieu, C., Kaga, M., Chilmonczyk, S., Zgrzebski, G., Yvore, P., and Pery, P., 1993, Cloning and characterisation of an *Eimeria acervulina* sporozoite gene homologous to aspartyl proteinases, *Mol. Biochem. Parasitol.* 62:303-312.
10. Richards, A.D., Roberts, R., Dunn, B.M., Graves, M.C., and Kay, J., 1989, Effective blocking of HIV-1 proteinase activity by characteristic inhibitors of aspartic proteinases, *FEBS Lett.* 247:113-117.
11. Runeberg-Roos, P., T rm angas, K., and Östman, A., 1991, Primary structure of a barley-grain aspartic proteinase, *Eur. J. Biochem.* 202:1021-1027.
12. Thomas, D.J., Richards, A.D., Jupp, R.A., Ueno, E., Yamamoto, K., Samloff, I.M., Dunn, B.M., and Kay, J., 1989, Stabilisation of cathepsin E by ATP, FEBS Lett. 243:145-148.
13. Valler, M.J., Kay, J., Aoyagi, T., and Dunn, B.M., 1985, The interaction of aspartic proteinases with naturally-occurring inhibitors from Actinomycetes and Ascaris lumbricoides, *J. Enz. Inhib.* 1:77-82.

32

TETANUS AND BOTULISM NEUROTOXINS

A NOVEL GROUP OF ZINC-ENDOPEPTIDASES

F. Tonello, S. Morante,[1] O. Rossetto, G. Schiavo, and C. Montecucco

Centro CNR Biomembrane and Dipartimento di Scienze Biomediche
Universita di Padovà
Via Trieste 75
35121 Padovà, Italy
[1] Dipartimento di Fisica
Universita di Roma Tor Vergata
Roma, Italy

ABSTRACT

Tetanus and botulinum neurotoxins are produced by bacteria of the genus *Clostridium* and cause the paralytic syndromes of tetanus and botulism with a persistent inhibition of neurotransmitter release at central and peripheral synapses, respectively. These neurotoxins consist of two disulfide-linked polypeptides: H (100 kDa) is responsible for neurospecific binding and cell penetration of L (50 kDa), a zinc-endopeptidase specific for three protein subunits of the neuroexocytosis apparatus. Tetanus neurotoxin and botulinum neurotoxins serotypes B, D, F and G cleave at single sites, which differ for each neurotoxin, VAMP/synaptobrevin, a membrane protein of the synaptic vesicles. Botulinum A and E neurotoxins cleave SNAP-25, a protein of the presynaptic membrane, at two different carboxyl-terminal peptide bonds. Serotype C cleaves specifically syntaxin, another protein of the nerve plasmalemma. The target specificity of these metallo-proteinases relies on a double recognition of their substrates based on interactions with the cleavage site and with a non contiguous segment that contains a structural motif common to VAMP, SNAP-25 and syntaxin.

INTRODUCTION

Tetanus and botulinum neurotoxins are responsible for all clinical symptoms of tetanus and botulism. They are produced by bacteria of the genus *Clostridium* in eight different type: one tetanus neurotoxin (TeNT) and seven different botulinum neurotoxins (BoNT/A, /B, /C, /D, /E, /F, and /G). Toxin producing *Clostridia* are widespread in the environment, mainly as spores, which can germinate under appropriate anaerobic conditions.

Intracellular Protein Catabolism, Edited by Koichi Suzuki and Judith Bond
Plenum Press, New York, 1996

Tetanus follows the contamination of necrotic wounds with spores of *Clostridium tetani*, bacterial proliferation and toxin production and release. Botulism derives from ingestion of uncooked anaerobic food contaminated with spores of *Clostridium botulinum*, or related species. After germination, the bacteria can produce a botulinum neurotoxin complexed with accessory proteins that help the neurotoxin to survive the passage through the stomach. More rarely, spores germinate in anaerobic areas of the newborn intestine giving rise to infant botulism (Payling-Wright, 1955; Arnon, 1980; Simpson, 1989). Anti-tetanus vaccine consists of paraformaldheyde-treated TeNT. This chemical modification inactivates the toxin, but preserves its immunogenic properties (Ramon and Descombey, 1925). Tetanus toxoid is the major product of the biotechnology industry. Botulinum neurotoxins are increasingly used in the treatment of several dystonias and of some forms of strabismus (Jankovic and Hallett, 1994).

These toxins bind specifically to the presynaptic terminals of neuronal cells, enter the cytosol and block neurotransmitter release (van Heyningen, 1968; Simpson, 1989). While TeNT acts mainly within the central nervous system, BoNTs act at the neuromuscular junction (NMJ) where they inhibit acetylcholine release and hence cause a flaccid paralysis (Burgen *et al.*, 1949). Despite the different sites of action of the toxins, both TeNT and BoNTs cause a prolonged impairment of neurotransamitter release. Neurospecificity and catalytic activity (see below) account for the fact that they are the most poisonous substances known (estimated lethal dose in humans is between 0.1 and 1 ng/Kg of body weight) (Payling-Wright, 1955).

Structure of Tetanus and Botulinum Neurotoxins

As shown in Fig. 1, clostridial neurotoxins are produced as inactive polypeptide chains of 150 kDa, which accumulate in the cell cytosol until the toxin is released by cell autolysis. The neurotoxin is then cleaved at a proteinase-sensitive loop (arrow in the left panel of Fig. 1) to generate an active di-chain toxin (DasGupta, 1989; Krieglstein *et al.*, 1991). A single disulfide bond joins the heavy chain (H, 100 kDa) and light chain (L, 50 kDa) and this is required for neuron intoxication (Schiavo *et al.*, 1990; de Paiva *et al.*, 1993). The disulfide bond is reduced inside the cell at a yet undefined passage during cell entry. BoNT/A, BoNT/C and the HC domain of TeNT have been crystallized (Stevens *et al.*, 1991;

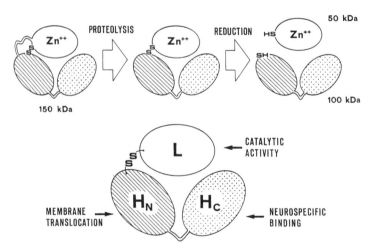

Figure 1. Scheme of the structure and mechanism of activation of tetanus and botulinum neurotoxins.

Anderson et al., 1993) and their analysis is under way. TeNT and BoNTs were suggested to fold into three distinct 50 kDa domains (Schiavo et al., 1993a), as shown in Fig. 1. This structural organization is functional to the accomplishment of the four distinct steps of the cell intoxication: binding, internalization, membrane translocation and intracellular catalytic activity (Montecucco et al., 1994). The amino-terminal 50 kDa H chain domain, termed HN, is implicated in membrane translocation, while the carboxy-terminal part of the H chain, termed HC, is mainly responsible for binding to neuronal cells and the L chain is responsible for the intracellular toxin activity.

These toxins are made as a 150 kDa single polypeptide chain composed of three 50 kDa domains, which play different roles in nerve cell intoxication (lower panel). The three domains are connected by protease-sensitive loops. The toxin becomes active upon selective proteolytic cleavage that generates two disulfide-linked chains. Hc is responsible for cell binding and Hn for membrane translocation. Reduction takes place inside the nerve cells and liberates the activity of the L chain, which blocks neuroexocytosis via a zinc-endopeptidase activity specific for three protein subunits of the neuroexocytosis apparatus.

Cell Binding, Internalization and Membrane Translocation

BoNTs and TeNT bind to the presynaptic membrane of the NMJ and are internalized into as yet undefined vesicles. The BoNTs remain within peripheral neurons, whereas TeNT is driven by its receptor inside intracellular vesicles that move retrogradely, along the motorneuron axon, to the soma. Eventually, TeNT is discharged in the intersynaptic space between the motorneuron and the inhibitory interneurons of the spinal cord (Schwab et al., 1979; Wellhoner, 1992; Halpern & Neale, 1995). Here it binds to unidentified receptors and is again internalized inside vesicles, wherefrom L enters the cytosol to block 4-aminobutyric acid (GABA) and glycine release. Possible mechanisms of penetration are discussed in details elsewhere (Montecucco et al., 1994; Montecucco & Schiavo, 1994).

The Zinc-Endopeptidase Activity of Tetanus and Botulism Neurotoxins

The L chains of TeNT and BoNTs are about 435 residues long and show few homologous segments, concentrated in the amino-terminal and central parts (Niemann, 1991; Minton, 1995). A highly conserved segment is present at the center of the L chain and contains the His-Glu-Xaa-Xaa-His zinc-binding motif of zinc-endopeptidases (Vallee and Auld, 1990; Schiavo et al., 1992a, c; Jiang and Bond, 1992; Bode et al., 1993). Atomic adsorption measurements of the zinc content of clostridial neurotoxins showed that they contain one atom of zinc per molecule of toxin, bound to the L chain (Schiavo et al., 1992a, c; 1993a). The sole exception is BoNT/C which coordinates two atoms of zinc (Schiavo et al., 1994). The affinity of zinc for TeNT and BoNT/A, /B and /E, measured by flow dialysis with $^{65}Zn^{2+}$, lies in the 50-150 nMolar range (Schiavo et al, 1992b).

Thermolysin-like enzymes bind zinc via two histidines and one carboxylate (Matthews et al., 1972; Pauptit et al., 1988; Thayer et al., 1991). At variance, astacin-like proteinases coordinate zinc via three histidines and one tyrosine, whereas adamalysin and collagenases via three histidines (Bode et al., 1992; Baumann et al., 1993; Gomis-Ruth et al., 1993; Lovejoy et al., 1994; Bode et al., 1993). In all cases, one molecule of water is an additional zinc ligand and is involved in the hydrolysis of the peptide bond.

The L chain of TeNT and BoNT/A, /B and /E were chemically modified with DEPC, a histidine specific reagent (Miles, 1977), in their holo and apo forms to determine if two or three additional histidines become available for modification upon zinc removal. Two additional histidines were modified with DEPC in the apo-toxin in each toxin tested (Schiavo et al., 1992a, b), as it is found with thermolysin. The apo forms of astacin, adamalysin and

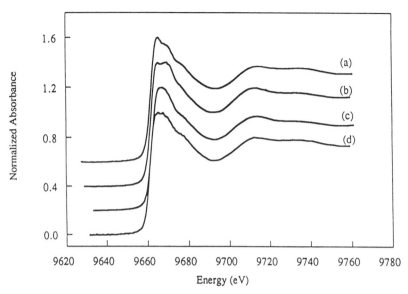

Figure 2. K-edge XANES of TeNT isolated L-chain, astacin, alkaline protease and thermolysin. The Zn K-edge XANES spectrum of TeNT isolated L-chain (a) is compared with those of astacin (b), alkaline protease (c) and thermolysin.

the alkaline protease from *Pseudomonas aeruginosa* aggregate when treated with DEPC, while their holo-forms do not (Tonello *et al.*, 1995). Zinc can be exchanged with other transition metals of similar size with preservation of toxic activity (Hohne-Zell *et al.*, 1993; Tonello *et al.*, in preparation). The cobalt-substituted TeNT shows an absorption spectrum different both from that of Co-thermolysisn and of Co-astacin (Tonello *et al.*, in preparation). These results indicate that TeNT coordinates zinc differently from known metallo-proteases.

Recently other information about the coordination of zinc in tetanus neurotoxin was obtained by X-ray absorption spectroscopy. We compared the zinc absorption spectrum of TeNT with those of structurally known zinc-endopeptidases: astacin, alkaline protease and thermolysin (Morante *et al.*, 1995). The XANES (X-ray Absorption Near Edge Structure) spectra reported in Fig. 2 show similar environments for the active site zinc of astacin and alkaline protease and of TeNT, which differ from that of thermolysin. This conclusion is reenforced by the zinc edge EXAFS (Extended X-ray Absorption Fine Structure) shown in Fig. 3. These results indicate that around the zinc atom of TeNT there are three, possibly four, groups that are able to scatter electrons as aromatic orbitals do (i.e., imidazole or phenyl or phenol or indole groups).

On the basis of these results and of an inspection of the sequence of the central part of the L chain of clostridial neurotoxins (see Fig. 4), we suggest that the zinc atom of these metallo-proteinases is bound via the two histidines of the motif, the water molecule bound to the Glu residue of the motif and to the conserved Tyr residue of the segment, which is part of the fully conserved tripeptide Leu-Tyr-Gly. We suggest that the polypeptide chain bends after the second His of the motif to bring the phenolic group near the zinc atom. It is noteworthy that this conserved Tyr occupies the same position of the third His ligand of astacin and alkaline protease. This proposal explains the loss of proteolytic activity of a tetanus neurotoxin L chain whose Tyr-243 was mutated to Phe (Yamasaki *et al.*, 1994a). If this novel mode of zinc coordination is definitively established, it will by itself identify this new group of zinc-endopeptidases (Montecucco & Schiavo, 1994).

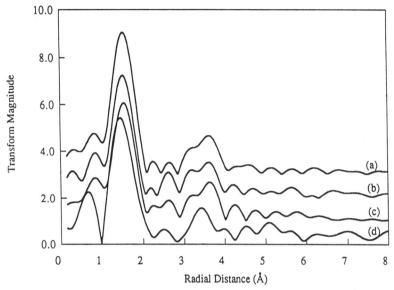

Figure 3. Fourier transforms of the experimental EXAFS signal $k^3(\aleph)$ (k) vs k for Zn K-edge. The four protein spectra are labeled as in Fig. 2.

TeNT and BoNTs were recently shown to be zinc-endopeptidases specific for protein components of the neuroexocytosis apparatus (Montecucco and Schiavo, 1993, 1994). Their cytosolic targets and their subcellular localization are schematically summarized in Fig. 5. TeNT and BoNT/B, /D, /F and /G cleave specifically VAMP (vesicle-associated membrane protein)/synaptobrevin, a membrane protein of synaptic vesicles (Schiavo *et al.*, 1992a, c, 1993b, c, 1994 a, 1994b; Yamasaki *et al.*, 1994b). VAMP consists of a Pro-rich amino terminal, not conserved among species and isoforms, of a highly conserved central part and of a membrane anchor (Trimble *et al.*, 1988). The clostridial neurotoxins act within the conserved part and remove from the synaptic vesicle cytosolic surface most of the VAMP mass (Fig. 5).

BoNT/A, /C and /E act on proteins of the presynaptic membrane: BoNT/A and /E cleave SNAP-25, while serotype C cleaves syntaxin (Blasi *et al.*, 1993, b; Schiavo *et al.*, 1993c, d, 1994; Binz *et al.*, 1994). SNAP-25 is a 205 residues protein bound to the cytosolic surface of the membrane and is required for axonal growth (Oyler *et al.*, 1989; Osen-Sand *et al.*, 1993). Syntaxin is anchored to the plasma membrane with most of its mass exposed to the cytosol and interacts with the calcium channels at the presynaptic terminal active zones (Bennett *et al.*, 1992).

Except for TeNT and BoNT/B, each one of the different clostridial neurotoxins catalyses the hydrolysis of different peptide bonds, as summarized in Table I. No common features of the amino acid sequences around the cleavage sites of the various neurotoxins in terms of amino acid lateral chain charge, hydrophilicity/hydrophobicity and volume can be identified.

Clostridial Neurotoxin Recognition of VAMP, SNAP-25 and Syntaxin

TeNT and BoNTs do not cleave short peptides containing the respective cleavage sites (Schiavo *et al.*, 1992a; Shone *et al.*, 1993), but these peptides bind and inhibit the

```
BoNT/A               D  P  A  V  T  L  A  H  E  L  I  H  A  G  H  R  L  Y  G
BoNT/A INFANT        D  P  A  V  T  L  A  H  E  L  I  H  A  E  H  R  L  Y  G
BoNT/B GPI           D  P  A  L  I  L  M  H  E  L  I  H  V  L  H  G  L  Y  G
BoNT/B GPII          D  P  A  L  I  L  M  H  E  L  I  H  V  L  H  G  L  Y  G
BoNT/C               D  P  I  L  I  L  M  H  E  L  N  H  A  M  H  N  L  Y  G
BoNT/D               D  P  V  I  A  L  M  H  E  L  T  H  S  L  H  Q  L  Y  G
BoNT/E               D  P  A  L  T  L  M  H  E  L  I  H  S  L  H  G  L  Y  G
BoNT/E BUTYRICUM     D  P  A  L  T  L  M  H  E  L  I  H  S  L  H  G  L  Y  G
BoNT/F               D  P  A  I  S  L  A  H  E  L  I  H  A  L  H  G  L  Y  G
BoNT/F BARATI        D  P  A  I  S  L  A  H  E  L  I  H  V  L  H  G  L  Y  G
BoNT/F LANGELAND     D  P  A  I  S  L  A  H  E  L  I  H  A  L  H  G  L  Y  G
BoNT/G               D  P  A  L  T  L  M  H  E  L  I  H  V  L  H  G  L  Y  G
TeNT                 D  P  A  L  L  L  M  H  E  L  I  H  V  L  H  G  L  Y  G
                     *  *           *     *  *  *     *        *     *  *  *

ZINCINS                               H  E  x  x  H
                                      *     *  *     *
METZINCINS                   x  h  x  H  E  x  h  H  x  h  G  h  x  H
```

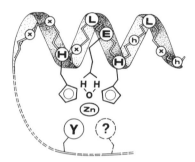

Figure 4. The central segment of the L chain of clostridial neurotoxin contains the zinc-endopeptidase motif. The upper panel reports the sequence of the longest conserved segment among the L chains of tetanus and botulinum neurotoxins. It contains the zinc-binding motif of metallo-proteinases (zincins) and it is dissimilar from the consensus sequence of the metzincins metallo-proteinase family. The lower panel depicts the possible mode of zinc coordination of clostridial neurotoxins, as deduced from recent results. Zinc is bound via the two histidines and the glutamate-bound water molecule of the motif arranged in a alpha-helical configuration. The conserved Tyr of the segment is the strongest candidate as fourth zinc ligand. The possibility of the existence of a fifth ligand cannot at the moment be discarded.

respective toxin. BoNT/B proteolyses VAMP peptides of forty residues or longer and maximal cleavage rates are only observed with the 33-96 segment (Shone et al, 1993). These proteinases cleave only one peptide bond out of the several identical bonds present in the sequence of their targets. Moreover, clostridial neurotoxins do not cleave proteins containing the same peptide bonds cleaved within the three targets discussed above. These results indicate that TeNT and BoNTs are proteases that recognize the tertiary structure of their targets and that residues around the cleaved peptide bond is not the sole determinant of the very specific recognition of VAMP, SNAP-25 and syntaxin.

As mentioned above, the L chains of clostridial neurotoxins have strong similarities, thus indicating that they derive from a common ancestral metalloprotease. This feature suggested that their three neuronal targets could share a region involved in toxin binding. An inspection of the sequences of VAMP, SNAP-25 and syntaxin revealed that these three proteins share a short motif, which is comprised within regions predicted to adopt an α-helical configuration (Rossetto et al., 1994). The motif, shows a face with three negative charges contiguous to a face formed by

Table 1. Target and peptide bond specificities
of tetanus and botulism neurotoxins

Toxin type	Target	Peptide bond cleaved
		P3- P2 - P1 - P1'-P2'- P3'
TeNT	VAMP	Ala-Ser-Gln——-Phe-Glu-Thr
BoNT/A	SNAP-25	Ala-Asn-Gln——-Arg-Ala-Thr
BoNT/B	VAMP	Ala-Ser-Gln——-Phe-Glu-Thr
BoNT/C	syntaxin	Thr-Lys-Lys——-Ala-Val-Lys
BoNT/D	VAMP	Asp-Gln-Lys——-Leu-Ser-Glu
BoNT/E	SNAP-25	Ile-Asp-Arg——-Ile-Met-Glu
BoNT/F	VAMP	Arg-Asp-Gln——-Lys-Leu-Ser
BoNT/G	VAMP	Thr-Ser-Ala——-Ala-Lys-Leu

hydrophobic residues. This motif is present in all VAMP, SNAP-25 and syntaxin isoforms known to be expressed in the nervous tissue of animals sensitive to the neurotoxins. Variations are present in VAMP and syntaxin of *Drosophila* and yeast, species which are resistant to these neurotoxins, and in syntaxin isoforms known not to be involved in exocytosis. There are two copies of the motif in VAMP, four copies in SNAP-25 and two copies in syntaxin. Peptides corresponding to the specific sequences of the motifs of each of the three neurotoxin target proteins inhibit *in vitro* and *in vivo* the activity of the neurotoxins, irrespectively of their origin and the neurotoxin type used. Antibodies raised versus these peptides cross-react among the three target proteins. These results indicate that the motif is exposed on the protein surface and adopts a similar configuration in each of the three neurotoxin targets. Moreover, the VAMP-specific, the SNAP-25-specific and the

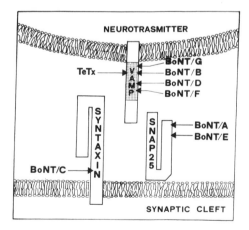

Figure 5. The targets of the intracellular zinc-endopeptidase activity of tetanus and botulinum neurotoxins and their sub-cellular localization. VAMP, SNAP-25 and syntaxin are essential parts of the 20S multi-protein complex which mediates synaptic vesicle docking and fusion with the presynaptic membrane. VAMP is located on the synaptic vesicle membrane, while SNAP-25 and syntaxin are located on the target membrane. Zinc-dependent proteolytic cleavage of VAMP by TeNT and BoNT/B, /D, /F or /G releases the amino-terminal part of VAMP in the cytosol and impairs the function of the neuroexocytosis apparatus. Each toxin cleaves a different peptide bond. BoNT/A and /E cleave SNAP-25 at the carboxyl terminus with the release of nine residues and twenty six residues respectively. BoNT/C cleaves the two syntaxin isoforms present in the nervous tissue at a single site near the carboxy-terminus. Thus, the action of TeNT and BoNT/B, /C, /D, /F and /G causes the release of a large portion of the cytosolic domain of VAMP and syntaxin. At variance, only a small part of SNAP-25 is released by the selective proteolysis of BoNT/A and /E.

syntaxin-specific neurotoxins cross-inhibit each other, namely they are able to compete for the same binding site, though are unable to cleave other targets than their specific one (Rossetto *et al.*, 1994). Experiments of replacement of the various residues of the motif are under way and preliminary results indicate that the negative charges of the carboxylate residues are essential for substrate hydrolysis (Pellizzari *et al.*, in preparation).

Taken together, these results indicate that TeNT and BoNTs recognize their protein substrates via a two site interaction with: a) a region that includes the peptide bond to be cleaved and b) a binding region closely similar among VAMP, SNAP-25 and syntaxin, which accounts for antibody cross-reactivity and cross-inhibition of the different neurotoxin types. The relative contribution of regions a) and b) to the specificity and strength of neurotoxin binding remains to be determined. It can be anticipated that hydrolysis of region a) leaves the toxin bound to its substrate only via its interaction with b) with a large decrease in binding strength, which is expected to lead to a rapid release of the hydrolysed substrate. Such a mode of binding would account for the high concentrations of peptides needed to inhibit the toxin protease activity (Rossetto *et al.*, 1994). A "double check" binding explains the high specificty of action of these neurotoxins as well as the available data on the minimal size of the target still cleaved by the neurotoxins (Shone *et al.*, 1993; Binz *et al.*, 1994; Yamasaki *et al.*, 1994a), because, in order to be cleaved, a peptide must include both region a) and region b).

CONCLUDING REMARKS

The neurotoxins produced by *Clostridia* responsible for tetanus and botulism form a new group of zinc-endopeptidases endowed with characteristic properties. They are produced as inactive precursors and are activated by specific proteolysis and disulfide reduction. Their amino acid sequence around the zinc binding motif does not correspond to any of the known families of zinc-endopeptidases. They act in the cell cytosol and are very specific in terms of both protein target and peptide bond cleaved. A future challenge is that of a better definition of the structural basis of the recognition and cleavage of their substrates. This is particularly important for the design of highly specific inhibitors of their zinc-endopeptidase activity to be evaluated as potential therapeutic agents for the treatment of tetanus and botulism. To be effective *in vivo* such inhibitors must be able to cross the nerve plasmalemma. Another important development will be the determination of their receptors, which is related to the possibility of using the H chain of these toxins as carriers of biological molecules and drugs inside nerve cells.

ACKNOWLEDGEMENTS

The authors are supported by CNR, Telethon-Italia grant 473 and MURST.

ABBREVIATIONS

BoNT, botulinum neurotoxins; DEPC, diethylpyrocarbonate; NGF, nerve growth factor; NMJ, neuromuscular junction; SNAP-25, synaptosomal associated protein of 25 kDa; VAMP-1, VAMP/synaptobrevin isoform 1; VAMP-2, VAMP/synaptobrevin isoform 2; TeNT, tetanus neurotoxin.

REFERENCES

Anderson, M.D., Fairweather. N., Charles, I.G., Emsley, P., Isaacs, N,W, and MacDermott, G. 1993, Crystal-lographic characterization of tetanus toxin fragment C. *J. Mol. Biol.*, 230, 673-674.

Baumann, U., Wu, S., Flaherty. K.M. and McKay. D.B. 1993, Three-dimensional structure of the alkaline protease of Pseudomonas aeruginosa: a two domain protein with a calcium binding parallel beta roll motif. *EMBO J.*, 12, 3357-3364.

Bennett, M.K., Calakos, N. and Scheller, R.H. 1992, Syntaxin: a synaptic protein implicated in docking of synaptic vesicles at presynaptic active zones. *Science*, 257, 255-259.

Binz, T., Blasi, J., Yamasaki, S., Baumeister, A., Link, E., Sudhof, T.C., Jahn, R. and Niemann, H. 1994, Proteolysis of SNAP-25 by types E and A botulinal neurotoxins. *J. Biol. Chem.*, 269, 1617-1620.

Blasi, J., Chapman, E.R., Yamasaki, S., Binz, T., Niemann, H. and Jahn, R. 1993, Botulinum neurotoxin C blocks neurotransmitter release by means of cleaving HPC-1/syntaxin. *EMBO J.*, 12, 4821-4828.

Bode, W., Gomis-Ruth, F.X., Huber, R., Zwilling, R. and Stcker, W. 1992, Structure of astacin and implication of astacins and zinc-ligation of collagenases. *Nature*, 358, 164-166.

Bode, W., Gomis-Ruth, F.X. and Stcker, W. 1993, Astacins, serralysin, snake venoms and matrix metallopro-teinases exhibit identical zinc-binding environments (HEXXHXXGXXH and Met-turn) and topolo-gies and should be grouped into a common family, the "metzincins". *FEBS Lett.*, 331, 134-140.

Burgen, A.S.V., Dickens, F. and Zatman, L.Y. 1949, The action of botulinum toxin on the neuromuscular junction. *J. Physiol.* (London), 109, 10-24.

DasGupta, B.R. 1989, The structure of botulinum neurotoxin. In Botulinum neurotoxins and tetanus toxin (ed L.L. Simpson), Academic Press, New York, pp. 53-67.

de Paiva, A., Poulain, B., Lawrence, G.W., Shone, C.C., Tauc, L. and Dolly, J.O. 1993, A role for the interchain disulfide or its participating thiols in the internalization of botulinum neurotoxin A revealed by a toxin derivative that binds to ecto-acceptors and inhibits transmitter release intracellularly. *J. Biol. Chem.*, 268, 20838-20844.

Gomis-Ruth, F.X., Kress, L.F. and Bode, W. 1993, First structure of a snake venom metalloproteinase: a prototype for matrix metalloproteinases/collagenases. *EMBO J.*, 12, 4151-4157.

Höhne-Zell, B., Stecher, B. and Gratzl, M. 1993, Functional characterization of the catalytic site of the tetanus toxin light chain using permeabilized adrenal chromaffin cells. *FEBS Lett.*, 336, 175-180.

Krieglstein, K., Henschen, A., Weller, U. and Habermann, E. 1991, Limited proteolysis of tetanus toxin: relation to activity and identification of cleavage sites. *Eur. J. Biochem.*, 202, 41-51.

Jankovic, J. and Hallett, M. 1994, (eds.) Therapy with botulinum toxin. Marcel Dekker, New York.

Jiang, W. and Bond, J.S. 1992, Families of metalloendopeptidases and their relationships. *FEBS Lett.*, 312, 110-114.

Lovejoy, B., Cleasby, A., Hassell, A.M., Longley, K., Luther, M.A., Weigl, D., McGeehan, G., McElroy, A.B., Drewry, D., Lambert M.H. and Jordan, S.R. 1994, Structure of the catalytic domain of fibroblast collagenase complexed with an inhibitor. *Science*, 263, 375-377.

Matthews, B.W., Jansonius, J.N. and Colman, P.M. 1972, Three-dimensional structure of thermolysin. *Nature New. Biol.*, 238, 37-41.

Miles, E.W. 1977, Modification of histidyl residues in proteins by diethylcarbonate. *Methods Enzymol.*, 47, 431-442.

Montecucco, C. and Schiavo, G. 1993, Tetanus and botulism neurotoxins: a new group of zinc proteases. *Trends Biochem. Sci.* 18, 324-327.

Montecucco, C. and Schiavo, G. 1994, Mechanism of action of tetanus and botulism neurotoxins. *Mol. Microbiol.* 13, 1-8.

Montecucco, C., Papini, E. and Schiavo, G. 1994, Bacterial protein toxins penetrate cells via a four-step mechanism. *FEBS Lett.* 346, 92-98.

Niemann, H. 1991, Molecular biology of clostridial neurotoxins. In A Sourcebook of Bacterial Protein Toxins (eds J.E. Alouf and J.H. Freer), Academic Press, London, pp. 303-348.

Osen-Sand, A., Catsicas, M., Staple, J.K., Jones, K.A., Ayala, G., Knowles, J., Grenningloh, G. and Catsicas, S. 1993, Inhibition of axonal growth by SNAP-25 antisense oligonucleotides in vitro and in vivo. *Nature*, 364, 445-448.

Oyler, G.A., Higgins, G.A., Hart, R.A., Battenberg, E., Billingsley, M., Bloom, F.E. and Wilson, M.C. 1989, The identification of a novel synaptosomal-associated protein, SNAP-25, differently expressed by neuronal subpopulations. *J. Cell. Biol. 109*, 3039-3052.

Pauptit, R.A., Karlsson, R., Picot, D., Jenkins, J.A., Niklaus-Reimer, A.S. and Jansonius, J.N. 1988, Crystal structure of neutral protease from Bacillus cereus refined at 3.0 A resolution and comparison with the homologous but more thermostable enzyme thermolysin. *J. Mol. Biol.*, 199, 525-537.

Payling-Wright, G. 1955, The neurotoxins of Clostridium botulinum and Clostridum tetani. *Pharmacol. Rev.*, 7, 413-465.

Ramon, G. and Descombey, P.A. 1925, Sur l'immunization antitetanique et sur la production de l'antitoxine tetanique. *Compt. Rend. Soc. Biol.*, 93, 508-598.

Schiavo, G., Papini, E., Genna, G. and Montecucco, C. 1990, An intact interchain disulfide bond is required for the neurotoxicity of tetanus toxin. *Infect. Immun.*, 58, 4136-4141.

Schiavo, G., Poulain, B., Rossetto, O., Benfenati, F., Tauc, L. and Montecucco, C. 1992a, Tetanus toxin is a zinc protein and its inhibition of neurotrasmitter release and protease activity depend on zinc. *EMBO J.*, 11, 3577-3583.

Schiavo, G., Benfenati, F., Poulain, B., Rossetto, O., Polverino de Laureto, P., DasGupta, B.R. and Montecucco, C. 1992b, Tetanus and botulinum-B neurotoxins block neurotransmitter release by a proteolytic cleavage of synaptobrevin. *Nature*, 359, 832-835.

Schiavo, G., Rossetto, O., Santucci, A., DasGupta, B.R. and Montecucco, C. 1992c, Botulinum neurotoxins are zinc proteins. *J. Biol. Chem.*, 267, 23479-23483.

Schiavo, G., Shone, C.C., Rossetto, O., Alexandre, F.C.G. and Montecucco, C. 1993a, Botulinum neurotoxin serotype F is a zinc endopeptidase specific for VAMP/synaptobrevin. *J. Biol. Chem.*, 268, 11516-11519.

Schiavo, G., Rossetto, O., Catsicas, S., Polverino de Laureto, P., DasGupta, B.R., Benfenati, F. and Montecucco, C. 1993b, Identification of the nerve-terminal targets of botulinum neurotoxins serotypes A, D and E. *J. Biol. Chem.*, 268, 23784-23787.

Schiavo, G., Santucci, A., DasGupta, B.R., Metha, P.P., Jontes, J., Benfenati, F., Wilson, M.C. and Montecucco, C. 1993c, Botulinum neurotoxins serotypes A and E cleave SNAP-25 at distinct COOH-terminal peptide bonds. *FEBS Lett.*, 335, 99-103.

Schiavo, G., Malizio, C., Trimble, W.S., Polverino de Laureto, P., Milan, G., Sugiyama, H., Johnson, E.A. and Montecucco, C. 1994a, Botulinum G neurotoxin cleaves VAMP/synaptobrevin at a single Ala-Ala peptide bond. *J. Biol. Chem.* 269, 20213-20216.

Schiavo, G., Rossetto, O. and Montecucco, C. 1994b, in "Zinc metalloproteases in Health and Disease", Hooper, N.M. Editor, Ellis Horwood, London, in press.

Schwab, M.E., Suda, K. and Thoenen, H. 1979, Selective retrograde trans-synaptic transfer of a protein, tetanus toxin, subsequent to its retrograde axonal transport. *J. Cell. Biol.*, 82, 798-810.

Simpson, L.L. 1989, (ed) Botulinum neurotoxins and tetanus toxin. Academic Press, New York.

Shone, C.C., Quinn, C.P., Wait, R., Hallis, B., Fooks, S.G. and Hambleton, P. 1993, Proteolytic cleavage of synthetic fragments of vesicle-associated membrane protein, isoform-2 by botulinum type B neurotoxin. *Eur. J. Biochem.* 217, 965-971.

Stevens, R.C., Evenson, M.L., Tepp, W. and DasGupta, B.R. 1991, Crystallization and preliminary X-ray analysis of botulinum neurotoxin type A. *J. Mol. Biol.* 222, 877-880.

Thayer, M.M., Flaherty, K.M. and McKay, D.B. 1991, Three-dimensional structure of the elastase of Pseudomonas aeruginosa at 1.5 A resolution. *J. Biol. Chem.*, 266, 2864-2871.

Trimble, W.S., Cowan, D.M. and Scheller, R.H. 1988, VAMP-1: a synaptic vesicle-associated integral membrane protein. *Proc. Natl. Acad. Sci. USA*, 85, 4538-4542.

Vallee, B.L. and Auld, D.S. 1990, Zinc coordination, function and structure of zinc enzymes and other proteins. *Biochemistry*, 29, 5647-5659.

Van Heyningen, W.E. (1968) Tetanus. *Sci. Am.*, 218, 69-77.

Yamasaki, S., Hu, Y., Binz, T., Kalkuhl, A., Kurazono, H., Tamura, T., Jahn, R., Kandel, E. & Niemann, H. 1994a, Synaptobrevin/VAMP of Aplisia californica: structure and proteolysis by tetanus and botulinal neurotoxins type D and F. *Proc. Natl. Acad. Sci. USA*, 91, 4688-4692.

Yamasaki, S., Baumeister, A., Binz, T., Blasi, J., Link, E., Cornille, F., Roques, B., Fykse, EM., Südhof. TC., Jahn, R., Niemann, H., 1994b, Cleavage of members of the synaptobrevin/VAMP family by types D and F botulinal neurotoxins and tetanus toxin. *J. Biol. Chem.*, 269, 12764-12772.

Wellhoner, H.H. 1992, Tetanus and botulinum neurotoxins. In Handbook of Experimental Pharmacology, vol. 102 (eds H. Herken and F. Hucho), Springer-Verlag, Berlin, pp. 357-417.

33

ENDOSOME-LYSOSOMES, UBIQUITIN AND NEURODEGENERATION

R. John Mayer, Carron Tipler, Jane Arnold, Lajos Laszlo,
Abdulaziz Al-Khedhairy, James Lowe,[1] and Michael Landon

[1] Departments of Biochemistry and Pathology
University of Nottingham Medical School
Queen's Medical Centre
Nottingham, NG7 2UH, United Kingdom

SUMMARY

Before the advent of ubiquitin immunocytochemistry and immunogold electron microscopy, there was no known intracellular molecular commonality between neurodegenerative diseases. The application of antibodies which primarily detect ubiquitin protein conjugates has shown that all of the human and animal idiopathic and transmissible chronic neurodegenerative diseases, (including Alzheimer's disease (AD), Lewy body disease (LBD), amyotrophic lateral sclerosis (ALS), Creutzfeldt-Jakob disease (CJD) and scrapie) are related by some form of intraneuronal inclusion which contains ubiquitin protein conjugates.

In addition, disorders such as Alzheimer's disease, CJD and sheep scrapie, are characterised by deposits of amyloid, arising through incomplete breakdown of membrane proteins which may be associated with cytoskeletal reorganisation. Although our knowledge about these diseases is increasing, they remain largely untreatable. Recently, attention has focussed on the mechanisms of production of different types of amyloid and the likely involvement within cells of the endosome-lysosome system, organelles which are immunopositive for ubiquitin protein conjugates. These organelles may be 'bioreactor' sites for the unfolding and partial degradation of membrane proteins to generate the amyloid materials or their precursors which subsequently become expelled from the cell, or are released from dead cells, and accumulate as pathological entities. Such common features of the disease processes give new direction to therapeutic intervention.

INTRODUCTION

The human neurodegenerative diseases can be categorised into idiopathic and transmissible disorders: the former having a far higher incidence and include Alzheimer's disease

(AD) and Lewy body dementia (LBD); the latter include Creutzfeldt Jakob disease (CJD), Gerstmann Straussler Scheinker syndrome (GSS) and fatal familial insomia (FFI).

AD and related disorders such as LBD exhibit the commonality of amyloid plaques. The extracellular amyloid in Alzheimer-related disorders contains, as a major component, aggregates of a fragment, Ab (previously referred to as b-protein or b/A4), of the Alzheimer amyloid precursor protein (APP). The functions of this transmembrane protein are unknown.

The major prion disorders in humans consist of CJD and GSS and in animals scrapie of sheep and bovine spongiform encephalopathy (BSE). There is evidence that these diseases have both genetic and infectious elements. Contrary to the central dogma of molecular biology, the evidence strongly indicates that replication of the causative agents, prions, occurs in the absence of a nucleic acid (Prusiner, 1991; Prusiner, 1992). Therefore site(s) for and mechanisms of replication of prions are the subjects of much current research into the disorders.

In some cases, prion diseases also exhibit extracellular amyloid, where the aggregates are composed of the prion protein (PrP) which accumulates predominantly in a non-truncated form (Prusiner, 1991; Prusiner, 1992). PrP is a glypiated molecule anchored to the cell membrane by a phosphatidyl inositol glycan tail. As with APP, the functions of PrP remain to be established; transgenic mice where the PrP gene has been deleted show no obvious physiological or behavioural defect (Bueler et al. 1992), suggesting that its functions can be substituted by another gene product.

The accumulating molecular cell biological evidence supports the notion that the processing events in the generation of amyloidogenic versions of the membrane-associated proteins APP and PrP occurs after endocytosis of the proteins in endosome-lysosomes (Borchelt et al., 1992; Haass et al. 1992; Laszlo et al., 1992; Mayer et al., 1992; McKinley et al., 1991).

IDIOPATHIC NEURODEGENERATIVE DISEASES

AD and LBD both exhibit extracellular amyloid plaques which contain as a major component Ab, a fragment of APP. The processing of APP has been the subject of intense study over the last few years.

The Processing of App and Catabolism of App

Different isoforms of APP are produced due to alternative splicing of the APP gene which resides on chromosome 21; the major form found in the nervous system is APP695. The amyloidogenic Ab fragments are the 39-43 amino acids derived from part of the single membrane spanning domain of APP plus amino terminal residues from the external region abutting onto the membrane. Cleavage of APP within the Ab domain by an incompletely characterised a-secretase releases a large amino-terminal non-amyloidogenic APP fragment (Esch et al., 1990), while a further enzyme(s) b-secretase(s) can cleave precisely at the amino terminus of Ab (Seubert et al., 1993). The enzymes cleaving at the carboxyl-terminus of Ab are not fastidious, releasing several fragments with variable carboxyl-termini. In vitro studies suggest that Ab molecules containing up to 40 residues are much less likely to form insoluble material than those which are two or three residues longer (Jarrett et al., 1993).

APP occurs in the endosome-lysosome system of cultured neurones (Benowitz et al., 1989) and presumably, therefore, in neurones in vivo. Increasingly, evidence suggests that a variety of Ab-related peptides (perhaps up to a dozen variants) can be found in such organelles (Haass et al., 1992; Siman et al. 1993). The cellular site(s) of an enzyme(s) cutting at the carboxyl-terminus of Ab is unknown, but this event could happen in the endosome-lysosome system. Recent studies indicate that cathepsin D-type enzymes may be involved in the generation of Ab-related peptides, while the thiol-cathepsins (e.g. cathepsins B, H, and L) may be involved in the degradation of frayed Ab-related molecules to amino acids (Siman *et al.* 1993).

Inclusion Bodies of Ad and Lbd

Although AD and LBD both exhibit extracellular amyloid plaques, differences are seen in the nature of the intraneuronal inclusions, ie. intraneuronal neurofibrillary tangles in AD and cortical Lewy bodies in LBD.

The advent of ubiquitin immunocytochemistry has revolutionised the detection of cortical Lewy bodies (Byrne *et al.*, 1991; Lennox *et al.*, 1989a; Lennox *et al.*, 1989b). Since cortical Lewy bodies, like neurofibrillary tangles of AD, are immunochemically ubiquitin-positive, they must contain some form(s) of currently unknown ubiquitinated proteins. Speculations have arisen based on the distribution of some molecular markers in neurofibrillary tangles and Lewy bodies (PGP 9.5, a ubiquitin carboxyl-terminal hydrolase; aB crystallin and gelsolin) that tangles represent "tombstone" inclusions in dying or dead cells. In contrast, Lewy bodies may form in neurones as part of an active process in which neurofilaments are used in an attempt to "cocoon" unwanted or lethal proteins or organelles prior to protein degradation by the ATP and ubiquitin-dependent non-lysosomal system or in the lysosomal apparatus (Lowe and Mayer, 1990; Lowe *et al.*, 1993; Mayer *et al.*, 1991). The presence of 20S components of the multi-complex enzyme, the proteosome, involved in the breakdown of ubiquitinated proteins in diffuse Lewy bodies of LBD provides further evidence of an active ubiquitin system in these inclusions (Kwak *et al.*, 1991). Immunohistochemical studies suggest that ubiquitin protein conjugates are additionally associated with the endosome-lysosome system (see later).

The Disruption of Membrane Proteins and Cytoskeletal Change

Any hypothesis set out to explain amyloid generation in AD and related disorders must also be able to explain, either directly or indirectly, the concurrent intraneuronal cytoskeletal changes, namely neurofibrillary tangles or cortical Lewy bodies; studies on Epstein-Barr virus (EBV)-transformed lymphoblastoid cells seem to provide some clues (Laszlo *et al.*, 1991). Immunocytochemistry shows that lymphocyte transformation by the virus or transfection of cells with the gene for the latent membrane protein (LMP, a multiple membrane-spanning protein of the virus that is responsible for tumourigenesis in susceptible animal cells) causes a reorganisation of the lymphocyte cytoskeleton to produce a vimentin-based filament inclusion which also contains LMP, a heat-shock protein 70 (hsp 70) and ubiquitin-protein conjugates (vimentin is the intermediate filament protein in cells of mesenchymal origin). The inclusions in part resemble Lewy bodies, which contain neuro-filaments (the intermediate filament protein in neurons) and ubiquitin-protein conjugates, together with several other proteins. Ultrastructural investigation by immunogold electron microscopy of EBV transformed cells shows a reorganisation of the endosome-lysosome system to orientate large lysosome-related organelles, containing LMP, hsp 70 and ubiquitin-protein conjugates, around the microtubule organising centre of the cells. Clearly, cytoskeletal reorganisation can occur acutely in response to loading of a single membrane viral protein into the cell surface. Similarly, problems in the handling of a single membrane protein in Alzheimer-related disorders may trigger cytoskeletal and endosome-lysosome rearrangements in neurones and other cell types. Evidence for such changes has been recently demonstrated by the transient transfection of COS-7 cells with the APP gene. Immunohistochemical studies have shown that in addition to increased levels of APP these cells also

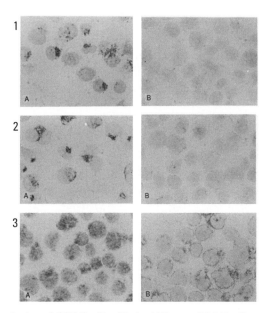

Figure 1. Transient transfection of COS-7 cells with the APP gene COS-7 cells were transiently transfected by the DEAE-dextran/chloroquine method with a vector that contained both an SV40 promoter and the APP gene. The cells were harvested 24 hours post-transfection and applied to glass microscope slides (cytospin) for immunohistochemical analysis. Mock transfected (panel A) and transfected (panel B) cells were stained with the following antisera: 1A and 1B - anti-APP antiserum AB54, (kindly donated by Dr D. Allsopp, SmithKline Beecham Pharmaceuticals, Molecular Neuropathology Department, Harlow, Essex, U.K.); dilution 1:500. 2A and 2B - anti-vimentin antiserum, (DAKO Ltd., Weybridge, Surrey, U.K.); dilution 1:1000 3A and 3B - anti-ubiquitin antiserum; dilution 1:1000. The immunoreactivity was subsequently detected by development with 3,3'-diaminobenzidine tetrahydrochloride.

show a cytoskeletal reorganisation similar to that seen in the EBV transformed cells, including aggregation of vimentin accompanied by increased levels of ubiquitin protein conjugates (see Figure 1, Al-Khedhairy and Mayer unpublished results).

Endosome-Lysosomes and Ad

Endosome-lysosome-like organelles can be detected at three neuronal sites by ubiquitin immunocytochemistry in Alzheimer-related disorders using the light microscope: in dystrophic neurites surrounding amyloid plaques; in axosomatic processes abutting onto neurones (or in autophagic granules within certain neurones e.g. those of the hippocampus) and as dot-like structures in the neuropil (Lowe *et al.* 1993). The dot-like structures in the neuropil are also a feature of the normal ageing brain in the absence of AD pathology (Dickson *et al.*, 1990). Experimental studies have shown that protein ubiquitination has unexpected roles in the maturation of both autophagic vacuoles (Gropper *et al.*, 1991) and endosome-lysosomes (Low *et al.*, 1993) and also in the response of some receptors to ligand binding (Cenciarelli *et al.*, 1992). This could account for the presence of ubiquitin-protein conjugates in predominantly endosome-like structures in both normal and diseased brains (Lowe *et al.* 1993). It is likely that the normal neuronal distribution of endosome-lysosomes becomes disrupted in those cells involved in APP fragmentation and amyloid formation, resulting in the observed distribution of ubiquitin-positive structures in diseased brain.

Confirmation of the importance of endosome-lysosomes in Alzheimer-related disorders is provided by the immunohistochemical observation of cathepsins distributed in a manner similar to that of ubiquitin-protein conjugates in or around dystrophic neurites surrounding amyloid plaques in AD (Cataldo *et al.*, 1991). The release of cathepsins from endosome-lysosomes could result in cell and tissue damage and be responsible, at least in part, for the inflammatory reactions in and around amyloid plaques in AD.

Apolipoprotein E4 in Late-Onset Ad

Apolipoproteins have been known for some time to be associated with amyloid plaques in AD (Namba *et. al.*, 1991). Recently there have been several reports that late-onset AD is genetically associated with an apolipoprotein E polymorphism, specifically copies of the apolipoprotein E4 allele (Corder *et al.*, 1993). The molecular pathological reason for this observation is still not known.

Internalised apolipoprotein E travels through the endosome-lysosome pathway after dissociation from the LDL receptor in early endosomes (Jensen *et al.,* 1994). It has recently been proposed that apoplipoprotein E may act as an 'amyloid promoting factor' in stimulating the polymerisation of Ab into amyloid filaments (Ma *et al.*, 1994). It could therefore be postulated that endosomes containing both APP and apolipoprotein E produce Ab amyloid more rapidly than those which contain Ab alone.

TRANSMISSIBLE NEURODEGENERATIVE DISEASES

The inherited prion diseases of man (familial CJD or GSS) are characterised by a series of point mutations and multiple base insertions in the PrP gene. The molecular genetic characterisation of this gene is required for every suspected case of prion disease because of varied and sometimes insignificant neuropathology of the diseased brain (Collinge *et al.*, 1990).

The Prion Protein

The normal form of PrP is designated PrP^C (where the superscript C indicates the normal cellular form) to distinguish it from the abnormal protein PrP^{Sc} (where Sc refers to the scrapie form of the protein or its equivalents in other species). PrP is coded in a single host exon excluding any possibility of alternative splicing of the primary transcript RNA to account for the alternative forms of the protein. The amino acid sequence of PrP^C and PrP^{Sc} is identical to the predicted amino acid sequence from the genomic DNA. No post-translational chemical modification has been identified that would distinguish PrP^C from PrP^{Sc} indicating the most likely method for the formation of PrP^{Sc} is due to the conformational manipulation of the normal isoform of the molecule (Prusiner, 1991; Prusiner, 1992).

The Infectious Agent

Despite continuing efforts to isolate a scrapie-specific nucleic acid, no material other than PrP has been found to be associated with the infectious agent, the prion (Prusiner, 1992). Recent findings that transgenic mice lacking the PrP gene cannot be infected with scrapie (Bueler *et al.*, 1993) further emphasise the key role of host PrP for agent replication. Transmission experiments between animal species led to the establishment of two concepts relating to infectivity; the species barrier and strains of agent. A seminal issue in replication of the infectious agent in transmissible encephalopathies is whether alternative strains can be generated by a protein-only mechanism.

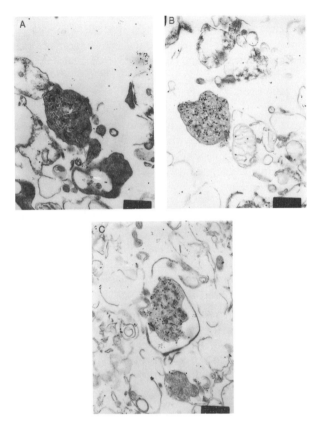

Figure 2. Two-sized immunogold electron microscopal analysis of subcellular fractions from terminal scrapie-infected mouse brain Homogenates of terminal ME7 scrapie-infected mouse brains were subjected to
subcellular fractionation on a discontinuous Nycodenz gradient (Arnold *et al.* 1995). Fractions were assayed
for several subcellular markers and two-sized immunogold electron microscopal analysis was performed on a
fraction found to be enriched in late endosomes as previously described (Arnold *et al.*, 1995). The following
primary antibodies were used all at a dilution of 1:50: A - anti-ubiquitin (10nm gold) and anti-mannose 6
phosphate receptor (20nm gold), (kindly donated by Dr. B. Hoflack, EMBL, Heidelberg, Germany, magnification X75400) B - anti-ubiquitin (10nm gold) and anti-PrP (20nm gold) (magnification X81200) C -
anti-ubiquitin (10nm gold) and anti-b-glucuronidase (20nm gold) (magnification X75400).

The Conversion of Prp^c to Prp^{sc}

Host PrP^C, internalised from the cell membrane by some form of endocytosis,
will enter the endosome-lysosome system for degradation. After experimental infection
of animals by direct injection into the brain, prions will be taken up by phagocytosis
into a variety of cells, including neurones. Recent studies with the chicken homologue
of PrP^C suggest that it is internalised in to the endocytic pathway via clathrin coated
pits; an unusual mode of entry for a GPI-anchored protein (Shyng *et al.*, 1994). In
this way prions could be introduced directly into the endosome-lysosome system, a
critically important fact relevant to the replication of a protein-only infectious agent.
Within the endosome-lysosome system PrP^C will be unfolded, this exposes protein
surfaces which through secondary structural interactions can form the intermolecular
b-sheet arrangements that are the basis of amyloid formation. This slow, but continuous

import of PrP^C into the endosome-lysosome system will result in the accumulation of PrP^{Sc}, which may be expelled from the cell following fusion of endosome-lysosomes with the cell membrane. Alternatively, the PrP^{Sc} which may be resistant to proteolysis in the endosome-lysosome system could accumulate to the point at which the organelle bursts, releasing prions and hydrolases, many of which are still active at neutral pH, into the cytoplasm. The hydrolases will cause cytolysis leading to cell death, spongiform change and host morbidity (Lowe et al., 1992).

Recent subcellular fractionation studies indicate that late multivesicular endosomes have a key role in the conversion of PrP^C into PrP^{Sc}. This is demonstrated by the immunogold electron microscopal co-localisation of PrP^{Sc}, the cation-independant mannose 6-phosphate receptor, ubiquitin protein conjugates and β-glucuronidase to the same type of organelle (see Figure 2) (Arnold et al., 1995). In scrapie-infected neuroblastoma cells a similar accumulation of PrP^{Sc} occurs in this family of organelles (Borchelt et al., 1992). At terminal stages of disease, electron microscopic evidence for intraneuronal spongiform lesions becomes apparent, with early spongiform change visible adjacent to dense endosome-lysosome-like multivesicular and tubulovesicular bodies (Laszlo et al., 1992; Lowe et al., 1992).

Generation of protein-only infectious particles in endosome-lysosomes can account for strains of scrapie-like agents without the need for nucleic acids. The denaturing environment of the organelles together with the complex mixture of proteases and other hydrolytic enzymes means that there may be alternative pathways for the unfolding and degradation of a single protein. The unfolding and hydrolytic pathways are likely to vary betwen cell types, depending on cell function (cf microglial cells and neurones) and, undoubtedly, will be influenced by the genotype of the infected animal. Given that different fragments of PrP can be produced by such means, then alternative products may nucleate the formation of varying types of PrP^{Sc}-containing amyloid by triggering one or other of a range of secondary structural interactions and intermolecular b-sheet conformations. This type of process could give rise to a thermodynamically-controlled, limited number of alternative strains of amyloids, which can faithfully reproduce themselves on reinfection of new animals of the same genotype (DeArmond 1993).

THE ROLE OF ENDOSOME-LYSOSOMES AS BIOREACTORS IN NEURODEGENERATIVE DISEASES

The combined studies on idiopathic and transmissible neurodegenerative diseases present observations which allocate a key role to the post-functional processing of membrane proteins within the endosome-lysosome system in the generation of amyloidogenic protein aggregates.

It could be argued that there is no cellular site other than the interior of an endosome-lysosome that has the capacity to catalyse the protein-protein interactions which eventually result in the genesis of amyloid. The organelles are designed to unfold and degrade macromolecules and, thus, are a site in which, over many years, the dismantling of membrane proteins could generate intermediates for amyloid formation. There is an exponential-like element to disease progression and neuropathology in transmissible neurodegenerative diseases. Such an exponential process would be predicted on the basis of repeated rounds of phagocytosis of prions, endosome-lysosome replication and release of prions by regurgitation or during cell death. Similar events could generate amyloid from APP in AD.

REFERENCES

1. Arnold JE, Tipler C, Laszlo L, Hope J, Landon M, Mayer RJ. The abnormal form of the prion protein accumulates in endosome/lysosome-like organelles in scrapie-infected mouse brain. *J. Pathol.* 1995; In press.

2. Benowitz LI, Rodriguez W, Paskevich P, Mufson EJ, Schenk D, Neve RL. The amyloid precursor protein is concentrated in neuronal lysosomes in normal and Alzheimer disease subjects. *Exp. Neurol.* 1989;106:237-250.

3. Borchelt DR, Taraboulos A, Prusiner SB. Evidence for synthesis of scrapie prion proteins in the endocytic pathway. *J. Biol. Chem.* 1992;267:16188-16199.

4. Bueler H, Aguzzi A, Sailer A, et al. Mice devoid of PrP are resistant to scrapie. *Cell* 1993;73:1339-1347.

5. Bueler H, Fischer M, Lang Y, et al. Normal development and behaviour of mice lacking the neuronal cell-surface PrP protein. *Nature* 1992;356:577-582.

6. Byrne EJ, Lennox GG, Godwin-Austen RB, et al. Dementia associated with cortical Lewy bodies: proposed clinical diagnostic criteria. *Dementia* 1991;2:283-284.

7. Cataldo AM, Paskevich PA, Kominami E, Nixon RA. Lysosomal hydrolases of different classes are abnormally distributed in brains of patients with Alzheimer disease. *Proc. Natl. Acad. Sci. USA* 1991;88:10998-11002.

8. Cenciarelli C, Hou D, Hsu KC, et al. Activation-induced ubiquitination of the T-cell antigen receptor. *Science* 1992;257:795-797.

9. Collinge J, Owen F, Poulter M, et al. Prion dementia without characteristic pathology. *Lancet* 1990;336:7-9.

10. Corder E H , Saunders A M , Strittmatter, W.J., et. al. Gene dose of apolipoprotein E type 4 allele and the risk of Alzheimer's disease in late-onset families. *Science* 1993;261:921-923

11. DeArmond SJ. Alzheimer's disease and Creutzfeldt-Jakob disease: overlap of pathogenic mechanisms. *Current Opinion in Neurology* 1993;6:872-881.

12. Dickson DW, Wertkin A, Kress Y, Ksiezak-Reding H, Yen S-H. Ubiquitin immunoreactive structures in normal human brains. *Lab. Invest.* 1990;63:87-99.

13. Esch FS, Keim PS, Beattie EC, et al. Cleavage of amyloid b-peptide during constitutive processing of its precursor. *Science* 1990;248:1122-1124.

14. Gropper R, Brandt RA, Elias S, et al. The ubiquitin-activating enzyme, E1, is required for stress-induced lysosomal degradation of cellular proteins. *J. Biol. Chem.* 1991;266:3602-3610.

15. Haass C, Koo EH, Mellon A, Hung AY, Selkoe DJ. Targeting of cell-surface ß-amyloid precursor protein to lysosomes: alternative processing into amyloid-bearing fragments. *Nature* 1992;357:500-503.

16. Jarrett JT, Berger EP, Lansbury PT. The carboxy terminus of the beta-amyloid protein is critical for the seeding of amyloid formation: implications for the pathogenesis of Alzheimer's disease. *Biochemistry* 1993;32:4693-4697.

17. Jensen T G , Roses A D and Jorgensen A L. Apolipoprotein E uptake and degradation via chloroquine-sensitive pathway in cultivated monkey cells overexpressing low density lipoprotein receptor. *Neurosci. Lett.* 1994;180:193-196

18. Kwak S , Masaki T , Ishiura S. and Sugita H. Multicatalytic proteinase is present in Lewy bodies in neurofibrillary tangles in diffuse Lewy body disease brains. *Neurosci. Lett.* 1991;128:21-24

19. Laszlo L, Lowe J, Self T, et al. Lysosomes are key organelles in the pathogensis of prion encephalopathies. *J. Pathol.* 1992;166:333-341.

20. Laszlo L, Tuckwell J, Self T, et al. The latent membrane protein-1 in Epstein-Barr virus-transformed lymphoblastoid cells is found with ubiquitin-protein conjugates and heat-shock protein-70 in lysosomes oriented around the microtubule organizing centre. *J. Pathol.* 1991;164:203-214.

21. Lennox G, Lowe J, Landon M, Byrne EJ, Mayer RJ, Godwin-Austen RB. Diffuse Lewy body disease: correlative neuropathology using anti-ubiquitin immunocytochemistry. *J.Neurol. Neurosurg. Psychiatr.* 1989a;52:1236-1247.

22. Lennox G, Lowe J, Morrell K, Landon M, Mayer RJ. Anti-ubiquitin immunocytochemistry is more sensitive than conventional techniques in the detection of diffuse Lewy body disease. *J.Neurol.Neurosurg.Psychiatr.* 1989b;52:67-71.

23. Low P, Doherty FJ, Sass M, Kovacs J, Mayer RJ, Laszlo L. Immunogold localisation of ubiquitin-protein conjugates in Sf9 insect cells: implications for the biogenesis of lysosome-related organelles. *FEBS Lett.* 1993;316:152-156.

24. Lowe J, Mayer RJ, Landon M. Ubiquitin in neurodegenerative diseases. *Brain Pathol.* 1993;3:55-65.

25. Lowe J, Fergusson J, Kenward N, et al. Immunoreactivity to ubiquitin-protein conjugates is present early in the disease process in the brains of scrapie-infected mice. *J. Pathol.* 1992;168:169-177.

26. Lowe J, Mayer RJ. Ubiquitin, cell stress and diseases of the nervous system. *Neuropathol. Appl. Neurobiol.* 1990;16:281-291.

27. Ma J., Yee A., Brewer H.B Jnr, Das S and Potter H. Amyloid-associated proteins a_1-antichymotrypsin and apolipoprotein E promote assembly of Alzheimer b- protein into filaments. *Nature* 1994;372:92-94

28. Mayer RJ, Landon M, Laszlo L, Lennox G, Lowe J. Protein processing in lysosomes - the new therapeutic target in neurodegenerative disease. *Lancet* 1992;340:156-159.

29. Mayer RJ, Arnold J, Laszlo L, Landon M, Lowe J. Ubiquitin in health and disease. *Biochim. Biophys. Acta* 1991;1089:141-157.

30. McKinley MP, Taraboulos A, Kenaga L, et al. Ultrastructural localisation of scrapie prions in cytoplasmic vesicles of infected cultured cells. *Lab. Invest.* 1991;65:622-630.

31. Namba Y, Tomonaga M., Kawasaki H , Otomo E and Ikeda K. Apolipoprotein E immunoreactivity in cerebral amyloid deposits and neurofibrillary tangles in Alzheimer's disease and Kuru plaque amyloid in Creutzfeldt-Jakob disease. *Brain Res.* 1991;541:163-166

32. Prusiner SB. Chemistry and biology of prions. *Biochemistry* 1992;31:12277-12288.

33. Prusiner SB. Molecular biology of prion diseases. *Science* 1991;252:1515-1522.

34. Seubert P, Oltersdorf T, Lee MG, et al. Secretion of b-amyloid precursor protein cleaved at the amino terminus of the b-amyloid peptide. *Nature* 1993;361:260-263.

35. Shyng S -L , Heuser J E and Harris D A. A glycolipid-anchored prion protein is endocytosed via clathrin coated pits. *J. Cell Biol.* 1994;125:1239-1250

36. Siman R, Mistretta S, Durkin JT, et al. Processing of the b-amyloid precursor. *J. Biol. Chem.* 1993;268:16602-16609.

ABNORMALITIES OF THE ENDOSOMAL-LYSOSOMAL SYSTEM IN ALZHEIMER'S DISEASE

Relationship To Disease Pathogenesis

Anne M. Cataldo,[1,2] Deborah J. Hamilton,[1] Jody L. Barnett,[1] Peter A. Paskevich,[1] and Ralph A. Nixon[1,2,3]

[1] Laboratories for Molecular Neuroscience
McLean Hospital
[2] Departments of Psychiatry and Neuropathology
[3] Program in Neuroscience
Harvard Medical School
Belmont, Maryland 02178

THE ENDOSOMAL-LYSOSOMAL SYSTEM

The lysosome is a major component of a dynamic and polymorphic system of acidic vacuolar compartments in the cell that is capable of degrading most large cellular molecules, i.e., nucleic acids, polysaccharides, proteins, and lipids, to low molecular weight products. As the major site of intracellular digestion, lysosomes contain dozens of hydrolytic enzymes that are housed within membrane-bound, low pH compartments (1, 2). The use of electron microscopy, enzyme and immunocytochemistry techniques have provided morphological criteria to identify lysosomes and to distinguish among different lysosomal stages and types. Lysosomal biogenesis begins within the endomembrane system of organelles. Synthesis and glycosylation take place in the endoplasmic reticulum followed by posttranslational modification and acquisition of phosphmannosyl residues within the Golgi apparatus. Mannose phosphate receptors interact with trans Golgi-derived coated vesicles containing nascent hydrolases and deliver these to acidic prelysosomal compartments, the late endosomes (3-5). The membrane-bound compartments that house both enzyme and the material to be digested are quite heterogeneous and include different morphological types of structures such as dense bodies, multivesicular bodies, and autophagic vacuoles. As digestion proceeds, a third type of lysosome-derived structure forms containing accumulating material that is resistant to degradation along with varying amounts of acid hydrolase activities. These residual granules, which are characterized by the color and autofluorescence of the accumulated material include lipofuscin, ceroid, and hemofuscin (1-5).

Intracellular Protein Catabolism, Edited by Koichi Suzuki and Judith Bond
Plenum Press, New York, 1996

Recent studies have revealed a greater complexity to the lysosomal system and its interaction with other acid vacuolar compartments in the cell. In addition to lysosomes, distinct early and late endosomal compartments traffic internalized materials to the lysosome for degradation and processing (6-11). Hydrolases present within lysosomal compartments are also abundant in late endosomes and, some acid proteases, such as cathepsins D and B, are also detectable in early endosomes where they are capable of performing limited proteolysis on selected proteins (6-11).

An association between lysosomal system disturbances and neurodegeneration has frequently been suggested, but the nature of the relationship is poorly understood. It is well known that the release of lysosomal constituents intracellularly or into the extracellular milieu can kill or injure cells, but in general, hydrolase release from cells appears to occur late in the degeneration process. In many forms of neurodegeneration, lysosome system participation is absent or minimal or may even represent the cell's attempt to repair or compensate for injury. In only a limited number of neurodegenerative states, however, do prominent morphologic or biochemical changes in the neuronal lysosomal system substantially precede late stage degenerative (atrophic/chromatolytic) changes. One of these conditions is Alzheimer's disease where a growing number of recent studies have been revealing abnormalities of the lysosomal system at both early and late stages of Alzheimer's disease, which may bear important relationships to known pathogenetic factors in the disease.

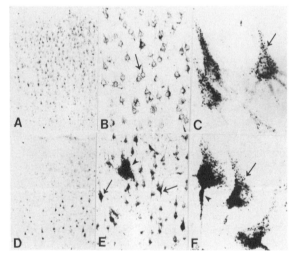

Figure 1. Increased expression of lysosomal hydrolases and lysosomal accumulation in pyramidal neurons of alzheimer brain. Lysosomal abnormalities are widespread in neurons of the prefrontal cortex of Alzheimer brains (D) when compared with the same neuronal populations from age-matched control brains (A). Higher magnification photomicrographs (B, C, E, F) demonstrate that cathepsin D-immunoreactive lysosomes are abundant within the neuronal perikarya and dendrites of control brains (B and C, arrows). In Alzheimer brains, immunostaining is increased in the basal pole of the soma (E and F, arrows) often extending into the axon hillock (F, arrowhead). A senile plaque in cortical laminae III is also intensely immunoreactive (E, arrowhead). (Panels B, C, E, and F reprinted with permission from Ref. 16).

THE ENDOSOMAL-LYSOSOMAL SYSTEM IN NORMAL HUMAN BRAIN

By using antibodies to lysosomal (acid) hydrolases to decorate all major lysosomal compartments in human pyramidal neurons, we observed that the lysosomal system in normal human pyramidal neurons is well-developed and highly variegated, consistent with its essential role in maintaining huge cytoplasmic volumes and large expanses of membrane surface area. Our studies using immunocytochemistry, immunoelectron microscopy, enzyme histochemistry and *in situ* hybridization showed that 10 antisera against different hydrolases specifically labeled intracellular lysosomal compartments in sections of prefrontal cortex and hippocampus from normal control brains (12-19). These hydrolases included all major cathepsins, D, B, L, and H and two glycosidases, ß-hexosaminidase and α-glycosidase, which intensely labeled lysosomes in neurons (Fig. 1). The hydrolase composition of lysosomes in neurons and glia differed markedly, indicating that the relative expression of acid hydrolases in different neuronal cell types may be tailored for the particular substrates encountered. Lysosome density was higher in neurons than glia. By contrast, the abundance of these hydrolases varied considerably in astrocytes. For example, lysosomes in astrocytes were darkly stained by antisera to cathepsin D and cathepsin L, but contained barely detectable hexosaminidase and cathepsin B immunoreactivities. Oligodendroglia stained weakly or not at all (12). Two other lysosomal hydrolases, cathepsin H and cathepsin G, were not detected in neurons. Cathepsin G immunoreactivity was not seen in any cell types in the brain, although the lysosomes of leukocytes in peripheral blood were intensely stained (12, 13).

ENDOSOMAL-LYSOSOMAL ABNORMALITIES IN ALZHEIMER'S DISEASE ARE EARLY MARKERS OF DYSFUNCTION

The same hydrolases detected in neuronal lysosomes from control brain tissue were also present in neurons from Alzheimer's disease brains. The number of lysosomes and their staining intensity, however, were greatly increased in many subpopulations of neurons in brains of Alzheimer patients (Fig. 1). In a group of Alzheimer's disease brains analyzed morphometrically, > 90% of the pyramidal neurons in layers III and V of the prefrontal cortex

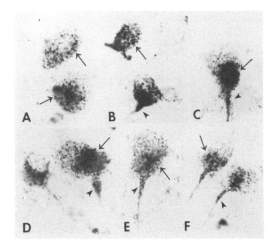

Figure 2. Widespread nature of lysosomal accumulation among neuronal populations in alzheimer disease. Less affected subcortical neurons from Alzheimer brain also displayed modest changes in the number and distribution of lysosomes detected here with an antiserum to cathepsin B. Neurons from the thalamus of control brain contain abundant cathepsin B-positive lysosomes (A, arrows). Corresponding populations of neurons from Alzheimer brain display alterations in cathepsin B immunoreactivity (B-F) within the cell body (arrows) and processes (arrowheads). (Reprinted with permission from Ref. 14).

displayed up to 8 fold higher numbers of hydrolase-positive lysosomes (20) than their normal counterparts in control brains. Most of these neurons exhibited no overt degenerative changes by histological and ultrastructural methods (14) indicating that these were the first evident intracellular manifestations of metabolic compromise in these neurons. Similar elevations were seen in neurons of other populations that are highly vulnerable in Alzheimer's disease, such as the CA1 and CA2 pyramids of the hippocampus. Moreover, populations of neurons affected only late in Alzheimer's disease and that are not lost in substantial numbers also show more modest lysosomal accumulations and redistributions (14). These were seen in subcortical areas of Alzheimer brain (Fig. 2) and include some Purkinje cells and their dendritic arborizations and many neurons within the brain stem, thalamus, and basal ganglia. Similar changes are seen in brains of patients with Down's syndrome at a stage before neurofibrillary pathology is evident in neurons (17). These findings provided direct evidence that altered endosomal-lysosomal function occurs at an early stage in Alzheimer's pathogenesis and may reflect early neuronal dysfunction. Our data suggests that the abnormal changes in acid hydrolase expression observed in Alzheimer brains involve several different compartments of the endosomal-lysosomal system reflecting an activation, rather than stasis, of the lysosomal system. Double-label immunocytochemistry with acid hydrolase antibodies combined with confocal microscopy confirmed that the vesicular compartments that accumulate are secondary lysosomal and, to a lesser extent, prelysosomal compartments (positive for the mannose-6-phosphate receptor). Lipofuscin, which is a relatively common non-specific concomitant of neurodegeneration, was not significantly increased at this stage.

Consistent with marked activation the lysosomal system in Alzheimer disease, we observed that the elevated levels of cathepsin D immunoreactive protein in at-risk neuronal populations is also associated with a substantial upregulation of cathepsin D gene expression in these same cell groups (21) (Fig. 3).

Hybridized tissue sections from the prefrontal cortices of control brains displayed a distribution of elevated specific cathepsin D mRNA expression in pyramidal neurons (Fig.

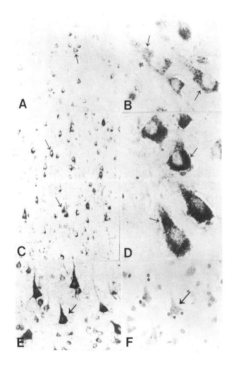

Figure 3. Early and widespread increased expression of Cathepsin D mRNA in Alzheimer brain. In the prefrontal cortex of control brains, pyramidal neurons in laminae III and V (A and B, arrows) displayed modest levels of cathepsin D mRNA within the soma and proximal dendrites. The sense cathepsin D RNA probe, which served as a negative control, yielded negligible hybridization signal (B, inset). In Alzheimer brains, the majority of neurons in the same populations exhibited increased levels of specific cathepsin D mRNA signal (C and D, arrows). Signal levels were the densest in the cell body and proximal dendrites with lower levels in the basal dendrites. Several of the same at-risk neurons (E and F, arrows) in Alzheimer brain containing dense cathepsin D mRNA signal expression (E) appear morphologically normal in a serial adjacent Nissl-stained section (F). (Reprinted with permission from Ref. 21).

3) paralleling that of immunoreactive cathepsin D seen in our cytochemical studies. To estimate the magnitude of the message increase and the proportion of the pyramidal cell population affected, we used computer-assisted microdensitometry to measure the density of cathepsin D mRNA signal in 125 individual, at-risk pyramidal neurons in neocortical laminae layer III and separately in lamina V from five Alzheimer brains and five control brains matched with respective postmortem interval and age of the individual. Greater than 94% of the pyramidal neurons in layer V and greater than 70% in layer III of the Alzheimer brains displayed abnormally high (1.5-5.0 fold) cathepsin D mRNA levels, averaging 2.5 fold higher than the mean value of the population of neurons analyzed in control cases (p < .001) (21). Most of these affected neurons displayed no evidence of atrophy or chromatolysis. Comparable increases in cathepsin D immunoreactivity were seen in the same cell populations analyzed by semiquantitative microdensitometry. By contrast, nonpyramidal neurons in cortical lamina IV, which are minimally involved in Alzheimer disease had relatively normal levels of cathepsin D mRNA and immunoreactivity (21). Neurons in advanced stages of degeneration contained massive amounts of cathepsin D immunoreactivity but, in many instances, exhibited lower than normal cathepsin D mRNA signals.

Preliminary studies in our laboratory have also shown that the lysosomal abnormalities in Alzheimer's disease revealed by acid hydrolase antibodies far exceeded those in a survey of other degenerative conditions (16). Relatively minor lysosomal accumulation was observed in affected neuronal populations associated with various disorders even as these neurons were degenerating. For example, neurons containing Pick bodies in Pick disease or atrophic neuronal populations in the caudate nucleus in late-stage Huntington's disease do not display elevated levels of cathepsin D immunoreactivity until the very end stages of cell death (21). These observations suggest that the massive accumulation of lysosomes seen in otherwise normal appearing at-risk populations in Alzheimer disease does not reflect a final common pathway of cell degeneration.

ADVANCED LYSOSOMAL SYSTEM ABNORMALITIES IN ALZHEIMER'S DISEASE

In the smaller proportion of neurons exhibiting neurofibrillary tangles or other signs of more advanced atrophy, hydrolase positive lysosomes massively accumulate throughout the perikaryon and proximal dendrite. Hydrolase-positive lipofuscin aggregates also become abundant in actively degenerating cells. Similar acid hydrolase-laden compartments and

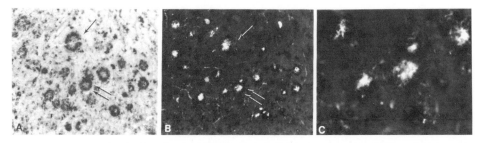

Figure 4. Immunocytochemical localization of acid hydrolase activity is associated with ß-amyloid deposition in senile plaques. Extracellular deposition of cathepsin D immunoreactivity (A, arrows) is prominent within a high number of senile plaques in the prefrontal cortex of Alzheimer brains. Counterstaining with thioflavin S (B, arrows and C) reveals histofluorescent ß-amyloid cores associated with most cathepsin D-positive plaques. (Panels A and B reprinted with permission from Ref. 16).

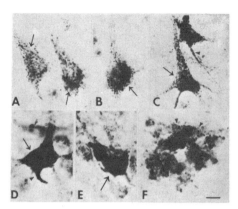

Figure 5. Progressive Lysosomal System Changes in Neurons of Alzheimer Brain. Pyramidal neurons from Alzheimer prefrontal cortex display various alterations in lysosomal hydrolase expression. Many neurons that are normal appearing show abnormally increased hydrolase immunoreactivity in the basal soma (A and B, arrows). Other neurons that display atrophic changes such as distended or attenuated processes, swelling in the axon hillock (C and D, arrowheads), or plasmalemmal alterations (E, arrowhead) contain greatly increased hydrolase immunoreactivity that frequently fills the perikaryon (C-E, arrows). Profiles of degenerating neurons similar to that in panel F (arrowheads) are often associated with hydrolase-positive plaques and are a major source of the extracellular lysosomal hydrolases in these lesions. (Reprinted with permission from Ref. 14).

lysosomal hydrolases are also found in focal deposits within the extracellular space in Alzheimer brain, suggesting that they are released as neurons or their processes degenerate (14). These compartments, containing a wide array of enzymatically competent hydrolases (15), persist in the extracellular space specifically in association with deposits of ß-amyloid in senile plaques (Fig. 4).

In the most cellular areas of the neocortex and hippocampus, and in the thalamus, an apparent progression from degenerating, intensely hydrolase-positive neurons (Fig. 5) to senile plaques was suggested by the presence of 'transitional' stages ranging from intact, intensely-stained neurons to senile plaques containing cluster of degenerating hydrolase-laden neurons, accumulations of immunoreactive lipofuscin, cell debris, and deposits of the ß-amyloid protein. The presence of acid hydrolases in focal extracellular deposits is a pathologic phenomenon that has only been observed in Alzheimer's pathogenesis and in other conditions where ß-amyloid is deposited (16, 22, 23). Consistent with its extracellular location within senile plaques, cathepsin D was found to be present in ventricular CSF of Alzheimer patients at levels averaging 4-fold higher than those in a control group of patients with neurodegenerative diseases (24).

Hydrolase expression may also be increased in astrocytes in Alzheimer disease (25), but histochemical, immunologic and ultrastructural evidence indicate that degenerating neurons and their processes are the principal sources of extracellular hydrolase in plaques of Alzheimer brain parenchyma (Fig. 6) (12, 26). More than a dozen acid hydrolases that accumulate in neuronal lysosomes in Alzheimer's disease brain are also abundant in senile plaques (12-16, 27) whereas cathepsins enriched in glial cells (cathepsin H) or neutrophils (cathepsin G), but not neurons, are not distributed extracellularly in senile plaques (13). Associated with the acid hydrolases in the plaques are various protease inhibitors including antithrombin III (28), α-antitrypsin (29), α-anti-chymotrypsin (30), and stefin A (31) and cystatin C (32). It has been suggested that some of these inhibitors may be released from non-neuronal cells to limit the deleterious actions of extracellular hydrolases.

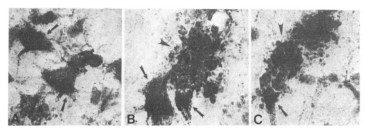

Figure 6. Contribution of Neuronal Lysosomal Hydrolases To Senile Plaque Development. Many neocortical neurons in more advanced stages of degeneration (A-C, arrows), which are intensely-stained with anti-HEX antiserum, are closely associated with or are constituents of hydrolase-positive senile plaques (A-C, arrowheads). (Reprinted with permission from Ref. 14).

Relationship of Lysosomal Abnormalities to Other Pathological Aspects of Alzheimer's disease

The close association between the intracellular and extracellular abnormalities of the endosomal-lysosomal system and conditions where ß-amyloid is deposited has tempted speculation that these two phenomena are related. That the association is more than coincidental is bolstered by a number of observations. Various mutations of the amyloid precursor protein (APP) gene have been shown to be the basis of several forms of familial Alzheimer's disease, attesting to the importance of this molecule, or derivatives of it, in Alzheimer pathogenesis (33). Lysosomes are normally responsible for degrading APP and, in the process, they generate intermediate "potentially amyloidogenic" fragments that encompass the entire ß-amyloid region (34), some of which may be neurotoxic (35). ß-Amyloid peptide itself is normally produced by cells in the endocytic pathway and possibly the secretory pathway (36, 37), both of which may harbor low levels of certain lysosomal hydrolases and perhaps higher levels when the system is upregulated. Several cathepsins, including cathepsin D and cathepsin B, satisfy important criteria for being candidate ß-amyloid generating enzymes, including a localization in appropriate compartments for Aß formation, abnormally upregulated expression in Alzheimer's disease, and appropriate bond specificity against APP and model peptides (38, 39). In addition, peptides that model a double amino acid substitution in APP present in individuals carrying the Swedish variant of familial Alzheimer's disease enhances cathepsin D cleavage at the ß-amyloid amino terminus by 10- to 70-fold (40). Other candidate proteases and scenarios for ß-amyloidogenesis have also been suggested and further investigation is needed to define the relative contributions of the endosomal-lysosomal system and other protease systems to APP metabolism and ß-amyloid formation.

Based on current knowledge of the various functions of the lysosomal system, its earlier upregulation in neurons in Alzheimer's disease could be interpreted as either compensatory and regenerative or a very early degenerative stage. It may be speculated that autophagy and endocytosis are responses expected in cells chronically attempting to regenerate and repair themselves by cannibalizing injured membranes and cytoplasmic proteins for new synthesis. Membrane dysfunction in Alzheimer's disease is suggested by many observations (41). In this regard, specific alleles of the apolipoprotein E (APO E) gene are considered to be either strong risk factors (ε4 allele) or a neuroprotective influence (ε2 allele) in late-onset Alzheimer's disease (42). APO E, which is not synthesized by neurons, is a cofactor in the transport of cholesterol into neurons and requires endosomal-lysosomal system activity for this function. APO E synthesis by astrocytes and uptake into neurons are markedly upregulated in response to various forms of neuronal injury, presumably to provide cholesterol for new membrane synthesis and regenerative processes (43). Another well-established risk factor for Alzheimer's disease, normal aging, is also accompanied by increased expression of some lysosomal proteases e.g., cathepsin D, cathepsin B (44, 45), which may superimpose on the disease-related changes. Thus, endosomal-lysosomal abnormalities seem to be at the cross-roads of various etiologic pathways, and their understanding may help to create a unified picture of Alzheimer pathogenesis.

REFERENCES

1. deDuve, C., Pressman, B.C., Gianetti, R., Wattiawux, R., and Appelman, F., 1955, Tissue fractionation studies. Intracellular distribution patterns of enzymes in rat liver tissues, *Biochem. J.* 60:604-617.
2. deDuve, C. and Wattiaux, R., 1966, Functions of lysosomes, *Annu. Rev. Physiol. 28:435-492.*
3. Dahms, N.M., Lobel, P., and Kornfeld, S., 1989, Mannose 6-phosphate receptors and lysosomal enzyme targeting, *J. Biol. Chem.* 264:12115-12118.
4. Geuze, H. J., Stoorvogel, W., Strous, G. J., Slot, J.W., Zuderhand- Bleekemolen, J., and Mellman, I., 1988, Sorting of mannose 6-phosphate receptors and lysosomal membrane proteins in endocytic vesicles, *J. Cell Biol.* 107:2491-2501.
5. Kornfeld, S., and Mellman, I., 1989, The biogenesis of lysosomes, *Annu. Rev. Cell Biol. 4:482-525.*
6. Bowser, R. and Murphy, R.F., 1990, Kinetics of hydrolysis of endocytosed substrates by mammalian cultured cells: early introduction of lysosomal enzymes into the endocytic pathway, *J. Cell Physiol.* 143:110-117.
7. Diment, S., Leech, M.S., and Stahl, P.D., 1988, Cathepsin D is membrane-associated in macrophage endosomes, *J. Biol. Chem.* 263:6901-6907.
8. Lemansky, P., Hasilik, A., VonFigura, K., Helmy, S., Fishman, J., Fine, R.E., Kerersha, N.L., and Rome, L.H., 1987, Lysosomal enzyme precursors in coated vesicles derived from the exocytic and endocytic pathways, *J. Cell Biol.* 104:1743-1748.
9. Rodman, J.S., Mercer, R. W., and Stahl, P.D., 1990, Endocytosis and transcytosis, *Curr. Opin. Cell Biol.* 2:664-672.
10. Erickson, A.H. and Blobel, G., 1983, Carboxyl-terminal proteolytic processing during biosynthesis of lysosomal enzymes beta-glucuronidase and cathepsin D, *Biochemistry* 22:52010-5205.
11. Riederer,. M.A., Soldati, T., Shapiro, A.D., Lin, J., and Pfeffer, S.R., 1994, Lysosome biogenesis requires Rab9 function and receptor recycling from endosomes to the trans-Golgi network, *J. Cell Biol.* 125:573-582.
12. Cataldo, A.M., Thayer, C.Y., Bird, E.D., Wheelock, T.R., and Nixon, R.A., 1990, Lysosomal proteinase antigens are prominently localized within senile plaques of a Alzheimer's disease: evidence for a neuronal origin, *Brain Res.* 513:181-192.
13. Cataldo, A.M., Paskevich, P.A., Kominami, E., and Nixon, R.A., 1991, Lysosomal hydrolases of different classes are abnormally distributed in brains of patients with Alzheimer's disease, *Proc. Natl. Acad. Sci. USA* 88:10998-11002.
14. Cataldo, A.M., Hamilton, D.J., and Nixon, R.A., 1994, Lysosomal abnormalities in degenerating neurons link neuronal compromise to senile plaque development in Alzheimer's disease, *Brain Res.* 640:68-80.
15. Cataldo, A.M. and Nixon, R.A., 1990, Enzymatically active lysosomal proteases are associated with amyloid deposits in Alzheimer's brain, *Proc. Natl. Acad. Sci. USA* 87:3861-3865.
16. Nixon, R.A., Cataldo, A.M., Paskevich, P.A., Hamilton, D.J., Wheelock, T.R., and Kanaley-Andrews, L., 1992, The lysosomal system in neurons: Involvement at multiple stages of Alzheimer's disease pathogenesis, *Ann. N. Y. Acad. Sci.* 674:65.
17. Nixon, R.A., Cataldo, A.M., Mann, D.M.A., Paskevich, P.A., Hamilton, D.J., and Wheelock, T.R., 1993, Abnormalities of lysosomal proteolysis in neurons in Alzheimer's disease and Down's syndrome: possible relationship to ß-amyloid deposition, in *Alzheimer's Disease: Advances in Clinical and Basic Research* B. Corain, K. Iqbal, K.M. Nicoline, B. Winblad, H. Wisniewski, and P. Zatta, Eds., John Wiley and Sons Ltd., New York, pp. 441-450.
18. Nixon, R.A. and Cataldo, A.M., 1991, Lysosomal proteolysis, in *Frontiers of Alzheimer's Research, T. Ishii, D. Allsop, and D. Selkoe, Eds. Elsevier, Amsterdam, pp. 133-146.*

19. Nixon, R.A. and Cataldo, A.M., 1993, The lysosomal system in neuronal cell death: a review, in *Markers of Neuronal Injury and Degeneration*, J.N. Johannessen, Ed., *Ann. N.Y. Acad. Sci.* 679:87-109.

20. Nixon, R.A., Cataldo, A.M., Hamilton, D.J., Barnett, J.L., and Paskevich., P.A., 1995, . In preparation.

21. Cataldo, A.M., Barnett, J.L., Berman, S.A., Li, J., Quarless, S., Bursztajn, S., Lippa, C., and Nixon, R.A., 1995, Gene expression and cellular content of cathepsin D in Alzheimer's disease brain: Evidence for early upregulation of the endosomal-lysosomal system, *Neuron* 14:1.

22. Bernstein, H.G., Kirschke, H. Wiederanders, B., Khudoerkov, R.M., Hinz, W., and Rinne, A., 1992, Lysosomal proteinases as putative diagnostic tools in human neuropathology: Alzheimer's disease (AD) and schizophrenia, *Acta Histochem.* Suppl. 42:19-24.

23. Schwagerl, A.L., Mohan, P.S., Cataldo, A.M., Vonsattel, J.P., Kowall, N.W., and Nixon, R.A., 1995, Elevated levels of the endosomal-lysosomal proteinase cathepsin D in Cerebrospinal fluid in Alzheimer's disease, *J. Neurochem.* 64:443.

24. Diedrich,. J.F., Minnigan, H., Carp, R.I., Whitaker, J.N., Race, R., Frey, W., 2d, and Haase, A.T., 1991, Neuropathological changes in scrapie and Alzheimer's disease are associated with increased expression of apolipoprotein E and cathepsin D in astrocytes, *J. Virol.* 65:4759-4768.

25. Nakamura, Y., Takeda, M., Suzuki, H. Hattori , H., Tada, K., Hariguchi, S., Hashimoto, S., and Nishimura, T., 1991, Abnormal distribution of cathepains in the brain of patients with Alzheimer's disease, *Neurosci. Lett.* 130:195-198.

26. Bernstein, H.G., Brusziz, S., Schmidt, D., Wiederanders, B., and Dorn, A., 1989, Immunodetection of cathepsin D in neuritic plaques found in brains of patients with dementia of Alzheimer type, *J. Hirnforsch.* 30:613-618.

27. Kalaria, R.N., Golde, T., Kroon, S.N., and Perry, G., 1993, Serine protease inhibitor antithrombin III and its messenger RNA in the pathogenesis of Alzheimer's disease, *Am. J. Pathol.* 143:886-893.

28. Gollin, P.A., Kalaria, R.N., Eikelenboom, P., Rosemuller, A., and Perry, G., 1992, Alpha 1-antitrypsin and alpha 1-antichymotrypsin are in the lesions of Alzheimer's disease, *Neuroreport* 3:201-203.

29. Abraham, C.R., Selkoe, D.J., and Potter, H., 1988, Immunochemical identification of the serine protease inhibitor alpha 1-antichymotrypsin in the brain amyloid deposits of Alzheimer's disease, *Cell* 52:487-501.

30. Bernstein, H.G., Rinne, R., Kirschke, H., Jarvinen., M., Knofel, B., and Rinne, A., 1994, Cystatin A-like immunoreactivity is widely distributed in human brain and accumulates in neuritic plaques of Alzheimer's disease subjects, *Brain Res. Bull.* 33:477.

31. Ii, K., Ito, H., Kominami, E., and Hirano, A., 1993, Abnormal distribution of cathepsin proteinases and endogenous inhibitors (cystatins) in the hippocampus of patients with Alzheimer's disease, parkinsonism dementia complex on Guam and senile dementia and in the aged, *Virchows Arch. A. Pathol. Anat. Histopathol.* 423:185-194.

32. Mullan, M. and Crawford, F., 1993, Genetic and molecular advance in Alzheimer's disease, *Trends Neurosci.* 16:398-403.

33. Estus, S., Golde, T., and Younkin, S., 1992, Normal processing of the Alzheimer's disease amyloid ß protein precursor generates potentially amyloidogenic carboxyl-terminal derivatives, *Ann. N. Y. Acad. Sci.* 674:138-148.

34. Neve, R.L., Kammesheidt, A., and Hohmann, C.F., 1992, Brain transplants of cells expressing the carboxyl terminal fragment of the Alzheimer amyloid protein precursor course specific neuropathology *in vivo*, *Proc. Natl. Acad. Sci. USA* 89:3448-3452.

35. Koo, E.H. and Squazzo, S.L., 1994, Evidence that production and release of amyloid ß-protein involves the endocytic pathway, *J. Biol. Chem.* 269:17386-17389.

36. Buscaglio, J., Gabuzda, D.H., Matsudaira, P., and Yankner, B.A., 1993, Generation of beta-amyloid in the secretory pathways in neuronal and non-neuronal cells, *Proc. Natl. Acad. Sci. USA* 90:2090-2096.

37. Tagawa, K., Maruyama, K., and Ishiura, S., 1992, Amyloid ß/A4 procursor protein (APP) processing in lysosomes, *Ann. N. Y. Acad. Sci.* 674:129-137.

38. Ladror, U.S., Snyder, S.W., Wang., G.T., Holzman, T.F., and Krafft, G.A., 1994, Cleavage at the amino and carboxyl termini of Alzheimer's amyloid-ß by cathepsin D, *J. Biol. Chem.* 269:18422-18428.

39. Dreyer, R.N., Bausch, K.M., Fracasso, P., Hammond, L.J., Wunderlich, D., Wirak, D.O., Davis, G., Brini, C.M., Buckholz, T.M., Konig, G., *et al.*, 1994, Processing of the pre-beta-amyloid protein by cathepsin D is enhanced by a familial Alzheimer's disease mutation, *Eur. J. Biochem.* 224:265-271.

40. Bosman, G.J., Bartholomeau, I.G.P., and DeGrip, W.J., 1991, Alzheimer's disease and cellular aging: membrane-related events as clues to primary mechanisms, *Gerontology* 37:95-112.

41. Corder, E.H., Saunders, A.M., Strittmatter, W.J., Schmechel, D.E., Gaskell, P.C., Small, G.W., Roses, A.D., Haines, J.L., and Pericak-Vance, M.A., 1993, Gene dose of apolipoprotein E type 4 allele and the risk of Alzheimer's disease in the latest onset families, *Science* 261:921-923.

42. Mahley, R.W., 1988, Apolipoprotein E: cholesterol transport protein with expanding role in cell biology, *Science* 240:622-630.
43. Matus, A. and Green, G.D.J., 1987, Age-related increase in α cathepsin D-like protease that degrades brain microtubule-associated proteins, *Biochemistry* 26:8083-8086.
44. Banay-Schwartz, M., DeGuzman, T., Kenessey, A., Palkovits, M., and Lajtha, A., 1992, The distribution of cathepsin D activity in adult and aging human brain regions, *J. Neurochem.* 58:2207-2211.

35

CATHEPSIN B EXPRESSION IN HUMAN TUMORS

Isabelle M. Berquin and Bonnie F. Sloane

Wayne State University
Department of Pharmacology
540 E. Canfield
Detroit, Michigan 48201

SUMMARY

Cathepsin B has been linked to tumor progression through observations that its activity, secretion or membrane association are increased. The most malignant tumors, and specifically the cells at the invasive edge of those tumors, express the highest activity. Cathepsin B may facilitate invasion directly by dissolving extracellular matrix barriers like the basement membrane, or indirectly by activating other proteases capable of digesting the extracellular matrix. Cathepsin B also might play a role in tumor growth and angiogenesis. Cathepsin B activity is the result of several levels of regulation: transcription, post-transcriptional processing, translation and glycosylation, maturation and trafficking, and inhibition. The majority of reports on cathepsin B expression in tumors have focused on measurements of activity or protein staining. In some tumors, e.g., gliomas, a correlation between the amounts of cathepsin B mRNA, protein and activity and tumor progression has been established. Regulation of cathepsin B at the transcriptional and post-transcriptional levels is still poorly understood. Although the putative promoter regions have characteristics of housekeeping-type promoters, cathepsin B mRNA expression varies depending on the cell type and state of differentiation. We have evidence that more than one promoter could direct expression of human cathepsin B. Multiple transcript species have been detected, resulting from alternative splicing in the 5'- and 3'-untranslated regions, and possibly the use of alternative promoter regions. The existence of transcript variants indicates a potential for post-transcriptional control of expression. In support of this, *ras*-transformation of MCF-10A human breast epithelial cells results in an increase in protein levels without a concomitant increase in mRNA levels. Cathepsin B mRNA species with distinct 5'-or 3'-untranslated regions may differ in their stability and translatability. Variations in the coding region may also alter cathepsin B properties. We and Frankfater's group have observed transcript species that would encode a truncated protein, lacking the prepeptide and about half of the propeptide. This truncated protein, if synthesized in cells, would be expected to be cytosolic; therefore its function is unclear. Once the several mechanisms of regulation of cathepsin B

Intracellular Protein Catabolism, Edited by Koichi Suzuki and Judith Bond
Plenum Press, New York, 1996

expression and activity are better understood, they could provide us with new strategies to specifically reduce cathepsin B activity in tumors.

CATHEPSIN B AND CANCER

Multiple lines of evidence suggest that the lysosomal cysteine protease cathepsin B is involved in tumor invasion and metastasis. Increases in expression at the mRNA and protein levels, increased activity and altered trafficking (localization and secretion) of cathepsin B have been found to correlate with malignancy of murine and human tumors. For a review of these correlative studies, see Sloane *et al.*, 1990; Keppler *et al.*, 1994; Sloane *et al.*, 1994a. Tumor progression is a multi-stage process which includes several steps that require proteolytic activity. Among these steps, the most representative are: local invasion, angiogenesis (necessary for tumor growth beyond a diameter of 2 mm), intravasation (when tumor cells enter the vascular system), and extravasation (when tumor cells exit the vascular system at the metastatic site). Our working hypothesis has been that the alterations in expression and trafficking of cathepsin B seen in malignant cells are responsible in part for

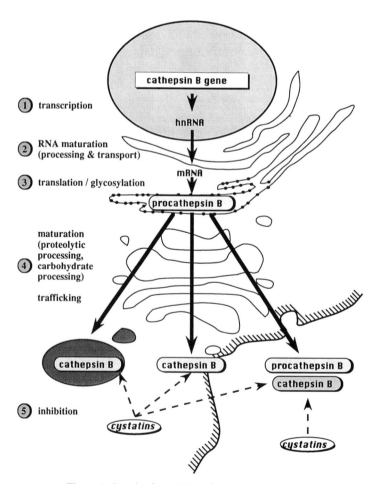

Figure 1. Levels of regulation of cathepsin B activity.

enhanced invasive capability of tumor cells. We have shown that both normal and tumor cathepsin B can degrade the extracellular matrix proteins laminin, fibronectin and type IV collagen at neutral as well as acidic pH (Lah *et al.*, 1989; Buck *et al.*, 1992). These are components of the basement membrane of blood vessels, a structure which forms a barrier to metastasizing tumor cells. Proteases from the four classes of endopeptidases – aspartic, cysteine, metallo and serine – have been linked to tumor progression, and dissolution of the basement membrane seems to be dependent on the sequential or simultaneous actions of proteases from more than one class. Indeed, several enzymes are secreted as inactive precursors which require activation by another endopeptidase. For instance, cathepsin B can activate prourokinase to urokinase (Kobayashi *et al.*, 1991), a serine protease activator of plasminogen, and in turn plasmin can activate the latent metalloprotease stromelysin (Matrisian and Bowden, 1990). Reciprocally, procathepsin B can be activated by serine proteases including urokinase (Dalet-Fumeron *et al.*, 1993). Thus, secreted cathepsin B might facilitate the local invasion and distant metastasis of tumor cells both directly and indirectly. Although proteases are frequently produced by tumor cells themselves, in some instances tumor cells induce the secretion of proteases by neighboring stromal cells. However, this has not yet been demonstrated for cathepsin B. Tumor cells and host cells exchange signals through cytokines and receptor-mediated contacts. Among such signals are growth factors stored in extracellular matrices and released by secreted endopeptidases (Flaumenhaft and Rifkin, 1992). In addition, staining for proteases (including cathepsin B) has been observed in endothelial cells of blood vessels irrigating tumors. This suggests that cathepsin B could participate in the invasive processes of angiogenesis. Hence, proteases, besides their obvious role in degradation of barriers to disseminating tumor cells, may favor tumor growth and progression by other means.

The gene is transcribed (1) as heterogeneous nuclear RNA (hnRNA) or primary transcript, then processed to messenger RNA (mRNA) and transported to the cytosol (2). Because of the presence of a signal peptide, preprocathepsin B synthesis (3) takes place on ribosomes associated with the endoplasmic reticulum. Maturation of cathepsin B (4) is coupled with intracellular trafficking and includes co-and post-transcriptional glycosylation and proteolytic processing events. Fully processed cathepsin B can be found in the lysosome, associated with the plasma membrane, or secreted. Procathepsin B can be secreted, perhaps through the default pathway of vesicular proteins. Endogenously, inhibition of cathepsin B activity (5) is mediated by cystatins.

LEVELS OF REGULATION OF CATHEPSIN B

The expression and activity of cathepsin B are regulated at multiple levels (Figure 1; Sloane *et al.*, 1994b). To date, all strategies to reduce cathepsin B activity in pathological states are aimed at the last step of regulation (step 5 in Figure 1), i.e., inhibition of proteolysis. Because they are proteins, the cystatins, endogenous low M_r inhibitors of cysteine proteases, are not optimal pharmacological agents. Although synthetic inhibitors for cathepsin B are an active area of research in biotechnical companies, they will not be discussed here. Other potential targets for controlling excessive cathepsin B activity are the factors involved in its biosynthesis and trafficking. As a lysosomal protease, cathepsin B is synthesized as a preproenzyme in the RER and co-translationally glycosylated (step 3 in Figure 1). Its maturation (step 4 in Figure 1) involves several post-translational steps (proteolysis and glycosylation) coupled with its movement in vesicular pathways to lysosomes. Some normal cells like osteoclasts and tumor cells secrete cathepsin B. In some cases, this may merely be a default pathway of trafficking, used when the lysosomal targeting pathways are saturated because of an overexpression of the protease. More than one intracellular trafficking route seems to be utilized, including the mannose 6-phosphate receptor pathway (Kornfeld, 1990) and the lysosomal proenzyme receptor pathway (McIntyre and Erickson, 1991). We have shown that in breast epithelial cells transformed with an activated *ras* oncogene (MCF-10A neoT), there is an increase in cathepsin B activity and protein associated with the plasma membrane and a more peripheral distribution of immunofluorescent and immunogold staining for cathepsin B (Sloane *et al.*, 1994c). In the same MCF-10A neoT cells, cathepsins B, L and D are located in different vesicular compartments (Sameni *et al.*, 1995; M. Sameni and B.F. Sloane, unpublished data). This raises the possibility that once the mechanisms of trafficking are understood, it may be possible to design specific agents that alter the intracellular movements or secretion of one lysosomal protease without affecting the others.

The first three steps in cathepsin B biosynthesis – transcription, RNA maturation and protein synthesis (Figure 1) – have not been thoroughly studied. There is evidence that cathepsin B is regulated at both the transcriptional and post-transcriptional levels. The next few sections of this review will focus on our current knowledge on the cathepsin B gene and its expression in normal and tumor cells.

THE CATHEPSIN B GENE

A single human cathepsin B gene has been identified, located on chromosome 8p22 (Wang *et al.*, 1988; Fong *et al.*, 1992). It is comprised of at least twelve exons (Gong *et al.*, 1993), spanning more than 27 kilobases (Berquin *et al.*, 1995). By comparison with the hepatoma and kidney composite cDNA sequence obtained by Chan and colleagues (1986), the translation initiation site is encoded in exon 3, so that exon 1, 2 and 25 bp of exon 3 are non-coding and constitute the 5'-untranslated region (UTR) of transcripts. We have recently identified two additional exons that can be alternatively spliced to become part of the 5'-UTR (see section on cathepsin B transcript variants). At the other end of the gene, exon 12 and most of exon 11 are non-coding and make up the 3'-UTR. The deduced amino acid sequence (Chan *et al.*, 1986) indicates that cathepsin B transcripts encode a preproenzyme, with a 17-residue prepeptide sequence (that directs vectorial protein synthesis into the endoplasmic reticulum), and a 62-residue propeptide region. The single-chain form of mature cathepsin B is encoded by 254 codons, and can be cleaved to the double-chain form by an internal cleavage at two sites, eliminating a dipeptide. A 6-residue extension at the COOH terminus is removed from both forms of mature cathepsin B.

The putative promoter region upstream of exon 1 of the human cathepsin B gene has characteristics of a housekeeping gene (Gong *et al.*, 1993). These characteristics include the absence of a TATA or CAAT box but a high GC content and the presence of several potential binding sites for the transcription factor Sp-1. The mouse cathepsin B gene had been characterized earlier (Ferrara *et al.*, 1990; Qian *et al.*, 1991) and a putative promoter region was found to be of the housekeeping type as well. Although the sequence and exon/intron junction points in the coding region are conserved between the mouse and human gene, there is more variability in the UTRs. Notably, one fewer exon has been identified in the mouse 5'-UTR as compared to the human, so that there is no direct correspondence of the putative promoter regions between the two species. The fact that human and murine cathepsin B genes have housekeeping-type promoter regions suggests that the expression of this protease is constitutive, as opposed to genes with regulated promoters containing TATA and CAAT boxes and enhancer elements. Nevertheless, as discussed below, there is considerable variability in cathepsin B mRNA levels among tissues or between normal and tumor tissues. The presence of housekeeping promoter characteristics does not preclude the possibility of transcriptional regulation, since certain genes with housekeeping-type promoters contain additional elements that enable them to be induced. This is the case for cathepsin D, a lysosomal aspartic protease with a mixed promoter (Cavaillès *et al.*, 1993). Estrogen induces expression of cathepsin D by acting through an atypical estrogen responsive element which activates a TATA box. Other transcription initiation sites are independent of this TATA box and function in the absence of estrogen induction. Such a dual strategy of regulation of expression could also exist for cathepsin B. In this way, the protease could serve both a maintenance role of intracellular protein degradation and tissue-specific roles like antigen presentation, osteoclastic bone resorption and macrophage-mediated degradation. Indeed, there is evidence that both the mouse and the human cathepsin B genes contain more than one promoter region (Rhaissi *et al.*, 1993; Berquin *et al.*, 1995). In the mouse, the two newly identified putative promoter regions contain a TATA box, with one also containing a CAAT box. If the existence of several promoters for cathepsin B is confirmed, each promoter might be regulated independently. In this context, the mechanism of the tissue-specific expression of cathepsin B could rely on the presence of transcription factors which vary with cell differentiation or microenvironment, and act differentially on each promoter.

TISSUE-SPECIFIC CATHEPSIN B EXPRESSION

Cathepsin B is a lysosomal endopeptidase which belongs to the cysteine protease class and exhibits an additional peptidyldipeptidase activity (Aronson and Barrett, 1978). Beside its ubiquitous role in intralysosomal protein degradation, it is likely that cathepsin B has specialized roles as its levels differ among tissue types. At the mRNA level, rat cathepsin B expression varies over 10-fold in various tissues: mRNA levels are highest in kidney, then progressively lower in spleen, lung, brain, heart, liver and thymus, with no mRNA detected in whole pancreas (San Segundo *et al.*, 1986). Murine cathepsin B transcripts are similarly distributed, with the most abundant levels in the kidney, and lower levels in decreasing order in lung, spleen, liver, heart, skin, brain and pancreas (Qian *et al.*, 1989). To our knowledge, there has not been a similar study of cathepsin B mRNA levels in normal human tissues. However, human cathepsin B activity levels have been determined and are highest in liver, thyroid, kidney and spleen. Intermediate activity is measured in heart, colon, adrenal and lung, whereas low activity is detected in prostate, testis, nerve, stomach, pancreas, brain, skeletal muscle, skin and breast (Shuja *et al.*, 1991). These activity measurements cannot directly be correlated to mRNA levels, because of the possible post-transcriptional modulation of cathepsin B expression and the variable contribution of endogenous inhibitors. Moreover, cathepsin B activity may not reflect total protein levels. The processing steps of cathepsin B biosynthesis could occur at different rates in distinct cell types and the turnover rate of the enzyme also may differ.

The above studies looking at cathepsin B expression in whole tissues are informative, but, as tissues are formed by more than one cell population, may not reflect all of the specialized roles of the protease. For example, in the study of rat cathepsin B mRNA levels described above (San Segundo *et al.*, 1986), mRNA could not be detected in whole pancreas, yet could be detected in islets of Langerhans and insulinomas. Thus, cathepsin B may have a specialized role in the endocrine function of the pancreas. Other studies indicate that cathepsin B mRNA and activity levels are altered during differentiation of certain cell lineages. Phorbol ester- or GM-CSF-induced differentiation of the human U937 promonocytic cell line is associated with an increase in cathepsin B mRNA and protein (Ward *et al.*, 1990). This is consistent with the observations that cathepsin B mRNA and activity are high in spleen. During osteogenic differentiation in the mouse, cathepsin B transcripts are expressed at higher levels (Friemert *et al.*, 1991) as are transcript levels and activity of cathepsin B during bovine myogenic differentiation (Béchet *et al.*, 1991). Several groups have reported a stimulatory or inhibitory effect of one or several cytokine(s) on cathepsin B

Table 1. Mammalian cathepsin B mRNA forms. Cathepsin B transcripts detected by northern blotting in rat, murine, human and bovine cells or tissues. Transcripts are listed from the most to the least abundant, with the most abundant transcript indicated in bold. kb, kilobase.

Source	Cell/Tissue Type	mRNA Forms (kb)	Reference
murine	myogenic cell line C2	**2.3-2.5**, 1.8	Colella *et al.*, 1986
murine	B16a melanoma	**2.2**, 4.0, 5.0	Qian *et al.*, 1989
murine	normal kidney, spleen, liver	**2.2**	Qian *et al.*, 1989
murine	normal liver, B16-F1 and B16a melanoma, Hepa cl 9 hepatoma	**2.2**, 4.1	Moin *et al.*, 1989
murine	radiation-induced osteosarcoma	**2.0**	Friemert *et al.*, 1991
rat	myogenic cell lines L$_6$, L$_8$	**2.3-2.5**, 1.8	Colella *et al.*, 1986
rat	various tissues	**2.3**	San Segundo *et al.*, 1986
rat	adult brain	**2.2**	Petanceska *et al.*, 1994
bovine	myoblasts, myotubes	**2.6**, 3.2	Béchet *et al.*, 1991
human	colorectal carcinoma	**2.2**, 4.0, 1.5, 3.0	Murnane *et al.*, 1991
human	colon carcinoma	**2.3**, 4.3	Corticchiato *et al.*, 1992
human	osteoclastomas	**2.1**, 1.9, 1.2	Page *et al.*, 1992
human	normal osteoblasts	**~2**, ~4	Oursler *et al.*, 1993
human	normal tissues, tumors	**2.3**, 4.0	Gong *et al.*, 1993
human	MCF-10A diploid breast epithelial cells	**2.2**, 4.1	Sloane *et al.*, 1994c

expression at the mRNA or activity level. This effect seems to be highly dependent upon the cell type under study. For instance, phorbol myristyl acetate decreases cathepsin B mRNA concentration in rat thyroid cells (Phillips *et al.*, 1989) and cathepsin B activity in Syrian hamster embryo cells (Nguyen-Ba *et al.*, 1994), whereas it increases cathepsin B mRNA and activity in promonocytes (Ward *et al.*, 1990) and stimulates secretion of procathepsin B from colon carcinoma cells (Keppler *et al.*, 1994). Similarly, dexamethasone decreases cathepsin B activity in Syrian hamster embryo cells (Nguyen-Ba *et al.*, 1994), but increases cathepsin B expression and secretion in human osteoblast-like cells (Oursler *et al.*, 1993). Interleukin-1 induces higher cathepsin B activity and intracellular staining in cultured synovial cells (Huet *et al.*, 1993), yet has no effect on cathepsin B expression in colon carcinoma cells (Keppler *et al.*, 1994). In the above studies, no evidence was gathered which would indicate a direct effect of a specific agent on cathepsin B transcription or synthesis. Rather, it seems more likely that the various factors studied act through signal transduction pathways which cause a change in the differentiation state of the cell.

CATHEPSIN B TRANSCRIPT VARIANTS

Several cathepsin B transcript species have been reported in various mammalian cells. Table 1 summarizes some of these reports for rat, murine, human and bovine cells and tissues. In rat, a transcript of ~2.3 kb is observed by all groups, with a minor 1.8 kb species reported in one study (Colella *et al.*, 1986). Likewise, an ~2.2 kb transcript is seen in all murine tissues examined, but additional minor transcripts are seen by different groups. Qian and colleagues (1989) detect 4.0 and 5.0 kb transcripts in murine tumor cells only, whereas Moin *et al.* (1989) detect a 4.0 kb message in both normal and tumor murine cells. Two transcripts of ~2.3 and 4.0 kb can be seen in most human tissues, with additional minor transcripts reported in colorectal carcinoma (Murnane *et al.*, 1992) and osteoclastoma (Page *et al.*, 1992). As only one human cathepsin B gene has been identified, all transcripts are likely to be created by the use of different initiation sites or by post-transcriptional processing events such as alternative splicing. One study of bovine myogenic cells indicates the presence of two transcripts of 2.6 and 3.2 kb (Béchet *et al.*, 1991). The slight differences in estimated mRNA sizes within one species are probably due to variations in the electrophoresis methods and markers used. Murine and human transcripts of ~4.0 kb contain the same coding region as the ~2.3 kb transcript, but longer 3'-UTRs (Qian *et al.*, 1991). The human 4-kb mRNA is produced by processing at a cryptic intron donor site in exon 11 and splicing to exon 12, resulting in a longer 3'-UTR (Gong *et al.*, 1993). Other minor transcripts detected in some cells could be artifactual or, more interestingly, may result from alternative transcriptional or post-transcriptional mechanisms in those cells. Whether these transcript variants play a specialized role in cathepsin B expression is not known.

Another type of transcript heterogeneity has been observed for human cathepsin B in cells and tissues. This involves small sequences in the 5'- and 3'-UTRs and is not detectable by a size shift on northern blots. Two groups describe a 10-nt insertion in the 3'-UTR of human cathepsin B cDNAs (Cao *et al.*, 1994; Tam *et al.*, 1994) which could form a stable stem-loop structure with potential implications for mRNA stability. Gong *et al.* (1993) have identified alternative splicing in the 5'-end of human transcripts, with one splicing species lacking exon 2 and one species lacking both exons 2 and 3 (figure 2). We have discovered two new exons in the human cathepsin B gene which map between exons 2 and 3 and thus have been denominated exons 2a and 2b (Berquin *et al.*, 1995). These two exons can be alternatively spliced and contribute to the diversity of the 5'-UTR of cathepsin B transcripts (Berquin *et al.*, 1995). In addition, we have evidence for other transcripts initiated from different sites. In a glioblastoma tumor sample, we found by 5'-RACE (rapid amplification of cDNA ends) that nearly half of the cathepsin B transcripts started in exon 3 and 25% in exon 4 (Berquin *et al.*, 1995). As the translation initiation codon is located in exon 3, both the transcripts lacking exons 2 and 3 described by Gong *et al.* and those initiated in exon 4 could encode a truncated form of procathepsin B, using an in-frame AUG (methionine) codon in exon 4 to initiate protein synthesis (Figure 2). Frankfater's group has reported in an abstract that such a truncated molecule could be expressed *in vitro* and refolded into a protein with enzymatic activity against a synthetic substrate (Gong and Frankfater, 1993). They proposed that

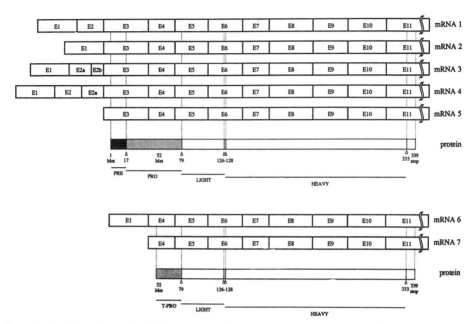

Figure 2. Possible cathepsin B mRNA variants and corresponding protein structure. Top panel: five mRNA species with different 5'-UTRs but which encode full-length preprocathepsin B. mRNA 1 contains exons 1 to 11 and corresponds to the first full-length human cathepsin B cDNAs that have been identified (Chan *et al.*, 1986); mRNA 2 lacks exon 2 and is found in cell lines and tissue samples (Cao *et al.*, 1994; Gong *et al.*, 1994; Berquin *et al.*, 1995); mRNA 3 lacks exon 2 but contains two additional exons, 2a and 2b, and was first seen in a gastric tumor cDNA (Berquin *et al.*, 1995); mRNA 4 contains exon 2 as well as exon 2a, and was first seen in a gastric tumor cDNA (Berquin *et al.*, 1995); mRNA 5 is initiated in exon 3 and was detected in a glioblastoma sample (Berquin *et al.*, 1995). Bottom panel: two mRNA species lacking exon 3, hence encoding a truncated procathepsin B protein. mRNA 6 is spliced between exons 1 and 4 and was identified by Gong *et al.* (1994) in tumor tissues and cell lines; mRNA 7 starts in exon 4 and was detected in a glioblastoma sample (Berquin *et al.*, 1995). Only part of exon 11 (886 nucleotides-long) is shown. The predicted size of the above transcripts ranges from ~1.7 to 2.1 kb. Note that for all of these mRNA species, alternative splicing may also occur in the 3'-UTR. In this case, the first 142 nucleotides of exon 1 are spliced to a 2.7-kb exon 12 (Gong *et al.*, 1994), forming ~3.6 to 4.0 kb mRNAs. Arrowheads under the protein structure indicate the sites of proteolytic processing, with corresponding amino acid residues. Prepeptide, propeptide, light and heavy chain domains are indicated. T-pro, truncated propeptide. Potential translation initiation (Met) sites and the stop codon are also shown

transcripts lacking exons 2 and 3 are tumor-specific, and that the resulting truncated protein may be expressed in tumors and contribute to the increased cathepsin B activity measured in many different animal and human tumors. The lack of a signal peptide, however, would prevent this protein from entering the vesicular pathway, and thus it would be expected to be cytoplasmic by default. What function a cytoplasmic form of cathepsin B could serve in tumor cells is unclear.

The role of multiple cathepsin B transcripts is not known. This multiplicity may allow for post-transcriptional regulation mechanisms. On one hand, different transcripts could vary in their inherent stability, resulting in the accumulation of different steady-state levels of cathepsin B mRNA. On the other hand, the rate of translation initiation from specific transcripts might differ. The relative abundance of cathepsin B mRNA species expressed in a cell at a given time may determine the amount of protease synthesized and its activity.

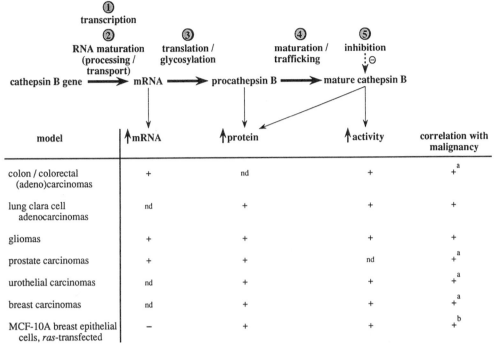

Figure 1. Correlation of cathepsin B mRNA, protein and activity levels with malignancy of human tumors. The five levels of regulation of cathepsin B expression (described in Figure 1) and their relation to increased mRNA, protein or activity are indicated at the top. Note that protein measurements include both pro and mature forms of cathepsin B. Increased mRNA was estimated by northern blotting or *in situ* hybridization, protein levels by immunoblotting or immunostaining of cells/tissues, and activity by assays with synthetic substrates. nd, not determined. [a]cathepsin B expression is increased in invasive edges of the tumor. [b]cathepsin B activity increases with the progression from a premalignant to a malignant lesion.

Further characterization of what species are expressed in different tissue types and in tumors may help elucidate some of the mechanisms by which cathepsin B expression is regulated.

CATHEPSIN B EXPRESSION IN TUMORS

Cathepsin B activity has been found to be elevated in a variety of animal and human tumors and tumor cell lines. In several cases, activity is correlated to malignant progression. Human tumor types overexpressing cathepsin B activity include carcinoma of the colon and rectum, breast, bladder, pancreas, lung, liver, esophagus, stomach, cervix and ovary. As depicted in Figure 3, cathepsin B activity is the end result of a complex biosynthetic pathway including **1)** transcription, **2)** post-transcriptional processing, **3)** translation and glycosylation, **4)** maturation and trafficking, and **5)** modulation by endogenous inhibitors. The majority of reports on cathepsin B expression in tumors have assessed only activity, so that the level(s) of regulation that is altered in each case is unknown. Figure 3 summarizes a few studies in human tumors where increased cathepsin B expression has been observed at more than one level (mRNA, protein and/or activity).

One example is human colorectal carcinoma, where cathepsin B mRNA levels are increased in tumors as compared to matched normal colorectal tissues (Murnane *et al.*, 1991;

E. Campo, M.R. Emmert-Buck and B.F. Sloane, unpublished data). In the former study, the highest increase among eight tumor samples was seen in Dukes' stages A and B (Murnane *et al.*, 1991). These early stages of progression represent tumors that are in the process of invading the bowel wall or local tissues, whereas the later stages C and D represent tumors which have already spread. In our more recent study looking at 47 matched normal/tumor samples, significantly ($p<0.05$) more metastatic tumors (Dukes' stages C+D) than localized tumors (Dukes' stages A+B) exhibited overexpression of cathepsin B mRNAs. The discordance between the two studies could be due to the difference in sample size (8 versus 47). Our mRNA studies confirm our immunohistochemical studies showing that cathepsin B staining increases with progression of colorectal carcinoma (Campo *et al.*, 1994). Staining for cathepsin B in colorectal carcinoma has prognostic value: among 69 cases, survival of patients which express high levels of the protease in tumor cells is significantly shorter (Campo *et al.*, 1994). In the same study, a redistribution of cathepsin B staining could be seen from perinuclear/apical in normal colonic epithelium to basal and/or diffuse cytoplasmic in colorectal carcinoma. This diffuse cytoplasmic staining may indicate the presence of truncated cathepsin B but this has not been confirmed. Overall, cathepsin B activity is elevated in colorectal carcinomas as compared to normal (Shuja *et al.*, 1991). Perhaps of more relevance to tumor invasion is the observation by Emmert-Buck and colleagues (1994) that both cathepsin B and gelatinase A activities are increased in invasive regions of colon tumors as compared to matched normal epithelial cells. For these analyses, a tissue microdissection technique was used to isolate tumor cells and matched normal epithelial cells. The same observation of increased cathepsin B activity in microdissected invasive tumor edges has been made for breast tumors (M.E. Emmert-Buck, E. Campo and B.F. Sloane, unpublished data).

Another striking example of increased expression of cathepsin B associated with tumor progression is found in human glioma. The abundance of cathepsin B transcripts and the intensity of staining for cathepsin B protein increase progressively in astrocytoma, anaplastic astrocytoma and malignant glioblastoma, and correlate with evidence of clinical invasion assessed by Magnetic Resonance Imaging (Rempel *et al.*, 1994). Thus, cathepsin B may serve as a diagnostic and prognostic marker for human glioma as well as for colon carcinoma. Levels of expression (mRNA, protein and activity) and altered trafficking of cathepsin B in glioblastoma cell lines parallel their ability to invade Matrigel *in vitro*. One of the hallmarks of gliomas is angiogenesis. Interestingly, cathepsin B staining is observed in the endothelial cells of the neovessels, especially in regions of tumor infiltration into adjacent normal brain, but not in the endothelial cells of the normal microvasculature of the brain (Mikkelsen *et al.*, 1995).

Proteolytic activity of an invading tumor is mostly required locally, i.e., near infiltrating cells or the invading front of a tumor mass. Such a localized expression of cathepsin B has been observed in urinary bladder carcinomas (Visscher *et al.*, 1994). Additionally, there is an inverse correlation between cathepsin B and type IV collagen staining, suggesting that cathepsin B participates in extracellular matrix degradation at the edge of bladder tumors (D.W. Visscher, M. Sameni and B.F. Sloane, unpublished results). In prostate tumors, cells at the invasive edge of the tumor express high amounts of cathepsin B mRNA (Sinha *et al.*, 1993) and protein (Sinha *et al.*, 1995a) as shown by *in situ* hybridization and immunohistochemistry, respectively. In addition, cathepsin B expression is detected in endothelial cells of microvessels within prostate hyperplasia and adenocarcinoma, with some leakage of cathepsin B into the stroma of neoplastic prostate (Sinha *et al.*, 1995b). This indicates that this enzyme may play a role in neovascularization in prostate carcinoma as well as in gliomas.

Since cathepsin B is thought to contribute to the invasive and metastatic phenotype of tumors, its expression could be expected to increase when tumor cells start to invade adjacent tissues. However, in at least one model of tumor progression, cathepsin B expression is altered at an earlier stage: the transition of a preneoplastic to a neoplastic lesion. This is the case for MCF-10 human breast epithelial cells, derived from a patient with fibrocystic breast disease. Diploid MCF-10A cells have undergone spontaneous immortalization in culture, but are non-tumorigenic. A variant cell line, MCF-10AneoT, has been created by transfecting MCF-10A cells with an activated *ras* oncogene. MCF-10AneoT exhibits a transformed phenotype *in vitro* and forms preneoplastic lesions in beige nude mice, which progress to neoplasias in 30% of the mice. This cell line expresses cathepsin B mRNA levels similar to those of the parental MCF-10A. However, it contains increased amounts of cathepsin B protein and activity, together with alterations in cathepsin B trafficking and its membrane association (Sloane *et al.*, 1994c). Note that *ras* transformation has different outcomes depending on the cell type and the type of *ras* used. For instance, transfection of HD6 human colon carcinoma cells with oncogenic c-Ki-*ras* increased the levels of cathepsin B mRNA, protein and activity (M. Ahram, J. Rozhin, B.F. Sloane and E.A. Friedman, unpublished results). Transforming murine 3T3 fibroblasts with *ras* can but does not always result in increased cathepsin B transcript levels (Chambers *et al.*, 1992; Zhang and Schultz, 1992).

The mechanism(s) responsible for the increased cathepsin B mRNA and protein levels observed in human colorectal carcinoma, glioma, prostate carcinoma and other tumors is unknown. There may be an increase in the rate of cathepsin B transcription caused by the presence in tumor cells of altered levels of some transcription factors, or the preferential use of one promoter region. In the case of murine melanoma, the increase in mRNA levels from low metastatic B16-F1 to intermediate B16-F10 to high metastatic B16a variants seems to be due to an increased transcription rate (Qian *et al.*, 1994). The same group also showed that a 240 bp promoter fragment (-155 to +85) including part of mouse cathepsin B exon 1 is sufficient to express a reporter gene at higher levels in B16a than in B16-F1 or F10. This differential expression among melanoma variants may be due to the presence of elevated amounts of a specific transcription factor in B16a cells (Qian *et al.*, 1994). Another possible mechanism that could account for increased cathepsin B mRNA levels in tumors is an alteration in post-transcriptional processing of the primary transcript. For example, a different mode of alternative splicing may produce mature mRNAs with greater inherent stability. In some instances, cathepsin B transcript levels are not increased, yet protein and activity are. This is the case in MCF-10A breast epithelial cells transfected with oncogenic *ras* (Sloane *et al.*, 1994c). Therefore, a post-transcriptional mechanism such as a higher rate of translation of the transcripts might be responsible for the high protein levels. A higher translation rate could derive from a different property of transcripts, like the removal of a loop structure which slows down ribosome progression along the mRNA chain, or the (un)masking of an RNA-binding protein recognition site on the transcript, mediated by alternative splicing or the use of a different transcription start site. On the other hand, the translation machinery in the transformed cells could be altered and favor cathepsin B protein synthesis, or the turnover rate of the protease could be slowed down. Many more studies will be required to fully elucidate the multi-level regulation of cathepsin B expression in normal cells and the alterations in this regulation which result in increased proteolytic activity in tumor cells.

ACKNOWLEDGEMENTS

This work was supported in part by USPHS Grants CA 36481 and CA 56586.

REFERENCES

Aronson, N.N., and Barrett, A.J., 1978, The specificity of cathepsin B, *Biochem. J.* 171:759-765.

Béchet, D.M., Ferrara, M.J., Mordier, S.B., Roux, M.-P., Deval, C.D., and Obled, A., 1991, Expression of lysosomal cathepsin B during calf myoblast-myotube differentiation, *J. Biol. Chem.* 266:14104-14112.

Berquin, I.M., Cao, L., Fong, D., and Sloane, B.F., 1995, Identification of two new exons and multiple transcription start points in the 5'-untranslated region of the human cathepsin B-encoding gene, *Gene* 159: 143-149.

Buck, M.R., Karustis, D.G., Day, N.A., Honn, K.V., and Sloane, B.F., 1992, Degradation of extracellular-matrix proteins by human cathepsin B from normal and tumour tissues, *Biochem. J.* 282:2273-2278.

Cao, L., Taggart, R.T., Berquin, I.M., Moin, K., Fong, D., and Sloane, B.F., 1994, Human gastric adenocarcinoma cathepsin B: isolation and sequencing of full-length cDNAs and polymorphisms of the gene. *Gene* 139:163-169.

Campo, E., Munoz, J., Miquel, R., Palacin, A., Cardesa, A., Sloane, B.F., and Emmert-Buck, M., 1994, Cathepsin B expression in colorectal carcinomas correlates with tumor progression and shortened patient survival, *Am. J. Pathol.* 145: 301-309.

Cavaillès, V., Augereau, P., and Rochefort, H., 1993, Cathepsin D gene is controlled by a mixed promoter, and estrogens stimulate only TATA-dependent transcription in breast cancer cells, *Proc. Natl. Acad. Sci. U.S.A.* 90:203-207.

Chambers, A.F., Colella, R., Denhardt, D., and Wilson, S.M., 1992, Increased expression of cathepsin-L and cathepsin-B and decreased activity of their inhibitors in metastatic, *ras*-transformed NIH 3T3 cells, *Mol. Carcinogen.* 5:238-245.

Chan, S.J., San Segundo, B., McCormick, M.B., and Steiner, D.F., 1986, Nucleotide and predicted amino acid sequences of cloned human and mouse preprocathepsin B cDNAs, *Proc. Natl. Acad. Sci. USA* 83:7721-7725.

Colella, R., Rosen, F.J., and Bird, J.W., 1986, mRNA levels of cathepsin B and D during myogenesis, *Biomed. Biochim. Acta* 45:11-12.

Corticchiato, O., Cajot, J.-F., Abrahamson, M., Chan, S.J., Keppler, D., and Sordat, B., 1992, Cystatin C and cathepsin B in human colon carcinoma: expression by cell lines and matrix degradation, *Intl. J. Cancer* 52:645-652.

Dalet-Fumeron, V., Guinec, N., and Pagano, M., 1993, High-performance liquid chromatographic method for the simultaneous purification of cathepsins B, H and L from human liver, *FEBS Lett.* 332:251-254.

Emmert-Buck, M.R., Roth, M.J., Zhuang, Z., Campo, E., Rozhin, J., Sloane, B.F., Liotta, L.A., and Stetler-Stevenson, W.G., 1994, Increased gelatinase A (MMP-2) and cathepsin B activity in microdissected human colon cancer samples, *Am. J. Pathol.* 145, 1285-1290.

Ferrara, M., Wojcik, F., Rhaissi, H., Mordier, S., Roux, M.-P., and Béchet, D., 1990, Gene structure of mouse cathepsin B, *FEBS Lett.* 273:195-199.

Flaumenhaft, R., and Rifkin, D.B., 1992, The extracellular regulation of growth factor action, *Mol. Biol. Cell* 3:1057-1065.

Fong, D., Chan, M.M.-Y., Hsieh, W.-T., Menninger, J.C., and Ward, D.C., 1992, Confirmation of the human cathepsin B gene (CTSB) assignment to chromosome 8, *Hum. Genet.* 89:10-12.

Friemert, C., Closs, E.I., Silbermann, M., Erfle, V., and Strauss, P.G., 1991, Isolation of a cathepsin B-encoding cDNA from murine osteogenic cells, *Gene* 103:259-261.

Gong, Q., Chan, S.J., Bajkowski, A.S., Steiner, D.F., and Frankfater, A., 1993, Characterization of the cathepsin B gene and multiple mRNAs in human tissues: evidence for alternative splicing of cathepsin B pre-mRNA, *DNA Cell Biol.* 12:299-309.

Gong, Q., and Frankfater, A., 1993, Alternative splicing of cathepsin B in human tumors produces a novel isoform which can be refolded into an active enzyme *in vitro*, *FASEB J.* 7:A1208.

Huet, G., Flipo, R.-M., Colin, C., Janin, A., Hemon, B., Collyn-d'Hooge, M., Lafyatis, R., Duquesnoy, B., and Degand, P., 1993, Stimulation of the secretion of latent cysteine proteinase activity by tumor necrosis factor α and interleukin-1, *Arthritis and Rheumatism* 36:772-780.

Keppler, D., Abrahamson, M., and Sordat, B., 1994, Proteases and cancer: secretion of cathepsin B and tumour invasion, *Biochem. Soc. Trans.* 22:43-49.

Keppler, D., Waridel, P., Abrahamson, M., Bachmann, D., Berdoz, J., and Sordat, B., 1994, Latency of cathepsin B secreted by human colon carcinoma cells is not linked to secretion of cystatin C and is relieved by neutrophil elastase, *Biochim. Biophys. Acta* 1226:117-125.

Kobayashi, H., Schmitt, M., Goretzki, L., Chucholowski, N., Calvete, J., Kramer, M., Gunzler, W.A., Janicke, F., and Graeff, H., 1991, Cathepsin B efficiently activates the soluble and the tumor cell receptor-bound form of the proenzyme urokinase-type plasminogen activator (pro-uPA), *J. Biol. Chem.* 266:5147-5152.

Kornfeld, S., 1990, Lysosomal enzyme targeting, *Biochem. Soc. Trans.* 18:367-374.

Lah, T.T., Buck, M.R., Honn, K.V., Crissman, J.D., Rao, N.C., Liotta, L.A., and Sloane, B.F., 1989, Degradation of laminin by human tumor cathepsin B, *Clin. Expl. Metastasis* 7:461-468.

Matrisian, L.M., and Bowden, G.T., 1990, Stromelysin/transin and tumor progression, *Semin. Cancer Biol.* 1:107-115.

McIntyre, G.F., and Erickson, A.H., 1991, Procathepsins L and D are membrane-bound in acidic microsomal vesicles, *J. Biol. Chem.* 266:15438-15445.

Mikkelsen, T., Yan, P.-S., Ho, K.-L., Sameni, M., Sloane, B.F., and Rosenblum, M.L., 1995, Immunolocalization of cathepsin B in human glioma: implications for human tumor invasion and angiogenesis, *J. Neurosurg.*,83: 285-290.

Moin, K., Rozhin, J., McKernan, T.B., Sanders, V.J., Fong, D., Honn, K.V., and Sloane, B.F., 1989, Enhanced levels of cathepsin B mRNA in murine tumors, *FEBS Lett.* 244: 61-64.

Murnane, M.J., Sheahan, K., Ozdemirli, M., and Shuja, S., 1991, Stage-specific increases in cathepsin B messenger RNA content in human colorectal carcinoma, *Cancer Res.* 51:1137-1142.

Nguyen-Ba, G., Robert, S., Dhalluin, S., Tapiero, H., and Hornebeck, W., 1994, Modulatory effects of dexamethasone on ornithine decarboxylase activity and gene expression: a possible post-transcriptional regulation by a neutral metalloprotease, *Cell Biochem. and Function* 12:121-128.

Oursler, M.J., Riggs, B.L., and Spelsberg, T.C., 1993, Glucocorticoid-induced activation of latent transforming growth factor-β by normal human osteoblast-like cells, *Endocrinology* 133:2187-2196.

Page, A.E., Warburton, M.J., Chambers, T.J., Pringle, J.A.S., and Hayman, A.R., 1992, Human osteoclastomas contain multiple forms of cathepsin B, *Biochim. Biophys. Acta* 1116:57-66.

Petanceska, S., Burke, S., Watson, S.J., and Devi, L., 1994, Differential distribution of messenger RNAs for cathepsin B, L and S in adult rat brain: an *in situ* hybridization study, *Neuroscience* 59:729-738.

Phillips, I.D., Black, E.G., Sheppard, M.C., and Docherty, K., 1989, Thyrotrophin, forskolin and ionomycin increase cathepsin B mRNA concentration in rat thyroid cells in culture, *J. Mol. Endocrinol.* 2:207-212.

Qian, F., Bajkowski, A.S., Steiner, D.F., Chan, S.J., and Frankfater, A., 1989, Expression of five cathepsins in murine melanomas of varying metastatic potential and normal tissues, *Cancer Res.* 49:4870-4875.

Qian, F., Frankfater, A., Chan, S.J., and Steiner, D.F., 1991a, The structure of the mouse cathepsin B gene and its putative promoter, *DNA Cell Biol.* 10:159-168.

Qian, F., Frankfater, A., Steiner, D.F., Bajkowski, A.S., and Chan, S.J., 1991b, Characterization of multiple cathepsin B mRNAs in murine B16a melanoma, *Anticancer Res.* 11:1445-1452.

Qian, F., Chan, S.J., Achkar, C., Steiner, D.F., and Frankfater, A., 1994, Transcriptional regulation of cathepsin B expression in B16 melanomas of varying metastatic potential, *Biochem. Biophys. Res. Comm.* 202:429-436.

Rempel, S.A., Rosenblum, M.L., Mikkelsen, T., Yan, P.S., Ellis, K.D., Golembieski, W.A., Sameni, M., Rozhin, J., Ziegler, G., and Sloane, B.F., 1994, Cathepsin B expression and localization in glioma progression and invasion. *Cancer Res.* 54:6027-6031.

Rhaissi, H., Béchet, D., and Ferrara, M., 1993, Multiple leader sequences for mouse cathepsin B mRNA? *Biochimie* 75:899-904.

Sameni, M., Elliott, E., Ziegler, G., Fortgens, P.H., Dennison, C., and Sloane, B.F., Cathepsins B and D are localized at surface of human breast cancer cells, *Pathol. Oncol. Res.*, in press.

San Segundo, B., Chan, S.J., and Steiner, D.F., 1986, Differences in cathepsin B mRNA levels in rat tissues suggest specialized functions, *FEBS Lett.* 201:251-256.

Shuja, S., Sheahan, K., and Murnane, M.J., 1991, Cysteine endopeptidase activity levels in normal human tissues, colorectal adenomas and carcinomas, *Intl. J. Cancer* 49:341-346.

Sinha, A.A., Gleason, D.F., DeLeon, O.F., Wilson, M.J., and Sloane, B.F., 1993, Localization of a biotinylated cathepsin B oligonucleotide probe in human prostate including invasive cells and invasive edges by in situ hybridization, *Anat. Rec.* 235:233-240.

Sinha, A.A., Wilson, M.J., Gleason, D.F., Reddy, P.K., Sameni, M., and Sloane, B.F., 1995a, Immunohistochemical localization of cathepsin B in neoplastic human prostate, *Prostate* 25, 26: 171-178.

Sinha, A.A., Gleason, D.F., Staley, N.A., Wilson, M.J., Sameni, M., and Sloane, B.F., 1995b, Cathepsin B in angiogenesis of human prostate: an immunohistochemical and immunoelectron microscope analysis, *Anat. Rec.*, 241: 353-262.

Sloane, B.F., Moin, K., Krepela, E., and Rozhin, J., 1990, Cathepsin B and its endogenous inhibitors: the role in tumor malignancy, *Cancer Metastasis Reviews* 9:333-352.

Sloane, B.F., Sameni, M., Cao, L., Berquin, I., Rozhin, J., Ziegler, G., Moin, K., and Day, N., 1994a, Alterations in processing and trafficking of cathepsin B during malignant progression, In Katunuma, N., Suzuki, K., Travis, J., and Fritz, H. (Eds) *Biological Functions of Proteases and Inhibitors*, Japan Scientific Societies Press, Tokyo, pp. 131-147.

Sloane, B.F., Moin, K., and Lah, T., 1994b, Regulation of lysosomal endopeptidases in malignant neoplasia, In Pretlow, T.G. and Pretlow, T.P. (Eds) *Biochemical and Molecular Aspects of Selected Cancers*, Vol. 2, Academic Press, San Diego, California, pp.411-454.

Sloane, B.F., Moin, K., Sameni, M., Tait, L.R., Rozhin, J., and Ziegler, G., 1994c, Membrane association of cathepsin B can be induced by transfection of human breast epithelial cells with c-Ha-*ras* oncogene, *J. Cell Science* 107:373-384.

Tam, S.W., Cote-Paulino, L.R., Peak, D.A., Sheahan, K., and Murnane, M.J., 1994, Human cathepsin B cDNAs: sequence variations in the 3'-untranslated region, *Gene* 139:171-176.

Visscher, D.W., Sloane, B.F., Sameni, M., Babiarz, J.W., Jacobson, J. and Crissman, J.D., 1994, Clinico-pathologic significance of cathepsin B immunostaining in transitional neoplasia. *Mod. Pathol.* 7: 76-81.

Wang, X., Chan, S.J., Eddy, R.L., Byers, M.G., Fukushima, Y., Henry, W.M., Haley, L.L., Steiner, D.F., and Shows, T.B., 1988, Chromosome assignment of cathepsin B (CTSB) to 8p22 and cathepsin H (CTSH) to 15q24-q25, *Cytogenet. Cell Genet.* 46:710-711.

Ward, C.J., Crocker, J., Chan, S.J., Stockley, R.A., and Burnett, D., 1990, Changes in the expression of elastase and cathepsin B with differentiation of U937 promonocytes by GMCSF, *Biochem. Biophys. Res. Comm.* 167:659-664.

Zhang, J.-Y., and Schultz, R.M., 1992, Fibroblasts transformed by different *ras* oncogenes show dissimilar patterns of protease gene expression and regulation, *Cancer Res.* 52:6682-6689.

36

TIMP-2 MEDIATES CELL SURFACE BINDING OF MMP-2

M. L. Corcoran, M. R. Emmert-Buck, J. L. McClanahan, M. Pelina-Parker, and W. G. Stetler-Stevenson

Extracellular Pathology Section
Laboratory of Pathology
National Cancer Institute
National Institutes of Health
Bethesda, Maryland

ABSTRACT

In order to understand the mechanism for neoplastic cell invasion, we utilized binding studies of TIMP-2, gelatinase A and the TIMP-2/gelatinase A complex to neoplastic cells and correlated these results with their capacity to invade a matrix substrate in a modified Boyden chamber assay. Binding studies were performed on malignant human breast cancer cells and fibrosarcoma cells with rTIMP-2, rGelatinase A, and TIMP-2/gelatinase A complex. Competition studies of the binding characteristics of these proteins indicated that gelatinase A and gelatinase A/TIMP-2 complex bound to the surface of cells via TIMP-2. Furthermore, the localization of either latent or active protease to the surface of MDA-MB-435 breast cancer cells facilitated the invasion of these neoplastic cells through a matrigel barrier. This suggests that in addition to binding this complex, these cells can activate this pro-enzyme-inhibitor complex and use this activity to facilitate cellular invasion. Moreover, their enhanced invasion was suppressed by exogenous additions of rTIMP-2. A working hypothesis and model for the role of gelatinase A/TIMP-2 complex in cellular invasion is presented.

INTRODUCTION

Metastasis formation is one of the major causes of the morbidity associated with cancer. The capacity of an neoplastic cell to proteolyze extracellular matrix is essential for metastasis formation. In order to control this process, the exact cascade of proteolytic events needs to be fully understood. In order for a cell to successfully invade through the matrix proteolytic degradation of the matrix must be regulated in a spatial and temporal fashion with respect to cell migration. The invasion of neoplastic cells through extracellular atrix matrix employs a dysfunctional and physiologically unregulated process of matrix, destruction (Barsky et al., 1984; Liotta & Stetler-Stevenson 1990). Although the four major groups of proteases maybe involved in this process, overexpression and activation of matrix metalloproteinases (Stetler-Stevenson et al., 1992) and suppression of tissue inhibitors of metalloproteinases expression in tumors strongly correlates with the development of an invasive (Campo et al., 1992) and metastatic phenotype (Hewitt et al., 1991; Sato et al., 1992). These zinc-atom dependent proteases are responsible for the matrix turnover that occurs extracellularly at neutral pH (Birkedal-Hansen et al., 1993). This family can be subdivided into three groups based on their substrate specificity. The interstitial collagenases, including the fibroblast (MMP-1) and polymorphonuclear leukocyte (MMP-8) collagenases, cleave fibrillar collagens, types I, II, III, VII, and X collagen resulting in a 3/4 amino terminal degradation fragment. Collagenase-3 (MMP-13), one of the newer members of this family, exclusively degrades type I collagen. The gelatinases, including gelatinase A (MMP-2) and gelatinase B (MMP-9), degrade elastin, fibronectin, gelatin as well as types IV, V, VII collagen. Because of their gelatinolytic activity, the proteases also finish degradation of fibrillar collagens which have been compromised by the action of interstitial collagenase. Lastly, the stromelysin subgroup, including stromelysins 1 (MMP-3), 2 (MMP-10), and 3 (MMP-11), and matrilysin (MMP-7) degrade a broader range of substrates including laminin, fibronectin, proteoglycans, and the non-helical regions of type IV collagen.

All the members of the MMP family possess conserved functional and structural domains which translate into similar mechanisms of proteolysis (Birkedal-Hansen, et al. 1993). The amino-terminal "activation locus" (PRCGXPDV) and the catalytic site that binds the zinc atom (VAAHEXGHXXGXXH) are the sites of greatest homology within this family. The carboxy-terminal region of all MMPs, except for matrilysin, possesses a hemopexin/vitronectin homology domain that aides in substrate specificity of interstitial collagenase and the stromelysins (Sanchez-Lopez et al., 1993). The gelatinases posses a fibronectin-like gelatin binding domain which also contributes to the substrate specificity (Strongin et al., 1993). Metal chelating agents (EDTA or o-phenanthroline) and peptide analogs designed to mimic the "activation locus" or substrate (Stetler-Stevenson et al., 1991; Van Ward & Birkedal-Hansen 1990) result in inhibition of MMP-mediated proteolysis.

Based on the work performed in a number of laboratories, the biochemical mechanism for activating the catalytic activity of mammalian MMP is currently believed to occur via a "cysteine switch" mechanism. In this theory, an unpaired cysteine residue in the PRCGXPD sequence of the pro-fragment of the enzyme maintains the latency of the zymogen by direct coordination with the active site zinc atom (Van Ward & Birkedal-Hansen 1990). Interrupting this interaction by physical or chemical means initiates a cascade of events that results in cleavage of the amino-terminus of this enzyme. Deletion of the R or C residues in the pro-sequence region of the MMPs disrupts the coordination of the zinc resulting in the auto-activation of the enzyme (Park et al., 1991; Sanchez-Lopez et al., 1988). Loss of the amino-terminal propeptide domain is evidence of, but not equivalent to, an active enzyme species. Moreover, synthetic peptides that mimic PRCGXPDV sequence have been previously shown to inhibit not only enzymatic activity (Nomura & Suzuki 1993; Stetler-Stevenson, et al. 1991) but also neoplastic cell invasion (Melchiori et al., 1992) and neurite outgrowth (Muir 1994). Although in vitro studies have aided in understanding the mechanism of action for these enzymes, the cellular or in vivo mechanism is under investigation.

Net proteolysis at a local micro-environment depends on the dynamic balance in the relative concentration of active MMPs and their inhibitors, namely tissue inhibitor of metalloproteinase or TIMPs. The production of both MMPs and TIMPs can be controlled at several levels including transcription, translation, and secretion of both these proteins. In addition the activation of the latent MMPs is an important step in controlling matrix turnover. The ability of TIMPs to inhibit the in vivo activity of MMPs is currently one of the parameters utilized to identify new members of the MMP family (Nagase et al., 1992). In addition to

TIMPs, serum derived d2 macroglobulin also non-specifically inhibits the action of a number of proteases including the MMPs. Currently, there are three well defined members of the TIMP family that share 40% amino acid homology. These are TIMP-1 (Docherty *et al.*, 1985), TIMP-2 (DeClerck *et al.*, 1989; Stetler-Stevenson *et al.*, 1989; Van Ward & Birkedal-Hansen 1990; Werb *et al.*, 1977) and TIMP-3 (Leco *et al.*, 1994; Pavloff *et al.*, 1992). The TIMPs bind with high affinity in a 1:1 molar ratio to active MMPs resulting in loss of proteolytic activity. All three proteins of this family contain 12 conserved cysteine residues which have been shown in TIMP-1(Williamson *et al.*, 1990) and TIMP-2 (DeClerck *et al.*, 1993) to form disulfide bridges resulting in six peptide loops and two peptide knots (Birkedal-Hansen, *et al.* 1993; DeClerck *et al.*, 1992; O'Shea *et al.*, 1992; Stetler-Stevenson *et al.*, 1992). TIMP-3 is a unique member of this family because it is tightly associated with the extracellular matrix and is not secreted free into conditioned media of cultured cells (Leco, *et al.* 1994; Pavloff, *et al.* 1992). Studies with truncated recombinant TIMP domains and proteolytically derived TIMP fragments (DeClerck, *et al.* 1993; Murphy *et al.*, 1991) have revealed that the conserved disulfide loop pattern in the amino-terminal domain (O'Shea, *et al.* 1992), plays a significant role in inhibiting MMP activity. Critical points in the evolution of an invasive/metastatic neoplastic phenotype is the induction and activation of MMPs and suppression of endogenous inhibitors. Thus, comprehending the mechanism of action of these enzymes and their inhibitors may prove important in therapeutic intervention of metastasis and the progression of cancer. Evidence exists that neoplastic cells possess a surface receptor that binds stromal cell derived gelatinase A (Emonard *et al.*, 1992) and could facilitate neoplastic cell invasion through the matrix. Evidence strongly supports a specific role for gelatinase A in most human tumors studied. A significant body of scientific evidence has been dedicated to correlating the degree of tumor progression with increase metalloproteinase production. Additional studies need to evaluate the degree of activated enzyme or if these enzymes are complexed to endogenous inhibitors. These studies could prove useful in identifying prognostic indicators for cancer progression. In the present study we assess the mechanism of cell surface binding and the contribution of the MMP-2/ TIMP-2 complex to cellular invasion.

MATERIALS AND METHODS

Binding Experiments

pro-MMP-2/TIMP-2 purified from A2058 conditioned media (Van Ward & Birkedal-Hansen 1990) and affinity purified rTIMP-2 kindly provided by R. Bird (Oncologics Inc., MD) were iodinated using the lactoperoxidase method to a specific activity of 4-5 x 10^6 dpm/μg. HT1080 cells were grown in 175 cm^2 flasks to 80-90% confluence in Dulbecco's modified Eagle's medium supplemented with 10% fetal bovine serum. Cells were washed 3 times with cold PBS containing 0.1% BSA (Buffer A) and incubated for 1 h in the same buffer. Cells were scraped off the flask, washed 3 times in buffer A, and resuspended in phosphate buffered saline containing 5 mM CaCl$_2$ and 0.1% BSA (Buffer B). In the binding saturation experiments, sequential dilutions of ^{125}I-labeled rTIMP-2 were added to the cells and incubated for 3 hours at 4°C either in the presence or absence of 100-fold excess unlabeled rTIMP-2. In the binding competition experiments, 100 fold excess unlabeled rTIMP-2, rTIMP-1 or pro-MMP-2/TIMP-2 was added to the cells for 15 minutes prior to the addition of labeled rTIMP-2 or pro-MMP-2/TIMP-2. Cells were washed 3 times with buffer B and counted in a gamma counter. Specific binding was calculated as the difference between the bound 125I-labeled rTIMP-2 or pro-MMP-2/TIMP-2 that occurred in the presence or absence of unlabeled pro-MMP-2/TIMP-2, rTIMP-2, or TIMP-1. For the analysis of binding data, best fit-curves were determined using the MLAB program with equations as previously described (Rodbard, 1973).

Binding assays for the MCF-7 cells were performed in 96-well culture plates (Nunc) coated with plasma membrane fractions prepared from MCF-7 cells (250 ng/well, overnight at 4°C). Just prior to binding assays, membrane preparations were saturated with PBS containing 0.1% bovine serum albumin (BSA). Labeled and unlabeled ligands were diluted in the same buffer. To determine the effect of time on ^{125}I-labeled rTIMP-2 binding, membrane preparations were incubated with 9 nM labeled ligand for various times up to 6 h at 22°C. Equilibrium binding of ^{125}I-labeled rTIMP-2 to membrane fractions was determined with binding assays for 3 h at 22°C, using three wells per point at various concentrations of labeled ligand (1.1 nM to 11 nM). Nonspecific binding was determined in the presence of 1,100 nM unlabeled rTIMP-2. In some experiments, 9 nM ^{125}I-labeled rTIMP-2 was incubated in the presence of various excesses of unlabeled TIMP-1 (70 nM to 3,500 nM). Binding assays were also performed with MCF-7 cells grown as a monolayer. Cells were first plated in a 24-well plate in Dulbecco's modified Eagle medium supplemented with 10% fetal calf serum (FCS). After 1 day of culture, culture medium was removed and cells were washed 3 times with DMEM containing 20 mM HEPES, 0.1% BSA. Saturation experiments were performed for 3 h at 22°C with various concentrations of iodinated ligand (0.2 nM to 20 nM) in the presence or absence of 2,000 nM unlabeled rTIMP-2. Both membrane fractions and cell monolayers were then washed 3 times with PBS-BSA (200 ml and 1 ml, respectively), lysed with 0.1 % sodium dodecyl sulfate in 0.5 M NaOH, and the bound radioactivity was measured with a gamma counter.

Matrigel Invasion Assay

The polycarbonate filters used in this modified Boyden chamber assay were coated with 5 μg type IV collagen followed by 25 μg/filter of Matrigel (a generous gift from Dr. Hynda Kleinman) as previously described (Albini et al., 1991). MDA-MB-435 cells were grown to 80-90% confluence in DMEM with 10% FBS. Cells were trypsinized and allowed to recover by gentle rocking at 30°C in DMEM with 10% FBS for 1 hour. The cells were then incubated at 37°C for 1 hour in serum free DMEM either in the absence or presence of MMP-2 or TIMP-2. The latent gelatinase A was activated in 1mM aminophenylmercuric acid as previously described (Kleiner et al., 1993). Prior to the invasion assay, the cells were washed twice in serum free DMEM. Equal amounts of cells were added to the top wells of the chamber and allowed to incubate for 6 hours at 37°C. The bottom wells contained chemotactic NIH 3T3 conditioned media. Cell invasion was quantitated by counting the H&E stained cells that bound the filter in at least 8 random fields and normalized to the number of cells that migrated in the untreated wells (Albini et al., 1991). Statistical analysis (student t test) were performed using the Statview program.

RESULTS

TIMP-2 Mediates Binding of Gelatinase A/TIMP-2 Complex to Neoplastic Cells

Specific and saturable binding of ^{125}I-labeled rTIMP-2 to HT-1080 cells in suspension was observed (Fig. 1).

Scatchard analysis of ^{125}I-labeled rTIMP-2 binding showed 30,000 sites per cell and KD of 2.5 nM in HT1080 cells and 25,000 sites per cell with a K_D=1.6 nM in breast

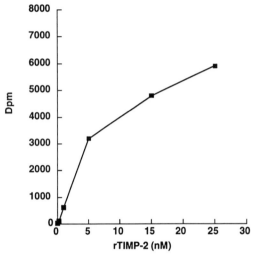

Figure 1. Saturable steady-state binding of ^{125}I-rTIMP-2 to HT-1080 cells.

adenocarcinoma, MCF-7 cells. MCF-7 and HT-1080 cells have similar TIMP-2 binding characteristics, but differ in the binding and activation of MMP-2 (Table 1).

Competition with 100 fold molar excess of rTIMP-1 did not affect rTIMP-2 binding. However, competition with 100-fold molar excess of pro-MMP-2/TIMP-2 complex completely abolished binding of rTIMP-2 to the cells (Fig. 2A). Although ^{125}I-labeled pro-rMMP-2 alone did not specifically bind to cells (data not shown), ^{125}I-labeled pro-MMP-2/TIMP-2 complex exhibited specific binding to HT-1080 cells in suspension (Fig. 2B).

Moreover, the binding of ^{125}I-labeled pro-MMP-2/TIMP-2 complex was not affected by the presence of 100-fold molar excess of TIMP-1, but was completely abolished in the presence of excess rTIMP-2. The binding of both components of the ^{125}I-labeled pro-MMP-2/TIMP-2 complex to the cells was confirmed by SDS-PAGE and autoradiography (data not shown).

Invasion Assays

The results from the binding assays indicated that TIMP-2 mediated the binding of gelatinase A and complex to the surface of these cells. Therefore, we evaluated if the localization of the protease to the surface of cells altered the invasive capacity of cells through a reconstituted basement membrane. For these assays we utilized breast adenocarcinoma cells, MDA-MB-435 which do not produce any gelatinase A when grown in culture.

Table 1. Scatchard Analysis of neoplastic cells

Cell Line	TIMP-2 Binding	
	KD	Sites/Cell
HT-1080	2.5 nM	30,000
MCF-7	1.6 nM	25,000

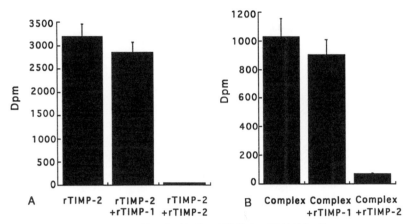

Figure 2. Competition for binding rTIMP-2 to HT-1080 cells.

Incubating these cells with either 25 or 50 ng preactivated (Fig. 3A) or latent gelatinase A (Fig. 3B) increased in a dose dependent manner the capacity of these cells to invade through the matrigel barrier.

However, this effect is saturable since addition of 100 ng active or latent gelatinase A did not further enhance their capacity to invade. Since active or latent gelatinase B did not enhance invasion (data not shown), these effects were specific for gelatinase A. These results suggest that the binding and localization of MMP-2 to the surface is via a saturable binding site. Moreover, the latent or activated gelatinase A which was previously shown to bind the cell surface via TIMP-2 binding site, increases the invasive capacity of these cells. Addition of excess rTIMP-2 to MDA-MB-435 (Fig. 4 A) or to HT0180 cells (Fig. 4B) suppressed the MMP-2 mediated enhanced invasion. The HT1080 cells have been previously demonstrated to produce gelatinases.

These results correlate with previous reports in the HT1080 fibrosarcoma cells (Albini *et al.*, 1991).

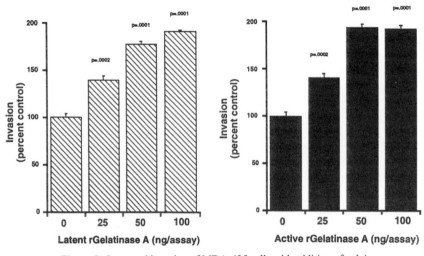

Figure 3. Increased invasion of MDA-435 cells with addition of gel A.

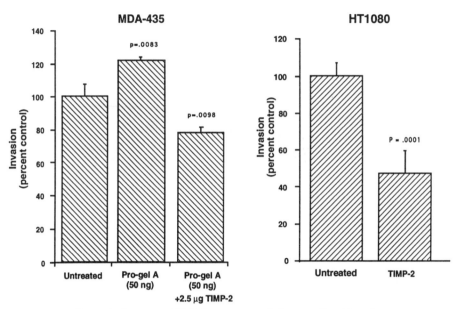

Figure 4. Exogenous TIMP-2 suppresses gelatinase mediated invasion.

DISCUSSION

Currently, two mechanisms for the activation of MMPs have been proposed. The first mechanism characterized for the activation of MMPs is the plasmin cascade which activates procollagenases, prostromelysins and progelatinase B. While some groups report that the plasmin cascade is also responsible for activation of gelatinase A (Keskioja *et al.*, 1992; Salo *et al.*, 1982), others believe that plasmin actually degrades this enzyme (Okada *et al.*, 1990). Activation of gelatinase A can occur in the presence of inhibitors of serine proteases and plasmin (Brown *et al.*, 1993; Murphy *et al.*, 1992; Ward *et al.*, 1991) suggesting that the plasmin cascade may not be the mechanism utilized *in vivo* for the activation of this enzyme. The cell-mediated activation of progelatinase A or progelatinase A/TIMP-2 complexes can be inhibited by chelating agents as well as by addition of exogenous free TIMP-2. Moreover, the carboxy-terminus on progelatinase A seems to be required for activation (Murphy *et al.* 1992). This domain is responsible for binding of TIMP-2 to the pro-form of gelatinase A (Fridman *et al.*, 1992). This observation is consistent with the requirement that TIMP-2 mediate binding of the enzyme to the cell surface. Evidence of a membrane associated activation of gelatinase A was reported by our laboratory to be induced by concanavalin A and phorbol esters (Brown *et al.*, 1993). Recently, a metalloproteinase which possesses a short membrane spanning domain was cloned and sequenced (Sato *et al.*, 1994). Currently, this protein is believed to be responsible for the cell associated activation only of gelatinase A. In addition to a transmembrane domain, the MT-MMP possesses the prosequence thought to be responsible for the latency of these enzymes.

In order for a neoplastic cell to metastasize, proteolysis of the matrix needs to be coordinated with cell migration. Here, we demonstrate that gelatinase A has the capacity to localize to the membrane of neoplastic cells. The localization of either latent or active gelatinase A to the cell surface presumably facilitates its activation by the MT-MMP activating protease. The localization to the membrane is shown in this report to be mediated

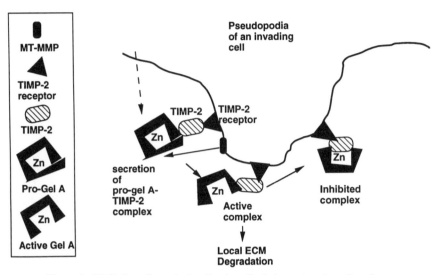

Figure 5. TIMP-2 mediates the localization of gelatinase A to the cell surface.

via TIMP-2 which binds a specific binding site on the surface of these cells. We hypothesize that once MT-MMP is activated, this activity is responsible for activation of progelatinase A/TIMP-2 complex that is either prebound or binds to the TIMP-2 binding site on the cell membrane. This cell surface binding and activation process strategically positions the gelatinase A activity at the invasion front.

The local gelatinase A activity at the cell surface is immediately controlled by this mechanism. The TIMP-2 binding site prevents the N-terminal metalloproteinase inhibitory domain of TIMP-2 from interacting with the active site of gelatinase A and blocking protease activity. Furthermore, dissociation from the cell surface allows free interaction between the inhibitor and the active gelatinase which results in enzyme inactivation. This allows precise cellular control of enzyme activity (Fig. 5). The exact nature of this mechanism is currently under intense investigation.

The results of this study suggest that cell surface localization of the gelatinase A is via complex formation with TIMP-2 and binding to a TIMP-2 receptor. The ability of saturating additions of TIMP-2 to suppresses invasion maybe due either to direct inhibition of the active gelatinase A, or dissociation of bound gelatinase A/TIMP-2 complex from the cell surface binding site, or both. These findings demonstrate that the invasive process is highly dependent on the dynamic balance between the gelatinase A and TIMP-2, and that cell invasion requires the action of both proteins.

REFERENCES

Albini, A., Melchiori, A., Santi, L., Liotta, L. A., Brown, P. D. and Stetler-Stevenson, W. G., 1991, Tumor cell invasion inhibited by TIMP-2 [see comments], *J. Natl. Cancer Inst.* 83: 775-9.

Barsky, S. J., Rao, C. N., Williams, J. E. and Liotta, L. A., 1984, Laminin molecular domains which alter metastsis in a murine model, *J. Clin. Invest.* 74: 843-848.

Birkedal-Hansen, H., Moore, W. G. I., Bodden, M. K., Windsor, L. J., Birkedal-Hansen, B., DeCarlo, A. and Engler, J. A., 1993, Matrix Metalloproteinases: A Review, *Crit. Rev. Oral Biol. Med.* 4: 197-250.

Brown, P. D., Bloxidge, R. E., Stuart, N. S. A., Gatter, K. C. and Carmichael, J., 1993, Association Between Expression of Activated 72-Kilodalton Gelatinase and Tumor Spread in Non-Small-Cell Lung Carcinoma, *J. Natl. Cancer Inst.* 85: 574-578.

Brown, P. D., Kleiner, D. E., Unsworth, E. J. and Stetler-Stevenson, W. G., 1993, Cellular activation of the 72 kDa type IV procollagenase/TIMP-2 complex, *Kidney Int.* 43: 163-170.

Campo, E., Merino, M. J., Liotta, L., Neumann, R. and Stetler-Stevenson, W., 1992, Distribution of the 72-kd type IV collagenase in nonneoplastic and neoplastic thyroid tissue, *Hum Pathol* 23: 1395-1401.

DeClerck, Y., Szpirer, C., Aly, M. S., Cassiman, J.-J., Eeckhout, Y. and Rousseau, G., 1992, The Gene for Tissue Inhibitor of Metalloproteinases-2 Is Localized on Human Chromosome Arm 17q25, *Genomics* 14: 782-784.

DeClerck, Y. A., Yean, T.-D., Ratzkin, B. J., Lu, H. S. and Langley, K. E., 1989, Purification and Characterization of Two Related but Distinct Metalloproteinase Inhibitors Secreted by Bovine Aortic Endothelial Cells, *J. Biol. Chem.* 264: 17445-17453.

DeClerck, Y. A., Yean, T. D., Lee, Y., Tomich, J. M. and Langley, K. E., 1993, Characterization of the Functional Domain of Tissue Inhibitor of Metalloproteinases-2 (TIMP-2), *Biochem. J.* 289: 65-69.

Docherty, A. J. P., Lyons, A., Smith, B. J., Wright, E. M., Stephens, P. E., Harris, T. J. R., Murphy, G. and Reynolds, J. J., 1985, Sequence of Human Tissue Inhibitor of Metalloproteinases and Its Identity to Erythroid-Potentiating Activity, *Nature* 318: 66-69.

Emonard, H. P., Remacle, A. G., No:el, A. C., Grimaud, J. A., Stetler-Stevenson, W. G. and Foidart, J. M., 1992, Tumor cell surface-associated binding site for the M(r) 72,000 type IV collagenase, *Cancer Res.* 52: 5845-8.

Fridman, R., Fuerst, T. R., Bird, R. E., Hoyhtya, M., Oelkuct, M., Kraus, S., Komarek, D., Liotta, L. A., Berman, M. L. and Stetler-Stevenson, W. G., 1992, Domain structure of human 72-kDa gelatinase/type IV collagenase. Characterization of proteolytic activity and identification of the tissue inhibitor of metalloproteinase-2 (TIMP-2) binding regions, *J. Biol. Chem.* 267: 15398-405.

Hewitt, R. E., Leach, I. H., Powe, D. G., Clark, I. M., Cawston, T. E. and Turner, D. R., 1991, Distribution of collagenase and tissue inhibitor of metalloproteinases (TIMP) in colorectal tumours, *Int. J. Cancer* 49: 666-672.

Keskioja, J., Lohi, J., Tuuttila, A., Tryggvason, K. and Vartio, T., 1992, Proteolytic Processing of the 72,000-Da Type-IV Collagenase by Urokinase Plasminogen Activator, *Exp. Cell Res.* 202: 471-476.

Kleiner, D., Jr., Tuuttila, A., Tryggvason, K. and Stetler-Stevenson, W. G., 1993, Stability analysis of latent and active 72-kDa type IV collagenase: the role of tissue inhibitor of metalloproteinases-2 (TIMP-2), *Biochemistry* 32: 1583-92.

Leco, K. J., Khokha, R., Pavloff, N., Hawkes, S. P. and Edwards, D. R., 1994, Tissue Inhibitor of Metalloproteinases-3 (TIMP-3) Is an Extracellular Matrix-associated Protein with a Distinctinve Pattern of Expression in Mouse Cells and Tissues, *J. Biol. Chem.* 269: 9352-9360.

Liotta, L. A. and Stetler-Stevenson, W. G., 1990, Metalloproteinases and cancer invasion, *Cancer Biology* 1: 99-106.

Melchiori, A., Albini, A., Ray, J. M. and Stetler-Stevenson, W. G., 1992, Inhibition of tumor cell invasion by a highly conserved peptide sequence from the matrix metalloproteinase enzyme prosegment, *Cancer Res.* 52: 2353-6.

Muir, D., 1994, Metalloproteinase-Dependent Neurite outgrowth within a synthetic extracellular matrix is induced by nerve growth factor, *Exp. Cell Res.* 210: 243-252.

Murphy, G., Houbrechts, A., Cockett, M. I., Williamson, R. A., O'Shea, M. and Docherty, A. J., 1991, The N-terminal domain of tissue inhibitor of metalloproteinases retains metalloproteinase inhibitory activity, *Biochemistry* 30: 8097-102.

Murphy, G., Willenbrock, F., Ward, R. V., Cockett, M. I., Eaton, D. and Docherty, A. J., 1992, The C-terminal domain of 72 kDa gelatinase A is not required for catalysis, but is essential for membrane activation and modulates interactions with tissue inhibitors of metalloproteinases, *Biochem. J.* 283: 637-641.

Nomura, K. and Suzuki, N., 1993, Stereo-specific inhibition of sea urchin envelysin (hatching enzyme) by a synthetic autoinhibitor peptide with a cysteine-switch consensus sequence., *FEBS Lett.* 321: 84-88.

O'Shea, M., Willenbrock, F., Williamson, R. A., Cockett, M. I., Freedman, R. B., Reynolds, J. J., Docherty, A. J. and Murphy, G., 1992, Site-Directed Mutations That Alter the Inhibitory Activity of the Tissue Inhibitor of Metalloproteinases-1: Importance of the N-Terminal Region Between Cysteine 3 and Cysteine 13, *Biochemistry* 31: 10146-10152.

Okada, Y., Morodomi, T., Enghild, J. J., Suzuki, K., Yasui, A., Nakanishi, I., Salvesen, G. and Nagase, H., 1990, Matrix metalloproteinase 2 from human rheumatoid synovial fibroblasts. Purification and activation of the precursor and enzymic properties, *Eur. J. Biochem.* 194: 721-730.

Park, A. J., Matrisian, L. M., Kells, A. F., Pearson, R., Yuan, Z. Y. and Navre, M., 1991, Mutational Analysis of the transin (rat stromelysin) autoinhibitor region demonstrates a role for residues surrounding the "cysteine switch", *J. Biol. Chem.* 266: 1584-1590.

Pavloff, N., Staskus, P. W., Kishnani, N. S. and Hawkes, S. P., 1992, A New Inhibitor of Metalloproteinases from Chicken: ChIMP-3. A Third Member of the TIMP Family, *J. Biol. Chem.*267: 17321-17326.

Rodbard, D., 1973, Mathematics of Horomonoe-receptor interaction. 1. Basic Principles, *Adv. Exp. Med. Biol.* 36: 289-326.

Salo, T., Liotta, L. A., Keski-Oja, J., Turpeennieme-Hujanen, T. and Tryggvason, K., 1982, Secretion of a Basement Membrane Collagen Degrading Enzyme and Plasminogen Activator by Transformed Cells-Role in Metastasis, *Int. J. Cancer* 30: 669-673.

Sanchez-Lopez, R., Alexander, C. M., Behrendtsen, O., Breathnach, R. and Werb, Z., 1993, Role of zinc-binding- and hemopexin domain-encoded sequences in the substrate specificity of collagenase and stromelysin-2 as revealed by chimeric proteins, *J. Biol. Chem.* 268: 7238-47.

Sanchez-Lopez, R., Nicholson, R., Gesnel, M.-C., Matrisian, L. M. and Breathnach, R., 1988, Structure-Function Relationships in the Collagenase Gene Family Member Transin, *J. Biol. Chem.* 263: 11892-11899.

Sato, H., Kida, Y., Mai, M., Endo, Y., Sasaki, T., Tanaka, J. and Seiki, M., 1992, Expression of genes encoding type IV collagen-degrading metalloproteinases and tissue inhibitors of metalloproteinases in various human tumor cells, *Oncogene* 7: 77-83.

Sato, H., Takahisa, T., Okada, Y., Cao, J., Shinagawa, A., Yamamoto, E. and Seiki, M., 1994, A matrix metalloproteinase expressed on the surface of invasive cells, *Nature* 370: 61-65.

Stetler-Stevenson, W. G., Bersch, N. and Golde, D. W., 1992, Tissue inhibitor of metalloproteinase-2 (TIMP-2) has erythroid-potentiating activity, *FEBS Lett..* 296: 231-4.

Stetler-Stevenson, W. G., Krutzsch, H. C. and Liotta, L. A., 1989, Tissue Inhibitor of Metalloproteinase (TIMP-2), *J. Biol. Chem.* 264: 17374-17378.

Stetler-Stevenson, W. G., Liotta, L. A. and Seldin, M. F., 1992, Linkage analysis demonstrates that the Timp-2 locus is on mouse chromosome 11, *Genomics* 14: 828-9.

Stetler-Stevenson, W. G., Talano, J. A., Gallagher, M. E., Krutzsch, H. C. and Liotta, L. A., 1991, Inhibition of human type IV collagenase by a highly conserved peptide sequence derived from its prosegment, *Amer. J. Med. Sci.* 302: 163-70.

Strongin, A. Y., Collier, I. E., Krasnov, P. A., Genrich, L. T., Marmer, B. L. and Goldberg, G. I., 1993, Human 92 kDa Type-IV Collagenase - Functional Analysis of Fibronectin and Carboxyl-End Domains, *Kidney Int.* 43: 158-162.

Van Ward, H. and Birkedal-Hansen, H., 1990, The cysteine switch: a principle of regulation of metalloproteinase activity with potential applicability to the entire matrix metalloproteinase gene family, *Proc. Natl. Acad. Sci. USA* 87: 5578-5582.

Ward, R. V., Atkinson, S. J., Slocombe, P. M., Docherty, A. J., Reynolds, J. J. and Murphy, G., 1991, Tissue inhibitor of metalloproteinases-2 inhibits the activation of 72 kDa progelatinase by fibroblast membranes, *Biochim. Biophys. Acta.* 1079: 242-6.

Werb, Z., Mainardi, C. L., Vater, C. and Harris, E. D., 1977, Endogenous Activation Of Latent Collagenase By Rheumatoid Synovial Cells. Evidence for a Role of Plasminogen Activator, *New Engl. J. Med.* 296: 1017-1023.

Williamson, R., Marston, F., Angal, S., Koklitis, P., Panico, M., Morris, H., Carne, A., Smith, B., Harris, T. and Freedman, R., 1990, Disulphide bond assignment in human tissue inhibitor of metalloproteinases (TIMP), *Biochem. J.* 268: 267-274.

INDEX